Engineering Geology

Christopher C. Mathewson
Texas A&M University

Charles E. Merrill Publishing Company
A Bell & Howell Company
Columbus Toronto London Sydney

Published by Charles E. Merrill Publishing Company
A Bell & Howell Company
Columbus, Ohio 43216

This book was set in Souvenir
Text Designer: Ann Mirels
Production Coordination: Jo Ellen Gohr
Cover Design Coordination: Will Chenoweth
Cover photograph by the author

Library of Congress Catalog Card Number: 80–84811

International Standard Book Number: 0–675–08032–0

Printed in the United States of America

1 2 3 4 5 6 7 8—85 84 83 82 81

To the memory of my father George,
who was my greatest teacher

foreword

This book fills a need for a text that both geology students and engineering students can use. It will be useful to geology students who want to know how to apply their science to civil works and to engineering students who want to know more about geologic hazards, construction site problems, and construction material sources. The book also is useful as a reference work for practicing engineering geologists who from time to time need to refresh their memories on a specific topic.

The text is organized into three parts: (1) earth materials (describing basic geologic concepts and engineering properties of rocks, soils, and fluids); (2) geologic processes and engineering geology (showing that many site-specific problems are related to the geologic process that formed the site); and (3) engineering geology in practice (dealing with the applications and practice of engineering geology, including ethics and registration).

Chapter 1 provides a perspective on the melding and interaction of geology and engineering. I would suggest that upper-division and graduate students then peruse Part III for an overview of the applications of geology to engineering practice.

Stream-wash alluvium and "d.g." (decomposed granite) may look and feel similar, but site preparation costs can be significantly different unless the engineering geologist understands the processes that created the site and describes the properties of the materials in terms useful to the engineer. Dr. Mathewson's book has gone a long way in explaining the essentials of engineering geology to the student and the practicing professional.

Richard J. Proctor
Former President,
Association of Engineering Geologists
Arcadia, California

preface

Engineering geology is a relatively young science by name, but it is an old profession. Many early workers, such as Karl Terzaghi, were both engineers and geologists. The specialization that comes with technological advancement resulted in the creation of the engineering geologist. Spectacular disasters like Vaiont and Teton dams, spectacular achievements like Hoover Dam, and the awakening of an awareness of the environment have all acted to emphasize the need for engineering geology.

The engineering geologist is a geologist who is also trained in the basics of engineering. The professional must know and understand geologic processes as they affect engineering works. Unlike the science of geology, where incorrect interpretations may be part of the advancement of the science, the engineering geologist's interpretation often has a direct impact on human life and property. In short, the engineering geologist is a geoscientist denied the scientific privilege of being wrong because an error can lead to the loss of life or property for which the courts may find the professional liable.

This text is written for civil engineering or geology students who wish to learn the fundamentals of engineering geology. To civil engineers, this text introduces the fundamentals of geology that apply directly to engineering: classification and properties of earth materials, surface processes, site investigation, and the application of geology to the solution of problems. To the geologist this text introduces the fundamentals of engineering: testing of earth materials, site-specific geology, engineering classification, and use of earth materials. The text assumes the student has no prior knowledge of soil, rock, or fluid mechanics and introduces them as part of the physical behavior of earth materials.

The primary objective of this text is to develop in the reader an awareness and understanding of the earth's surface and the geologic systems active on that surface. The engineer and geologist will view this text from opposite sides and will find a common meeting ground in engineering geology. The text has been written for three types of university classes: (1) geology for civil engineers or building construction at the sophomore level,

(2) engineering geology for geologists at the junior level, and (3) engineering geology for engineers and geologists at the junior or senior level. At the sophomore-level course "geology for engineers," the text provides a broad base for advanced engineering courses in rock and soil mechanics and in hydrology and a solid demonstration of the application of basic engineering sciences to the earth sciences. For the geologist, the text provides a basic understanding and introduction to a "unique" application of geology—that of geology *in* engineering. At the combined class level, the text acts to provide a common ground of knowledge and understanding from which the instructor can build the technical and professional skills and knowledge necessary to practice engineering geology.

ACKNOWLEDGEMENTS

This project would never have been started or completed without the support and encouragement of many. My wife Janet and our children Heather and Glenn were most understanding of Daddy's strange behavior as he cloistered himself in his office. Without the enthusiasm shown by my professors while I was a student, this book would not exist. Willard C. Lacy and the late Richard Sloane were professors that simply demanded that I think out a problem and conceptualize a solution before I wasted my time. To them I owe sincere thanks for my professional development.

My graduate students were my greatest and most consistent critics and sources of opinion. My many professional friends in the Association of Engineering Geologists provided hours of discussions about the practice of engineering geology. Field trip leaders, who organized interesting and exciting field trips, made it possible to build my slide collection from which I have selected most of the photographs used in this text. I am also indebted to the professionals who have provided some of the text's cases in point.

Richard J. Proctor; Allen W. Hatheway; James L. Kennedy III; Terry R. West, Purdue University; and Stanley P. Fisher, Ohio University, provided comprehensive critical reviews of the manuscript and made many valuable suggestions. Gary Robbins, who assisted me in teaching from this text, provided many new insights and valuable suggestions.

Christopher C. Mathewson
College Station, Texas

contents

one

part I

Earth Materials

two

three

four

five

Sedimentary Rocks 55

six

Metamorphic Rocks 75

seven

Soil Materials 89

eleven

twelve

thirteen

fourteen

sixteen

part 

Engineering Geology in Practice

one

Engineering Geology

ENGINEERING GEOLOGY

2

Even though engineering geology can be traced back to our ancestors, we must define its terms. Engineering geology, by its very name, invites many definitions. How does one mix engineering, a profession of widely differing fields, with a science having its own specialized fields? A quick analysis of the name reveals that *geology* is the noun and *engineering* the adjective. To begin, we shall define the science—or, more specifically, the geoscience—of geology.

Geology is defined in the *Dictionary of Geological Terms* as "the science which treats the history of the earth, the rocks of which it is composed, and the changes which it has undergone or is undergoing." This definition leaves us with the impression that geology is "the science of rocks and earth processes." *Webster's New World Dictionary* defines geology as "the science dealing with the physical nature and history of the earth, including the structure and development of its crust, the composition of its interior, individual rock types, the forms of life found as fossils, etc." From this definition it appears that the "physical nature and history of the earth" is a primary focus for the scientist who practices geology.

Engineering, on the other hand, is defined by Webster as "the science concerned with putting scientific knowledge to practical uses, divided into different branches, as civil, electrical, mechanical or chemical engineering." To many practicing professional engineers, this definition will appear incomplete because it does not include economics, which is a significant component of any practical use. A definition by the late Professor R. L. Sloane, a civil engineer from the University of Arizona, seems most explicit: "An engineer is a person who can do for one dollar what any fool can do for two."

Combining these definitions makes an engineering geologist "a scientist who can put geologic knowledge to practical use." Geologic knowledge concerning the earth's physical nature and history—that is, earth processes—directly affects at least three branches of engineering: civil engineering, where the engineer must economically and safely build on or in the earth's crust; petroleum engineering, where the engineer must economically design systems to recover and extract petroleum products from rocks; and mining engineering, where the engineer must design and develop methods to extract mineral resources from the crust economically and safely. Thus, there are actually three types of engineering geologists: civil engineering geologists, petroleum engineering geologists, and mining engineering geologists. The most common type is the "civil" engineering geologist, defined by the *Dictionary of Geological Terms* as one who applies the geological sciences to engineering practice "for the purpose of assuring that the geologic factors affecting the location, design, construction, operation and maintenance of engineering works are recognized and adequately provided for."

In this context, one role of an engineering geologist is that of a translator and interpreter. He or she translates the scientific facts, observed or measured, that describe the uniqueness of the physical character of the earth's crust into geologic information and then translates this information

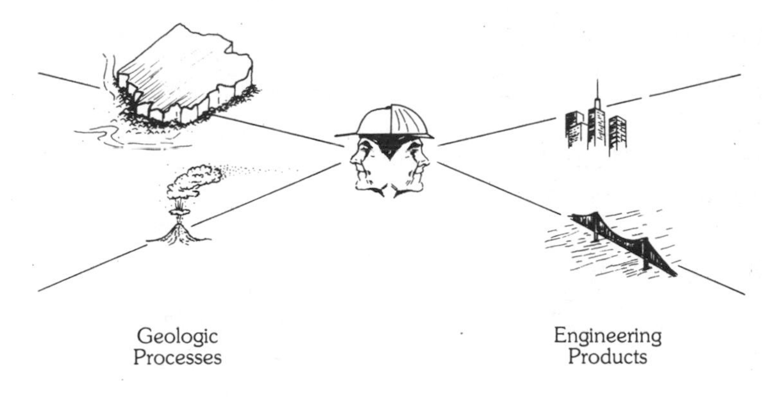

FIGURE 1–1 Portrait of an engineering geologist looking back at geologic processes and forward to engineering products. (From C. Mathewson, Academic preparation for the engineering geologist, in *AGI/AEG short course lecture notes: Engineering geology for geologists*, ed. A. F. Geiger and A. W. Hatheway, 1978. Reproduced by permission of the American Geological Institute.)

into engineering data to identify areas of significant physical constraint that will adversely affect the design, construction, and maintenance of any intended engineering project. In this role, the engineering geologist is identifying the geologic characteristics of a site that will *increase the cost* if they are adverse conditions or *reduce the cost* if they are favorable conditions. Thus, we can look at an engineering geologist as a "two-faced" professional looking back at geologic processes and forward to engineering products (Figure 1–1).

The engineering geologist is a fully competent geologist who is versed in the basic theory and practice of engineering and is aware of the complex natural systems that interact to produce the physical environment. How will these systems respond when people modify or act on them? Will a road cut require a high-level, expensive maintenance program due to unstable geologic conditions? Can it be redesigned to reduce the maintenance and still be economical? What geologic risk is associated with a particular dam? Unlike the geologist, the engineering geologist is denied the scientific privilege of being wrong because an error can lead to the loss of life or property for which the courts may find him or her liable.

ENGINEERING GEOLOGY AND CIVIL ENGINEERING

Since the engineering geologist is frequently associated with the civil engineer and is charged with the responsibility of interpreting the geologic conditions of a site into useful engineering data, the question of responsibility arises. The California State Board of Registration for Geologists and Geophysicists released the results of a report prepared by a commitee of geologists and civil engineers on April 5, 1973. This report defined the fields of expertise of both professions and then listed the responsibilities of each (Tables 1–1 and 1–2). It is a well-organized demonstration of the scope of the profession of engineering geology.

From Table 1–1, the student interested in engineering geology or civil engineering can examine what might be included in university programs in these two professions. The responsibilities of the two professions are usually separated as shown in Table 1–2.

TABLE 1-1 Fields of expertise

Engineering Geologist	Both	Civil Engineer
Classification and Physical Properties		
Rock description	Soil description	Soil testing Strength characteristics
Rock Mechanics		
Descriptive rock mechanics Rock structure Performance Configuration	*In situ* studies	Rock testing Stability analysis Stress distribution
Slope Stability		
Slope stability Interpretative Geologic analyses—geometrics	Grading	Slope stability analysis and testing
Mapping		
Geologic mapping Aerial photography Air photo interpretation Landforms Subsurface configurations	Soil mapping	Topographic mapping Surveying
Project Planning		
Development of geologic parameters Geologic feasibility		Urban planning Design Material analysis Economics
Surface Waters		
	Volume of runoff Stream description Silting potential Erosion potential Source of base flow Sedimentary processes Source of material	Design of control works Coastal-river engineering Design of development Hydrology
Groundwater		
Hydrogeology Occurrence Structural controls Direction of movement Underflow studies	Drainage Well design Subsidence Field permeability Transmissibility	Engineering hydrology Mathematical treatment of well systems Development concepts Regulation of supply

TABLE 1–1 (continued)

Engineering Geologist	Both	Civil Engineer
Storage computation	Specific yield	Economic considerations
Characteristics of water-bearing and non-water-bearing materials		Laboratory permeability
Earthquakes		
Seismicity	Seismic consideration	Seismicity
Location of faults	Earthquake probability	Response of soil and rock materials to seismic activity
Evaluation of active and inactive faults		Aseismic design of structures
Historic record of earthquakes		
Exploration		
	Planning	
	Supervision	
	Observation	
Design and Supervision		
	Grouting	
	Tunnel	
	Conduit	
Engineering Geophysics		
Soil and rock hardness		Engineering application
Mechanical properties		
Depth determinations		

The relationship between the engineer, the geologist, and the engineering geologist is summarized in Figure 1–2. At the foundation of the diagram is the science of geology, which is directed toward understanding the history of the earth. Its primary objective is to describe or explain how geologic processes in the past or within the earth have produced the physical environment. The engineering geologist is a geoscientist who uses the geologic foundation plus mechanics of earth materials to *predict.* These predictions may be directed toward the development of petroleum resources (petroleum geology), the extraction of mineral resources (mining geology), or the construction of safe and economic structures (engineering geology). The foundation of engineering is *conceptualization,* the development of a concept of how to solve a particular problem. The concept, of course, is based on scientific principles and experience in problem solving. Once a concept has been developed, the engineer then

TABLE 1–2 Professional responsibilities

1. *The Engineering Geologist*
   a. Description of the geologic environment pertinent to the engineering project.
   b. Description of earth materials, their distribution, and general physical/chemical characteristics.
   c. Deduction of the history of pertinent events affecting the earth materials.
   d. Forecasts of future events and conditions that may develop.
   e. Recommendation of materials for representative sampling and testing.
   f. Recommendation of ways to handle and treat various earth materials and processes; recommending or providing criteria for excavation design, particularly angle of cut slopes, in materials where engineering testing is inappropriate or where geologic elements control stability.
   g. Inspection during construction to confirm conditions.

2. *The Civil Engineer*
   a. Direction and coordination of team efforts where engineering is a predominant factor.
   b. Controlling the project in terms of time and money requirements and degree of safety desired.
   c. Engineering testing and analysis.
   d. Reviewing and evaluating data, conclusions, and recommendations of team members.
   e. Deciding on optimum procedures.
   f. Developing designs consistent with data and recommendations of team members.
   g. Inspection during construction to assure compliance.
   h. Making judgments on economy and safety matters.

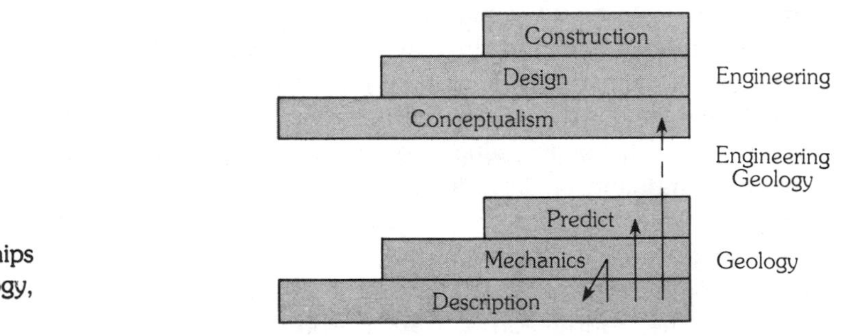

FIGURE 1–2 Functional relationships between geology, engineering geology, and engineering.

designs a solution to the problem and implements the design into the final product.

Engineering geology is a unique mix of geology and engineering which calls upon a broad background of information for every problem. A general list of reference texts and a selected bibliography are therefore given in Appendix 4 to assist the student in studying beyond the level of this general text. Appendix 4 also lists pertinent technical societies that publish journals with many useful case histories and the results of current research.

part I

Earth Materials

Physical Properties of Earth Materials

two

Every material has certain physical properties that control its behavior when it is subjected to some external force. Under different pressure and temperature conditions earth materials can be a solid, liquid, or gas. For example, water when frozen acts as a solid material; at 10°C it acts as a liquid or as water vapor. The engineering geologist is concerned with solid earth materials (rock and soil) and fluid earth materials (water and gases) and their interactions.

Solid materials have a definite shape and volume and some fundamental strength. Solid earth materials are composed of grains or crystals of one or more minerals. These minerals may be crystalline solids with ordered or regularly arranged atomic structures, or supercooled liquids with an amorphous structure composed of a random arrangement of atoms. Crystalline solids are the most common component of solid earth materials. Since these materials are composed of grains or crystals of one or more mineral types, the physical and mechanical properties of solid earth materials will be controlled by both the properties of the individual grains or crystals *(constituent properties)* and the properties of the entire mass *(bulk properties)*.

A fluid material does not have a definite shape but takes on the shape of its container. Fluids have no fundamental strength and deform by flowing when subjected to any shear stress, regardless of how small the stress may be. In most engineering geology cases, we assume most liquids to be incompressible and most gases to be compressible.

SOLID EARTH MATERIALS

CONSTITUENT PROPERTIES

The constituent properties of any solid earth material include the density or unit weight, specific gravity, grain size and shape, crystal size and shape, and composition of the individual grains in the material. **Density** is the mass of a substance per volume of substance and in geology is usually given in grams per cubic centimeter (g/cm^3). **Unit weight,** a term used in engineering, means the same as density. **Specific gravity** is the ratio of the density of the substance to the density of water at 4°C. Table 2–1 gives the density, unit weight, and specific gravity of common earth materials.

Grain size is the measure of the diameter of a spherical grain with the same volume as the natural grain regardless of its shape. Grain size classification systems vary between geologic and engineering professions; what is fine grain size to one profession may be medium to the other. Figure 2–1 compares the commonly used grain size classification systems. Grain size is frequently given in millimeters (mm) or phi (ϕ) sizes in geology, and in millimeters (mm) or inches (in.) in engineering.

Grain shape is a descriptive term used to differentiate between particles and to provide a means of describing shapes of particles that are not spherical. Grain shape terms are shown in Figure 2–2. In sedimentary rocks, which are composed of grains that have been transported and redeposited, grain shape is also a valuable indication of the amount and agent of transport. Well-rounded grains have been transported over longer distances than angular particles because mechanical processes break off the sharp corners of the angular grains.

Crystal size and **shape** are established during the crystallization of a molten mass as it cools. The size of the crystals is controlled by the cooling rate; slow cooling yields large crystals, while rapid cooling yields fine crystals.

TABLE 2–1 Weight properties of common earth materials

	Material	Density or Unit Weight (dry/saturated)		
		g/cm^3	lb/ft^3	kN/m^3
SOILS	Water @ 4°C	1.0	62.4	10
	Salt water	1.03	64.2	10.03
	Uniform loose sand	1.4/1.9	90/118	14/19
	Mixed loose sand	1.5/2.0	99/124	15/20
	Uniform dense sand	1.7/2.1	109/130	17/21
	Mixed dense sand	1.9/2.2	116/135	19/21
	Silt	1.6/2.2	100/135	16/21
	Soft clay	1.0/1.6	60/100	10/16
	Stiff clay	1.7/2.1	105/130	17/20
	Very stiff clay	1.8/2.3	115/145	18/23
ROCKS	Hard igneous	2.6/3.0	160–190*	26–30
	Broken hard igneous	1.8/2.0	110/125	18/20
	Metamorphic	2.6–2.9	160–180*	26–29
	Broken metamorphic	1.8/1.9	110/120	18/19
	Cemented sedimentary	2.4–2.9	150–180*	24–29
	Uncemented sedimentary	1.8/2.4	110/150	18/24

*The difference in unit weight between dry and saturated rocks is negligible.

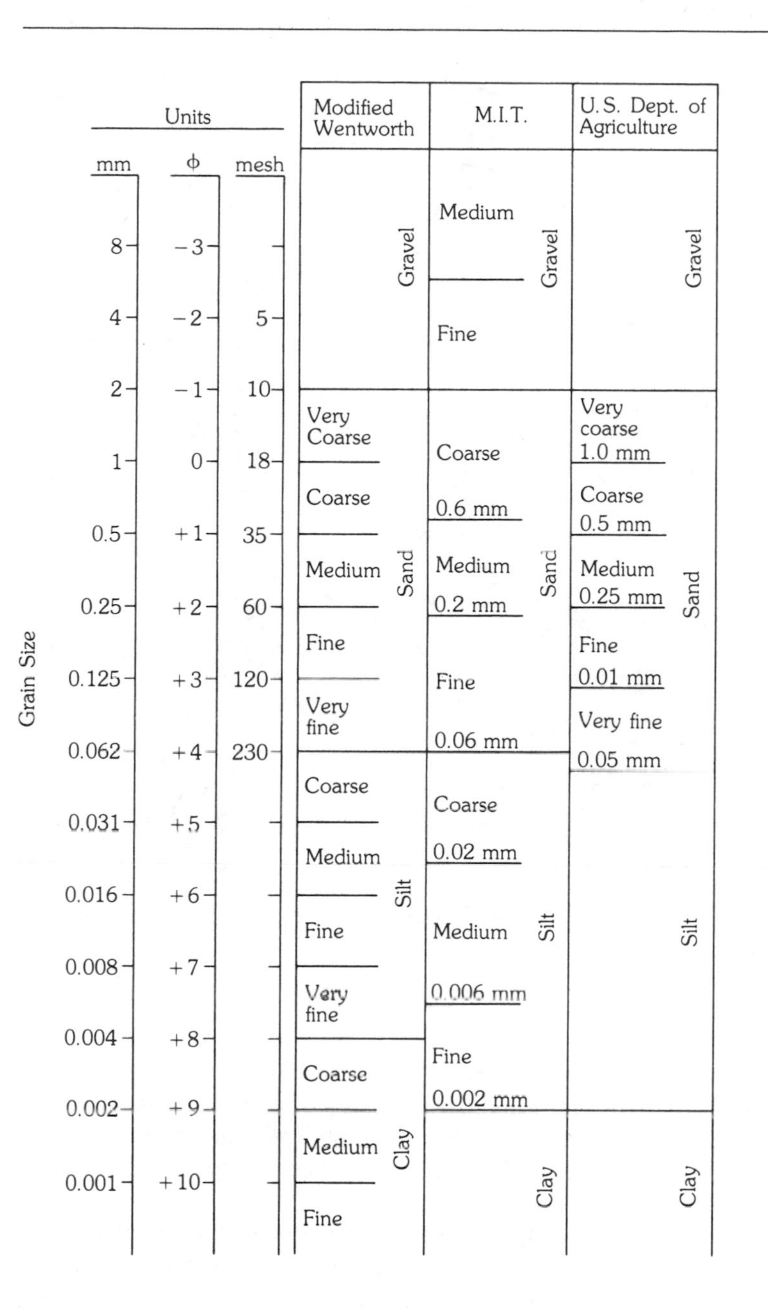

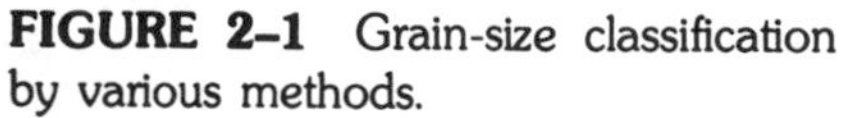
FIGURE 2–1 Grain-size classification by various methods.

Crystal shape is established by the arrangement of the atoms that make up the crystal and the space available for the crystals to grow. Free-growing crystals can take on the common forms shown in Figure 2–3.

Composition is one of the most important constituent properties of any solid earth material because the bulk properties of the material are affected by the composition of the constituents. Rock classification is based

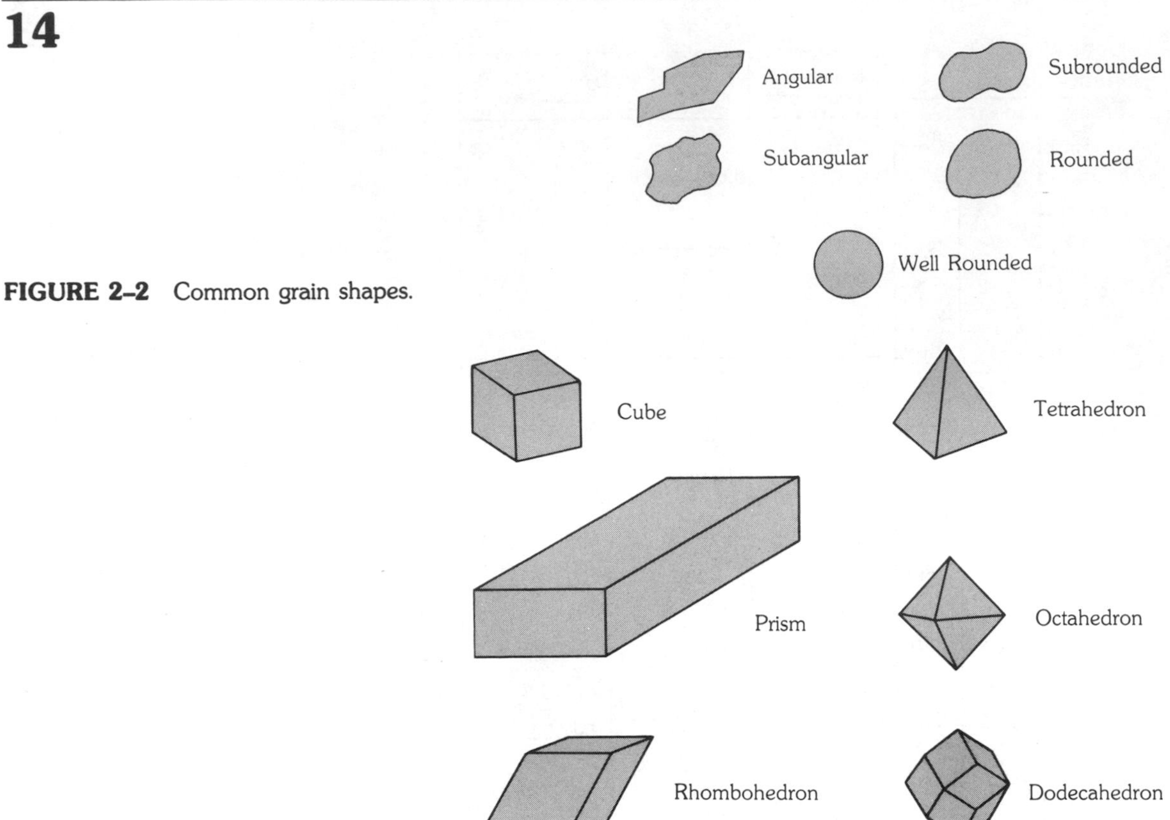

FIGURE 2–2 Common grain shapes.

FIGURE 2–3 Common crystal forms.

on the composition of the mineral *aggregates,* or crystals, or source rock grains that make up the rock.

BULK PROPERTIES

The engineering geologist is more concerned with the physical and mechanical properties of the entire bulk (mass) of earth material than with the characteristics of a specific grain or crystal. The bulk properties of an earth material are controlled or at least affected by the constituent properties.

Density and unit weight of solid earth materials are determined just as they are for the constituents except that the bulk density or total unit weight is the *average* of the densities or unit weights of the constituent components and void spaces. For example, a rock composed of grains with a density of 2.75 g/cm^3 may have a bulk density of only 2.30 g/cm^3 because of large void spaces between the grains.

Grain packing, or the arrangement of the grains in a solid material, is classified as shown in Figure 2–4. Grain packing plays a significant role in establishing the physical and mechanical properties of solid earth materials. Tighter packing increases bulk density and unit weight, increases shear strength, and decreases permeability and hydraulic conductivity. **Cementation** and **void filling** are processes that occur after an aggregate of grains is deposited. Cementation, frequently with calcium carbonate or silica cements, increases the bulk density and unit weight, increases shear strength, and decreases permeability and hydraulic conductivity.

Grain-size distribution is the measure of variation in grain sizes in a solid material. The term **sorting** is used by geologists to describe grain-size distribution: a sample composed of grains of the same size is *well sorted* and one composed of a greater mixture of sizes is *poorly sorted.* The engineer, on the other hand, uses **grading** or *gradation* to describe grain-size distribution. A sample that has uniform gradation or a mixture of all sizes is *well graded* while one that has nonuniform gradation or all particles of the same size is *poorly graded.* A comparison of grain-size distribution graphs is shown in Concept 2–1. Grain-size distribution affects the bulk density and unit weight of a sample, because an increase in the variability of grain sizes in a sample allows for smaller particles to fit the spaces left between larger particles, thereby increasing bulk density and unit weight.

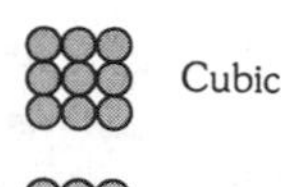

FIGURE 2–4 Cubic and rhombohedral packing of spherical grains. Notice that the porosity of the material decreases as the degree of packing increases. The addition of fine-grained materials further decreases the porosity.

CONCEPT 2–1

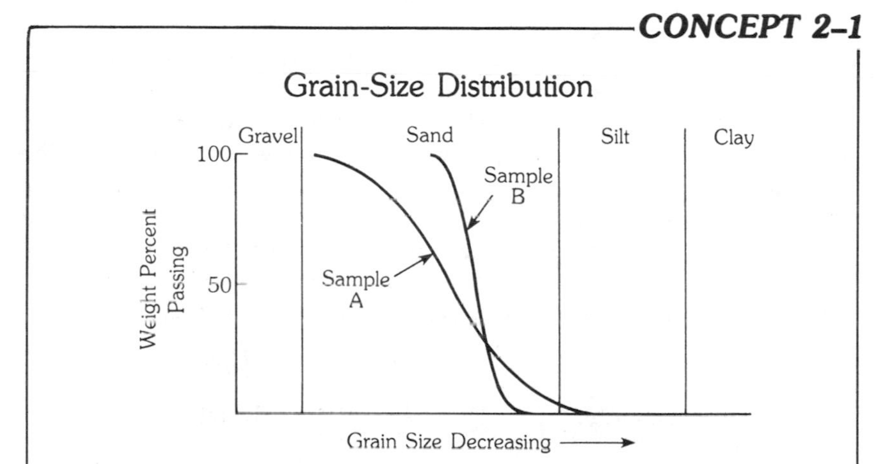

The terms used to describe grain-size distribution differ between geologists and engineers. **Geologists use sorting,** while **engineers use grading.** You can remember these differences by thinking of the philosophy of the two professions: geologists study past processes—thus a granular material has been *sorted*—while engineers are concerned with the material for some use—thus the material is *graded.* Sample A in the diagram shows a uniform distribution of all grain sizes; thus the sand is *poorly sorted* to a geologist and *well graded* (gradual gradation of sizes) to an engineer. Grains in sample B are nearly the same size; thus it is *well sorted* to a geologist but *poorly graded* to the engineer.

?

Crystal size, shape, and **composition** are significant in establishing the bulk physical and mechanical properties of crystalline materials. Many solid earth materials, as you will see in following chapters, are composed of a suite of different crystals. Materials composed of very fine crystals have higher strengths than materials composed of large crystals. Interlocking crystals produce a material with similar strengths in all directions, while the strength properties of a material composed of oriented, layered crystals depend upon the direction of loading. Crystal composition affects the strength and deformation character of the bulk material.

PHASE RELATIONSHIPS Solid earth materials composed of aggregates of grains can contain three phases of material: *solids,* containing just the grains; *liquids,* containing water or other liquids in the spaces between the grains; and *gases,* composed of the air or other gases filling the remaining spaces between the grains. Phase relationships are important physical properties of any solid earth material because many mechanical properties of the material relate to the existing phase relationships of the material. Just as geologists and engineers use different descriptive terms for grain-size distribution, so do they in phase relationships. The geologist uses the *total weight* or *total volume* as the basis of any ratio, while the engineer uses *weight of solids* or *volume of solids* as the basis of any ratio. These two basic differences reflect the basic philosophy of the two professions: the geologist is interested in the material as it occurs, and the engineer is interested in the material as it is used or as it responds to a load. Concept 2–2 describes phase relationships as they are defined by the geologist and the engineer.

A property of the bulk material that is independent of the density or viscosity of the fluid is **permeability. Hydraulic conductivity** is the flow velocity of a fluid through a porous media and is proportional to the hydraulic gradient. The hydraulic conductivity decreases as the density of the fluid decreases and viscosity of the fluid increases; the permeability is a constant. Engineers frequently use *permeability* in place of *hydraulic conductivity* to refer to the permeability of a material to water, oil, or gas. The relationship between permeability and hydraulic conductivity is:

	Equation	Units
Permeability:	$k = \frac{K\mu}{\rho g}$	(length)^2
Hydraulic conductivity:	$K = k\frac{\rho g}{\mu}$	length/time

where

μ = viscosity of the fluid
ρ = density of the fluid
g = gravitational acceleration

A property of weakly to uncemented fine-grained (clay- and silt-size) earth materials that changes as a function of moisture content of the material is **plasticity.** Plastic materials undergo a change from solid to liquid as the

CONCEPT 2–2

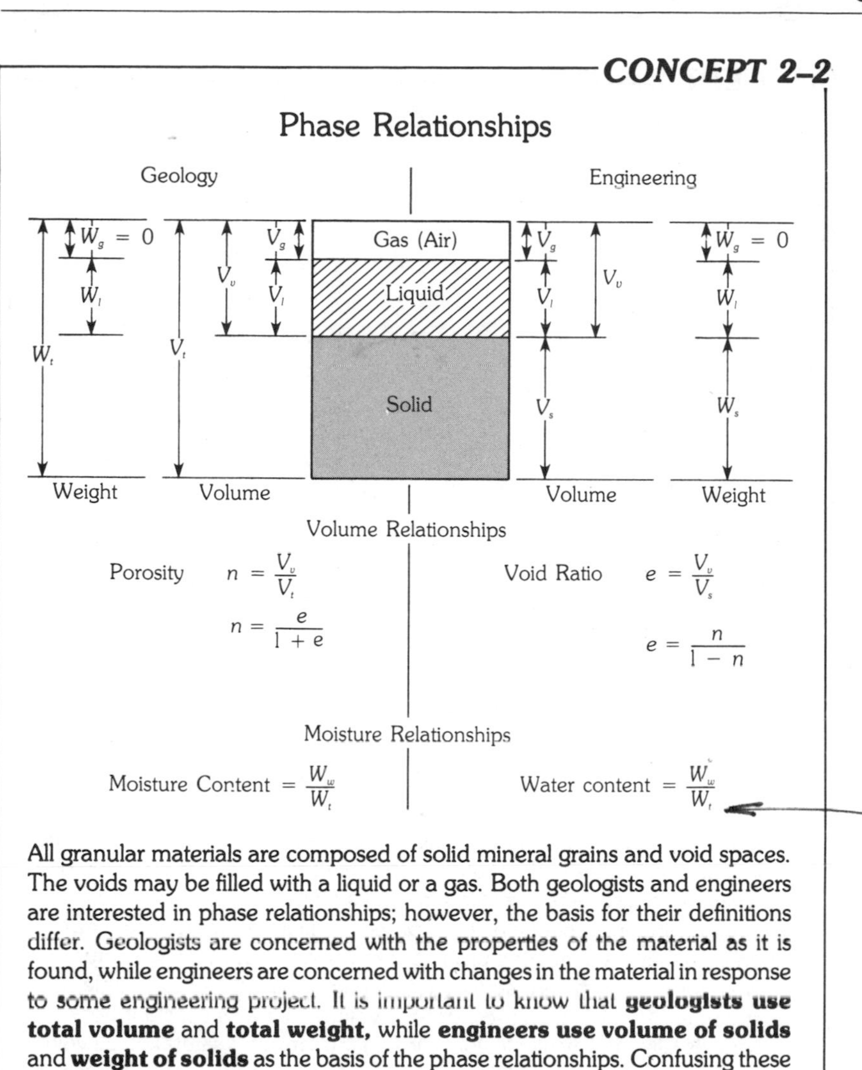

All granular materials are composed of solid mineral grains and void spaces. The voids may be filled with a liquid or a gas. Both geologists and engineers are interested in phase relationships; however, the basis for their definitions differ. Geologists are concerned with the properties of the material as it is found, while engineers are concerned with changes in the material in response to some engineering project. It is important to know that **geologists use total volume** and **total weight,** while **engineers use volume of solids** and **weight of solids** as the basis of the phase relationships. Confusing these two can make a significant difference: the geologist's porosity ($n = V_v/V_t$) and moisture content (W_w/W_t) range from zero to 1.00, while the engineer's void ratio ($e = V_v/V_s$) and water content (W_w/W_s) can range from zero to infinity. For example, a material with a moisture content of 50 percent has a water content of 100 percent.

moisture content increases. Concept 2–3 describes the relationship between total volume and phase as a function of increasing water content.

DEFORMATION CHARACTERISTICS As any solid earth material is subjected to an external load (F) or stress (force/area), the material deforms. The character of this deformation, or *strain* (change in length/original length), is controlled by the material's mechanical properties. Common stress-strain diagrams and mechanical models representing these diagrams

CONCEPT 2–3

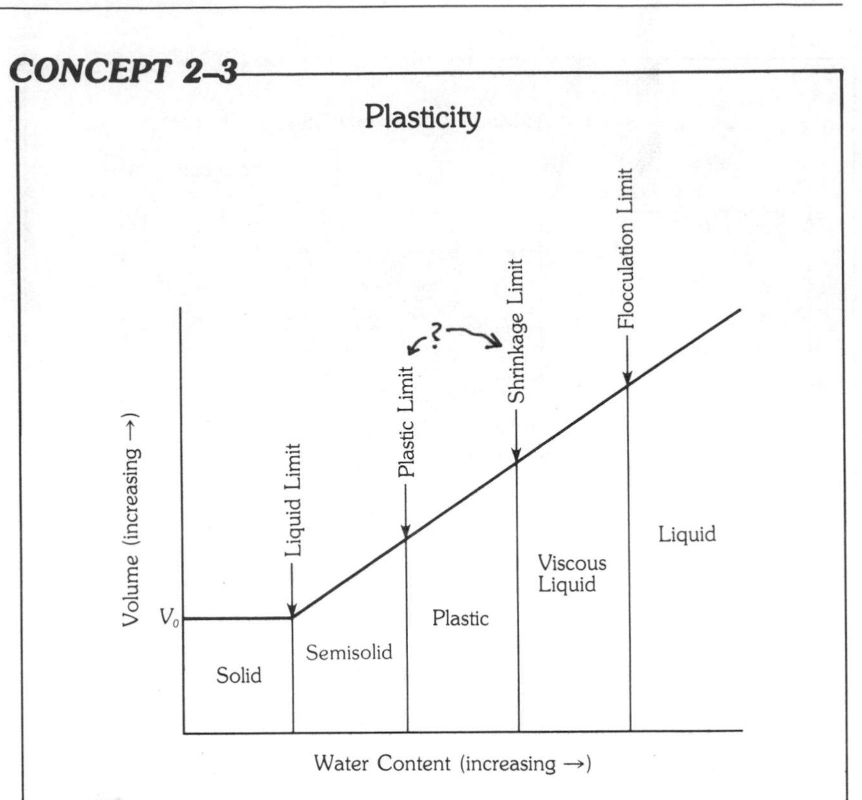

As water is added to a fine-grained, clay- and silt-sized material, the engineering properties of the material change. (This process is the same one that every child discovers in making mud pies.) If the material is air-dry, it will have some minimum volume (V_o) composed of solids and void space. The dry material will have engineering properties of a solid; that is, it will fracture or crush along discrete failure surfaces. As water is added to the sample, the water content (W_w/W_s) increases, but the volume remains constant because the water is filling the available voids. When the sample is saturated, all voids are filled. The total volume of the sample must increase if the water content is to increase. The sample is at the **shrinkage limit** when it is saturated and is at its smallest volume. At any water content that is at or above the shrinkage limit, the sample is saturated. As water is continually added, the volume increases and the sample eventually behaves as a plastic material. The water content at which the sample becomes a true plastic material is the **plastic limit.** Additional water eventually causes the sample to behave as a viscous fluid. The **liquid limit** is the water content of the sample when it has a shear strength of 1 g/cm^2. The range in water content between the liquid limit and the plastic limit is known as the **plasticity index.** A material is defined as a low plastic material if the plasticity index is low and a high plastic material if the plasticity index is high. When more water is added, the particles are separated by water, the sample behaves as a true liquid, and shear strength equals zero. This water content is called the **flocculation limit.**

are shown in Figure 2–5. The external load or stress can produce one of three basic stress conditions acting on a unit cube of earth material: **compressive, shear,** and **tensile.** *Compressive stresses* develop when external stresses are directly opposite and are directed toward each other. When the external stresses are offset, parallel, and directed in the opposite direction, thereby creating a force couple, *shear stress* develops. *Tensile stress* is a pulling agent, or the opposite of compressive stress (Figure 2–6). A general stress field can be a vector divided into three principal stress directions: *maximum, intermediate,* and *minimum principal stress* (Figure 2–7). Notice that these three stress vectors are mutually perpendicular and do not include shear stress.

Shear stress develops as a result of a stress differential (maximum principal stress − minimum principal stress), as we will learn later. Shear strength of a solid earth material will be first developed in a simple shear case *(direct shear)* in which only two stresses exist: the shear stress and normal

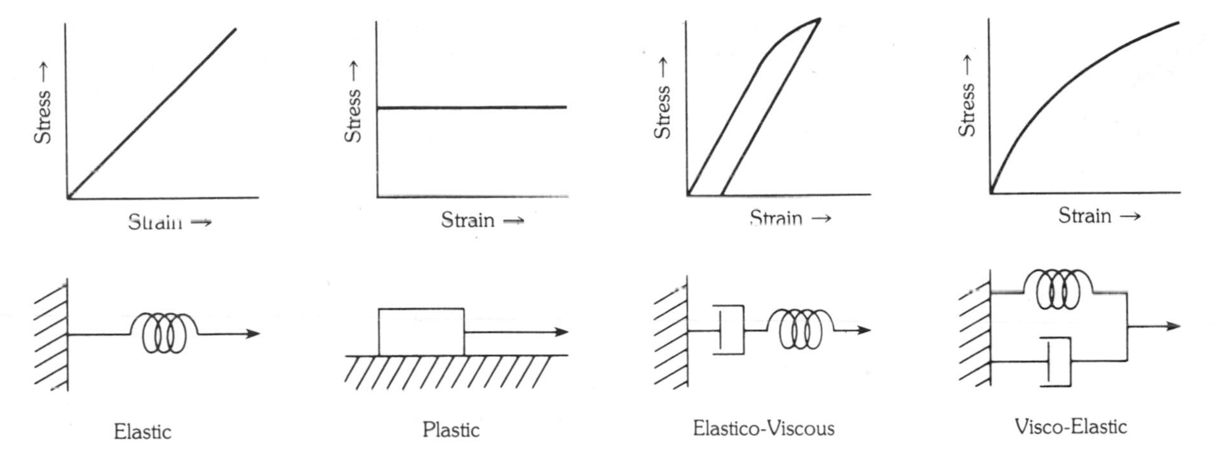

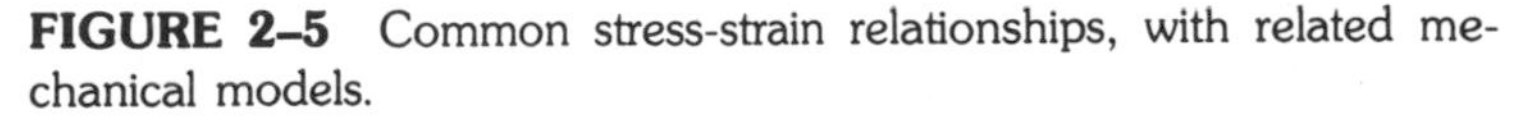

FIGURE 2–5 Common stress-strain relationships, with related mechanical models.

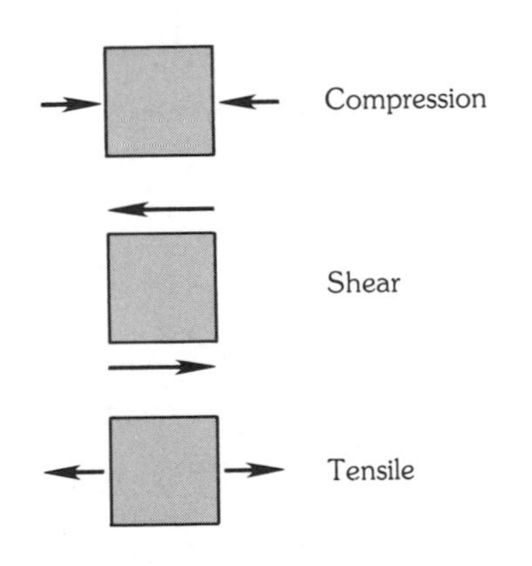

FIGURE 2–6 General stress conditions that act on any material.

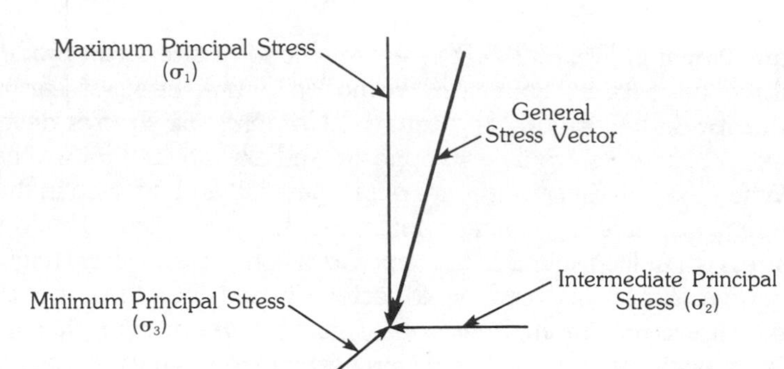

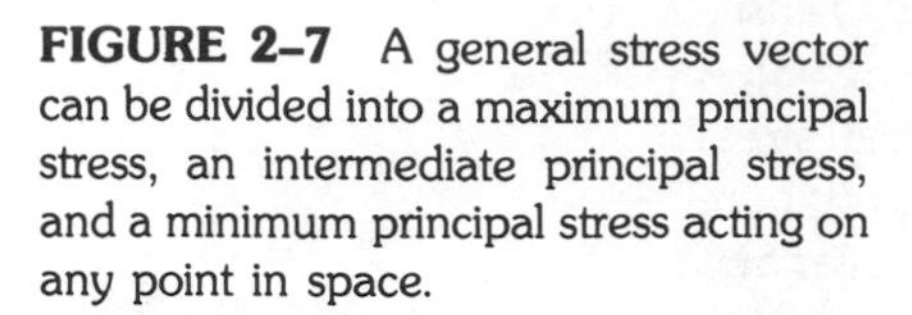
FIGURE 2–7 A general stress vector can be divided into a maximum principal stress, an intermediate principal stress, and a minimum principal stress acting on any point in space.

stress (see Concept 2–4). Notice the linear relationship between shear stress and normal stress in the diagram in Concept 2–4. Two material properties of solid earth materials are defined: **cohesion,** which is the shear strength when the normal stress is zero, and the **angle of internal friction** (friction angle), which is the angle between the shear strength–normal stress line and the horizontal axis. The coefficient of linear proportionality between shear strength and normal stress ($\Delta\tau/\Delta\sigma_n$) is the tangent of the angle of internal friction,

$$\frac{\Delta\tau}{\Delta\sigma_n} = \tan \phi$$

This angle is frequently referred to as the *phi angle* (ϕ). The shear strength of any solid earth material is the sum of these two basic physical properties of the material: cohesion and the friction angle. The normal stress and loading conditions are established by the material's external environment.

CONCEPT 2–4

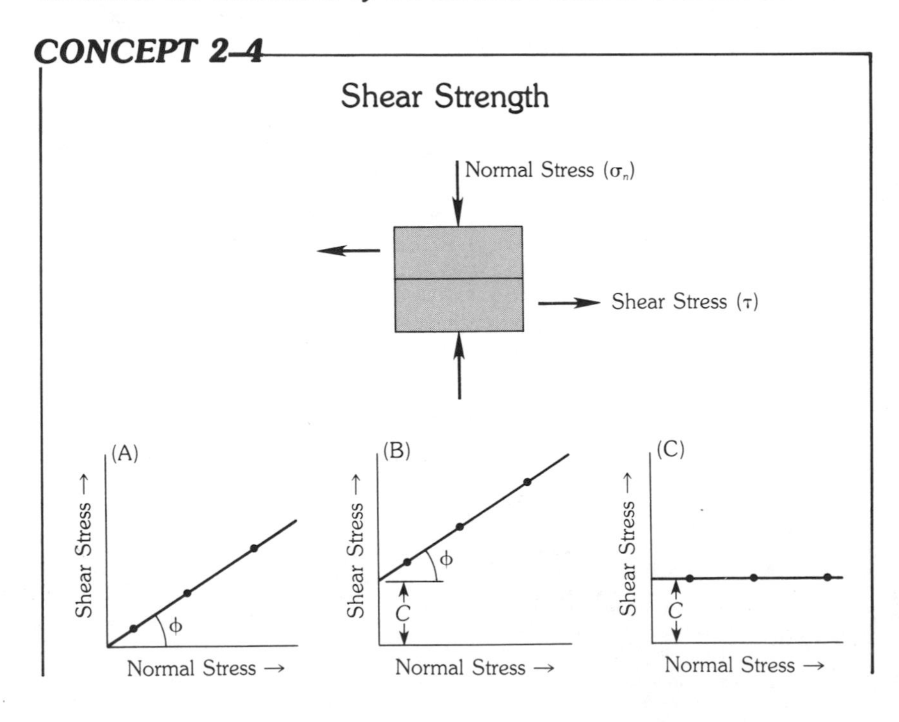

When a sample of earth material is loaded as shown in the diagram—a normal stress acting to hold the material together and a shear stress acting to slide (shear) the sample—the relationship between normal and shear stress at failure generally takes on one of three forms: A, B, or C. In case A, the material has no shear strength when the normal stress is zero and therefore will not hold together unless it is confined; it is noncohesive, (or *cohesionless*). If the material has a shear strength when the normal stress is zero (case B), the material *coheres* (is *cohesive*). In case C, the shear strength of the material does not increase with increasing normal stress; therefore, some process must be acting to resist the effect of normal stress. The pore fluid pressure increases to equal the normal stress; thus the shear strength is independent of normal stress in case C.

These three shear-strength tests can be related through the shear-strength equation:

$$\tau = C + (\sigma_n - u)\tan\phi$$

where

τ = shear strength

C = cohesion (shear strength when normal stress is zero)

σ_n = normal stress

u = pore fluid pressure which acts against σ_n

ϕ = friction angle

In case A, C and u equal zero:

$$\tau = \sigma_n \tan\phi$$

In case B, u is zero or less than σ_n:

$$\tau = C + (\sigma_n - u)\tan\phi$$

In case C, u equals σ_n as would occur in a confined sand or rapidly loaded clay:

$$\tau = C + 0$$

In a material that liquifies, the pore pressure increases such that the shear strength decreases to near-zero and the material behaves as a liquid.

The relationship between the direct shear conditions in a sample and the maximum and minimum principal stresses (σ_1, σ_3) on the sample is shown in Figure 2–8. Notice that as the minimum principal stress (σ_3) increases, both the maximum principal stress and shear strength increase because a vector component of the minimum principal stress increases the normal stress. Notice also that as the angle between the maximum or minimum principal stress and the shear plane changes, the shear stress conditions will also change. This relationship is graphically demonstrated by a Mohr diagram, as shown in Concept 2–5. The Mohr diagram can be used to construct a graphic solution for the state of stress along any plane in the sample.

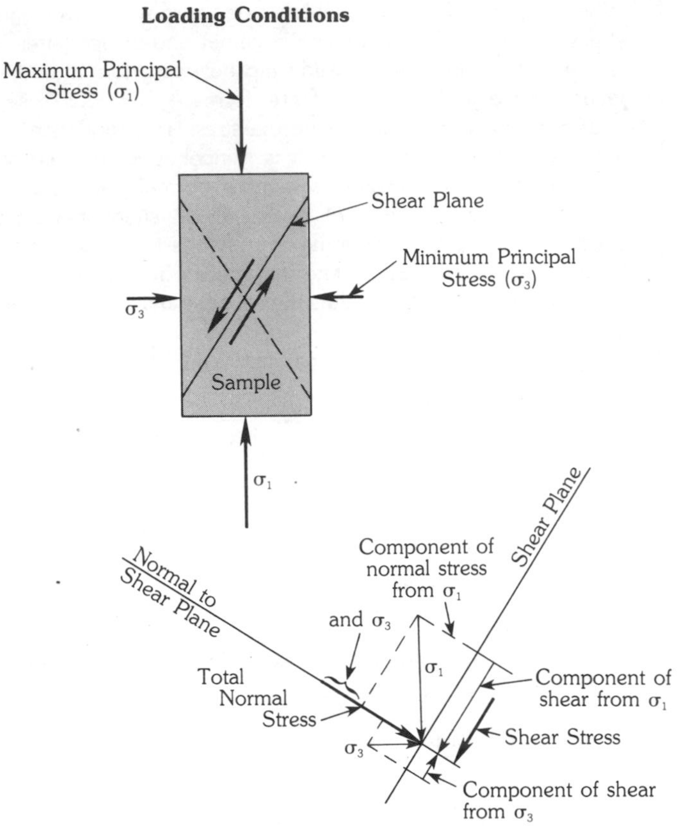

FIGURE 2–8 State of stress along any plane in a rock specimen. The stress conditions can be divided into normal and shear components by resolving the vectors.

Tensile strength becomes an important material property in engineering geology whenever earth materials are subjected to bending forces, as in the back (roof) of an underground opening. The plane perpendicular to the minimum principal stress (σ_3) is subjected to a tensile stress when σ_3 is zero. Changing the loading conditions from a piston to a point or line load will result in the formation of high tensile stresses along this plane. The Brazilian tensile test is performed on a circular sample and is the most common means of determining a rock's tensile properties.

Compressive stresses can cause the failure of grains or crystals because of crushing at points of contact, or they may cause a material to become more dense by squeezing it together. One important mechanical property of a material is its compressibility. **Compressibility** is most important in low bulk density, porous materials, because placing a foundation load on these materials will result in settlement of the structure. Settlement causes no problems if it is considered in the design; however, differential settlement can have disastrous consequences.

The compressibility of massive crystalline earth materials is an important consideration in the design of large, heavy structures or arch dams where the materials must support high stresses with little deformation.

CONCEPT 2–5

The Mohr Diagram

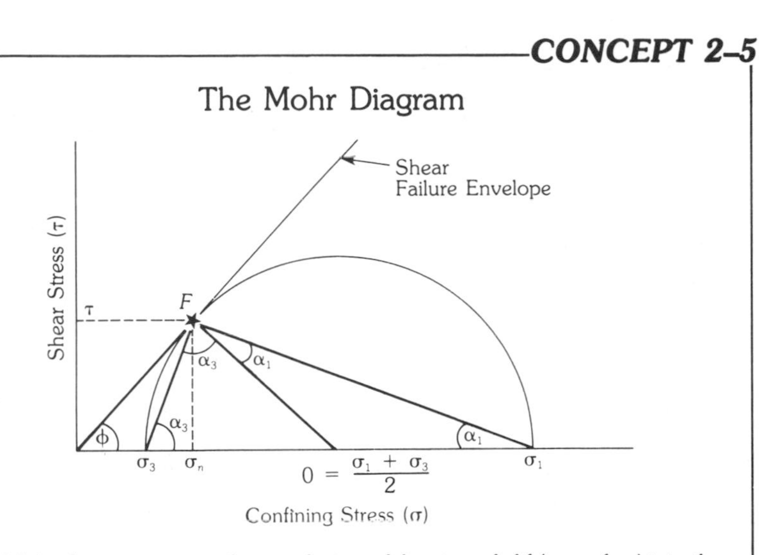

The Mohr diagram is a graphic resolution of the stress field (σ_1 and σ_3) into the shear stress (τ) and normal stress σ_n on any plane. In the diagram, notice that:

1. The lines σ_3–F and σ_1–F form a right angle at F and therefore have the same orientation as the σ_1 and σ_3 stress vectors in Figure 2–8.
2. As the angle of the shear plane changes, the influence of σ_1 and σ_3 on σ_n and τ also changes.
3. The shear failure envelope is the same as the normal stress–shear stress line developed in Concept 2–4.
4. The line O–F is normal to the failure envelope line and is the radius of the circle; therefore, lines O–F, O–σ_3, and O–σ_1 are all equal, and $\alpha_3 + \alpha_1 = 90°$.

If a mathematical relationship can be developed that relates α_1, α_3, and ϕ, then the Mohr diagram is a graphic solution similar to the vector resolution in Figure 2–8.

To solve for α_3, use the triangle ϕ–σ_3–F:

$$\phi + (180 - \alpha_3) + (90 - \alpha_3) = 180$$
$$\phi - 2\alpha_3 = -90$$
$$\alpha_3 = 45 + \frac{\phi}{2}$$

To solve for α_1, use triangle ϕ–F–α_1:

$$\phi + (90 + \alpha_1) + \alpha_1 = 180$$
$$2\alpha_1 = 90 - \phi$$
$$\alpha_1 = 45 - \frac{\phi}{2}$$

Notice that the α angles are the angles between the principal stress direction and the shear plane. The angle between the maximum principal stress direction (σ_1) and the shear surface (α_1) is most commonly used in geology ($\alpha_1 = 45 - \phi/2$), while the angle between the minimum principal stress direction (α_3) is frequently used in engineering ($\alpha_3 = 45 + \phi/2$).

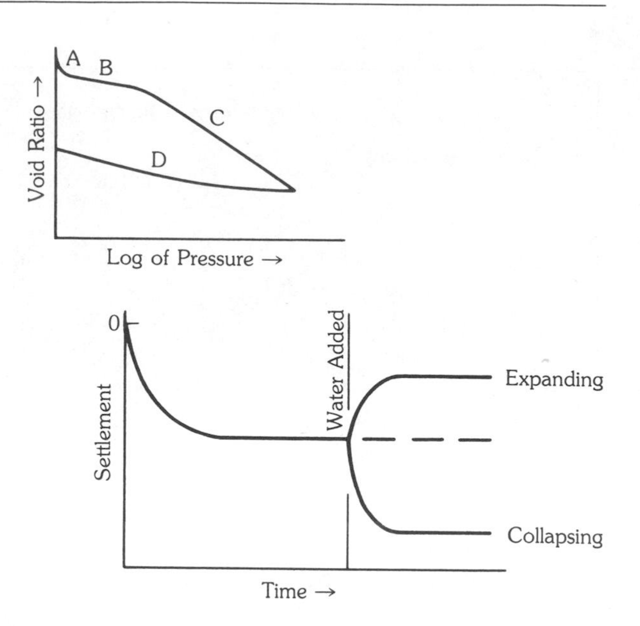

FIGURE 2–9 Void ratio versus the log of pressure. A is the initial compaction as the soil is loaded; B is the recompression of the soil to the past pressure; C is the virgin compression where the soil is compressed beyond the level of pressure in the ground; and D is the rebound as the soil responds to the removal of the pressure.

FIGURE 2–10 Time-settlement curves showing the effect of water on expanding or collapsing soils.

Compressive deformation of a material can be either elastic or nonelastic (permanent deformation), although some elastic rebound is expected in all materials when they are unloaded as shown in Figure 2–9.

Many physical and mechanical behavioral processes of solid earth materials are less common than the ones we have just discussed. However, the engineering geologist should be aware of them and determine whether or not they present problems to the engineering project. **Liquefaction** is the process by which a seemingly solid material looses all shear strength and behaves as a liquid. *Quick clays or sensitive clays* are clay materials deposited with a metastable fabric that may abruptly collapse, leading to the formation of a liquid mass of mud. Low-density sands are susceptible to liquefaction, particularly when subjected to vibrating loads, earthquakes, or high water pressures. Materials prone to **expansion** and **collapse** may either expand in volume when water is added or may abruptly collapse when saturated (Figure 2–10). Expansive materials are the single largest cause of damages to engineering structures and are discussed in detail in Chapter 7. Soils prone to collapse are common problems in arid regions where rapidly deposited silt and clay materials are left in a metastable fabric. The fine-grained clay particles tend to establish bonding forces to hold the coarser grains. Upon saturation, these bonding forces are removed, and the fabric collapses.

FLUID EARTH MATERIALS

The most important fluid earth material to the engineering geologist is water. Not only is it an important natural resource, but the presence of water can often mean the difference between a simple construction project and a disaster. Water, like any earth material, has *density* or *unit weight* and is the

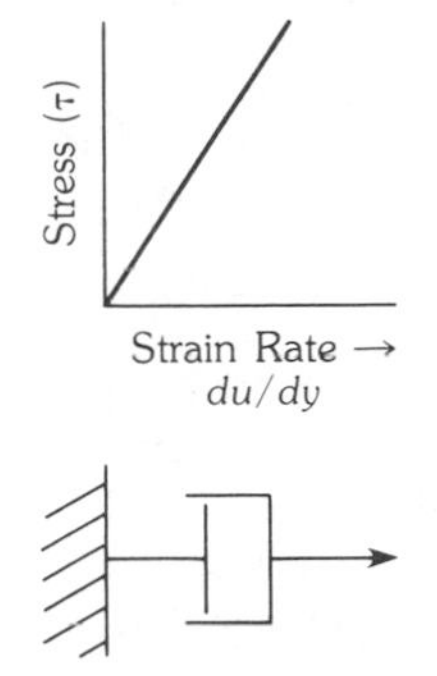

FIGURE 2–11 Viscosity of a fluid. The stress that the fluid can carry is related to the strain rate. Viscosity is defined as the slope of the stress-strain rate line. A dash pot is a mechanical model of a viscous material.

common unit for determining the *specific gravity* of all other earth materials. The density of any fluid varies as pressure or temperature conditions vary.

Viscosity (μ), a fundamental fluid property, offers resistance to the shear stress by changing the rate of deformation. Newton's law of viscosity states that for any rate of deformation of a fluid, the shear stress (τ) is directly proportional to the viscosity (μ):

$$\tau = \frac{du}{dy}$$

where du/dy = the deformation rate. (See Figure 2-11.)

Kinematic viscosity (ν) is a term used to relate the viscosity of one fluid condition to another and is expressed as

$$\nu = \frac{\mu}{\rho}$$

where ρ = the density of the reference fluid.

Fluid pressure is similar to stress in solid earth materials and is equal to force per area.

FLUID STATICS AND DYNAMICS When a fluid is at rest, it exerts a pressure on any surface (container) that is below the surface of the fluid. The pressure exerted is equal to the product of the height of the fluid column, or the **head** (h), and the density of the fluid:

$$p = \rho h$$

This pressure is often referred to as the **hydrostatic pressure.** Any material submerged in or floating on a fluid has a buoyant force acting upward on the object. The buoyant force is equal to the product of the volume of fluid displaced and the density of the fluid.

Fundamental laws of physics, conservation of mass, and conservation of energy are maintained anytime a fluid is placed into motion. The **continuity equation** requires the conservation of mass for fluid materials and states that the mass per unit time passing any two points must be equal if fluid is not lost or gained. Thus, the following relationship must be maintained in any fluid system:

$$\rho_1 V_1 A_1 = \rho_2 V_2 A_2$$

where

ρ_1 and ρ_2 are the fluid densities at points 1 and 2
V_1 and V_2 are the velocities of the fluid at points 1 and 2
A_1 and A_2 are the cross-sectional areas at points 1 and 2

If $\rho_1 = \rho_2$, the fluid is incompressible.

The continuity equation simply states that for an incompressible fluid, the quantity of fluid (volume/time $= Q$) must be constant in any fluid system if fluid is not lost or gained. Notice that the velocity (V) multiplied by the cross-sectional area (A) is the volume per time (Q).

Bernoulli's equation represents conservation of energy in fluid materials. Any fluid can be subjected to three basic energy states: (1) *pressure* is energy input from external sources; (2) *potential energy* is that energy gained from pressure head or elevation above some datum, and (3) *kinetic energy* is the dynamic energy of a moving fluid mass. In steady-state conditions where energy is not lost to heating, friction or turbulence, the sum of these three basic forms of energy must be constant in any incompressible fluid. Bernoulli's equation is

$$P_1 + Z_1 + \frac{1}{2}\left(\frac{V_1^2}{g}\right) = P_2 + Z_2 + \frac{1}{2}\left(\frac{V_2^2}{g}\right)$$

where

P_1 and P_2 are external pressures at points 1 and 2
Z_1 and Z_2 are potential energies at points 1 and 2
$\frac{1}{2}\left(\frac{V_1^2}{g}\right)$ and $\frac{1}{2}\left(\frac{V_2^2}{g}\right)$ are kinetic energies at points 1 and 2

CONCEPT 2–6

Continuity and Bernoulli's Equations

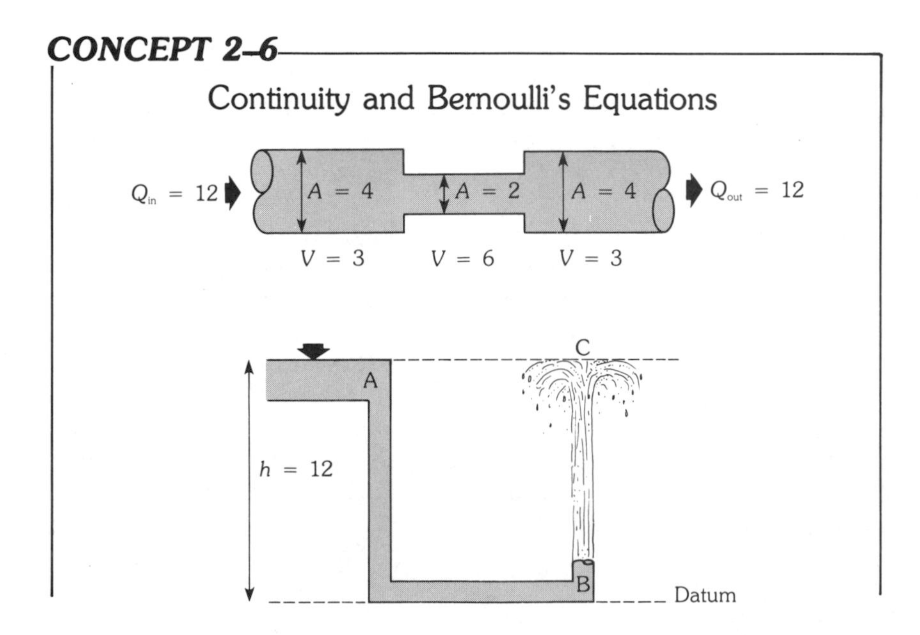

The continuity equation for an incompressible fluid states that the quantity of fluid moving past any point per unit time must be constant. In the pipe below, 12 units of fluid are flowing through the system. The flow velocity in the 4-unit area section is 3 velocity units; however, for continuity the velocity must increase to 6 units when the area is reduced from 4 to 2.

$$Q = 12 \times av = 4 \times 3 = 2 \times 6 = 4 \times 3 = 12$$

Bernoulli's equation states that energy cannot be added or taken from a fluid system. The energy in the fluid at points A and C is the same and is potential energy ($\gamma \times h$). At point B it is kinetic energy [$\frac{1}{2}(V^2/g)$]. It is the kinetic energy at B that forces the plume of water up to C. If the flow was stopped by a valve at B, the energy at B would become pressure energy equal to the potential energy at A.

Notice that the potential energy is equal to the hydrostatic pressure at each point. Concept 2–6 shows the relationships of the continuity equation to Bernoulli's equation.

Boundary shear is a measure of fluid friction between a moving fluid and the walls of the container (channel), as explained in Concept 2–7. Without boundary shear, the flow velocity in any river would soon reach infinity, since the system would be frictionless. Boundary shear along the bed of a stream causes the fluid velocity to reduce to zero at or just below the boundary (Figure 2–12).

CONCEPT 2–7

Boundary Shear

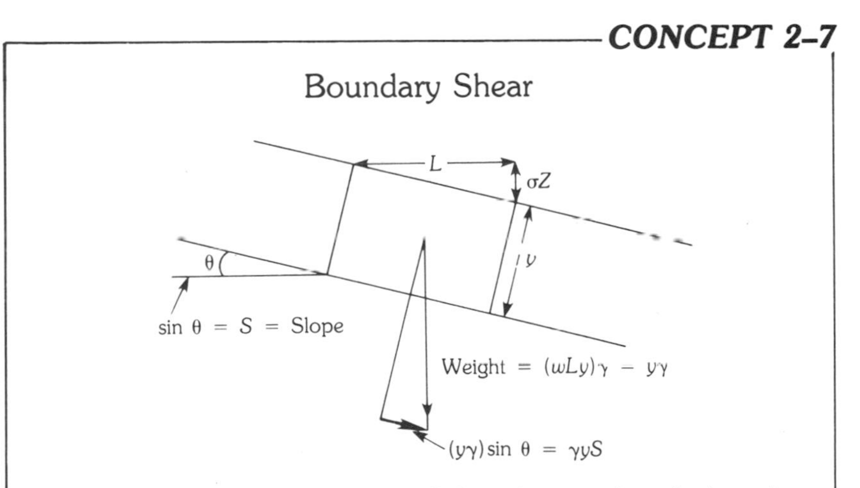

Boundary shear is the shear (frictional) force that acts along the boundary between a fluid and the channel. The boundary shear must be equal to the down-slope component of gravity whenever a fluid is flowing at a constant velocity (steady-state flow conditions). The down-slope component of gravity of a fluid with density γ and a volume (width $\times$ length $\times$ depth) is expressed as

$$\gamma(wLy) \sin \theta$$

where θ = slope angle.

If a unit width and unit length are selected, this becomes

$$\gamma y \sin \theta$$

In most channels the slope angle θ is small, and the slope ($S = \Delta Z/L$) can be substituted for $\sin \theta$. This substitution yields

$$\text{Boundary shear} = \gamma y S$$

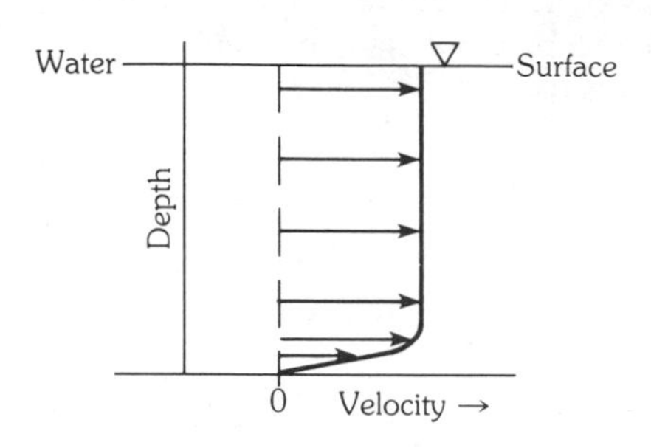

FIGURE 2–12 Velocity of flow versus depth in an open channel.

Rock Materials

three

"Between a *rock* and a hard place," the "*Rock* of Ages," "Own a Piece of the *Rock*," "Those who live in glass houses should not throw *stones*," and "Build your house upon the *rock*." Such common sayings and slogans all carry an implied definition of rock. Rocks are hard, permanent, strong earth materials. *Rock*, however, has many definitions depending upon the profession using the term. To an architect, rock is a type of building material, dimension stone (Figure 3–1). An engineer sees rock as a hard or brittle material that requires blasting to excavate or as a permanent, durable material for erosion control or other engineering uses (Figure 3–2). To a geologist, rock is an earth material produced by one or more natural, rock-forming processes (Figure 3–3). In the field of agronomy, rocks are the parent material from which natural processes produce soils (Figure 3–4).

Engineering geologists must view rocks from two different perspectives: the geologic view of rock as a material produced by the rock-forming processes, and the engineering view of a naturally occurring, hard, permanent material. Some things that are rocks to a geologist are not considered in the engineering field of rock mechanics, because the mechanical (engineering) properties of these materials are "soillike"; that is, they can be easily disaggregated. Regardless of the rock-forming process, all rocks have a common geologic basis and therefore a common geologic definition: **Rock** is any naturally formed aggregate or mass of mineral matter, whether or not coherent, constituting an essential and appreciable part of the earth's crust, or, more simply stated, rocks are natural earth materials composed of aggregates of one or more minerals, regardless of the mechanical properties of the material. Since the mineral content is a natural component of any rock, rocks are classified and identified by the mineral components and the processes that formed the minerals or accumulated the mineral aggregates and established their distribution within the rock.

FIGURE 3–1 Architectural rock, selected to enhance beauty or to project an image.

FIGURE 3–2 Engineering rock, selected for specific engineering properties to meet some design need.

FIGURE 3–3 Geologist's rock, a naturally occurring earth material formed by the rock-forming processes.

FIGURE 3–4 Agricultural rock, the parent material for soils.

ROCK-FORMING PROCESSES

There are three basic rock-forming processes: (1) cooling of molten material, (2) settling, depositional, or precipitation processes, and (3) heating or squeezing processes. These three processes form the basis for rock classification systems and are also significant factors in establishing the mechanical properties of rocks.

Rocks derived from molten material are **igneous** rocks, from the Latin *igneus* meaning fire. These rocks are usually hard and crystalline in character and are composed of minerals that formed as a molten mass cooled. Igneous rocks make up about 95 percent of the volume of the earth's crust. Settling, depositional, or precipitation processes form **sedimentary** rocks, from the Latin word *sedimentum* meaning to sit. These rocks are composed of materials that have been transported and then deposited, materials that have been precipitated from marine waters, or remains of organisms. Their mechanical character may range from very soft to very hard. Sedimentary rocks are the most widespread rock type on the earth's surface. The character of many sedimentary rocks may change over short distances. Heating to near melting temperatures, along with high pressures, causes chemical and/or physical changes in earth materials forming **metamorphic** rocks, from the Greek word *metamorphosis* meaning to transform. Metamorphic rocks are often complex, containing many minerals and having varying engineering properties.

The classification of rocks for purposes of engineering geology must be solidly based upon an understanding of how rock-forming processes control the rock's engineering properties. An igneous rock, formed from a molten mass, implies different characteristics in three-dimensional space than does a sedimentary rock that has been deposited by a river.

The dangers of extrapolating the engineering significance of a rock sample obtained in a core to the entire construction site cannot be overemphasized. The core sample is your clue; your understanding of the processes that formed the sample and its significance to the project is your challenge, because the core represents a very small fraction of the entire rock mass hidden below the earth's surface.

ENGINEERING CLASSIFICATION OF ROCKS

To the engineer, who must view rocks as an engineering material, any classification of rocks must relate to their engineering properties. There are two main engineering properties of a rock: (1) the properties of the intact, unfractured rock specimen (**rock substance**), and (2) the characteristics of the entire rock body, including fractures and other discontinuities (**rock mass**). A very strong rock material that is broken into pieces, the largest of which is only 10 cm, will behave entirely differently from the same rock in a solid mass.

ROCK SUBSTANCE CLASSIFICATION

Numerous attempts have been made to establish a simple classification system for the engineering properties of a rock substance. As you will see in Chapter 7 on soil materials, several two-letter or alpha-numerical classification schemes have been developed for soils. The problem with rock classification systems is that rock materials are highly variable and therefore the same geological rock type may have widely differing engineering properties. Rock characteristics, such as unconfined compressive strength, stress-strain relationships, failure behavior, or some modulus value could provide engineering information. In any classification system that relates to engineering properties, the properties must be measured and determined by simple diagnostic techniques. Elaborate laboratory testing procedures are expensive and difficult to use under adverse field conditions.

The simplest test for rock substance properties is the unconfined compression test or a variation of it. The unconfined compression test, however, requires careful preparation of a right cylinder sample with surface lapped ends. As a result, this test is not possible in the field unless a laboratory is available. Nevertheless, its simplicity and usual reliability have led to its general acceptance by the profession. Figure 3–5 shows a schematic diagram of the unconfined compression test.

Field tests of some compression characteristic of the rock substance that can be related to the unconfined compressive strength are being developed. The simplest one is a type of impact or rebound procedure that requires no sample preparation. Even a pick in the hand of an experienced engineering geologist is a highly valuable simple testing tool. The "rock hammer" test relates the sound, rebound, and impact marks of a hammer blow to the general strength of the rock material (Table 3–1). The necessary experience to "calibrate" the hammer simply requires some effort plus field observation of the hammer's responses compared to laboratory results of the unconfined compression test.

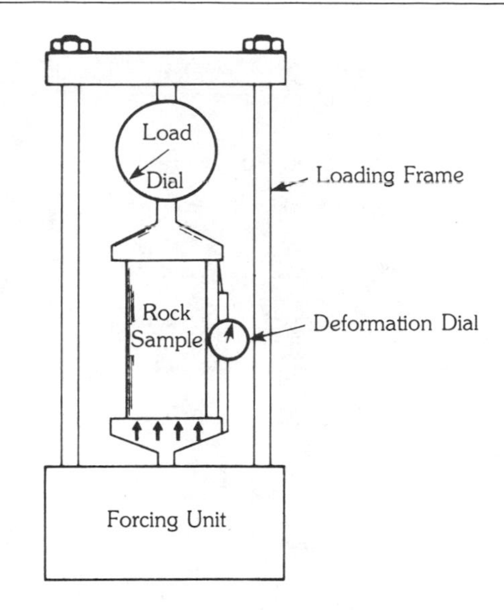

FIGURE 3–5 Diagram of an unconfined testing unit.

TABLE 3–1 Rock hammer test

Observations	Strength
Solid ring, clean rebound, no mark	Very high strength
Solid, thud, slight mark	High strength
Thud, no rebound, mark and fracture	Moderate strength
Thud, imprint of hammer, fractures	Low strength
Bury hammer, fracture	Very low strength

Impact testing, using a spring-loaded calibrated hammer, provides a more uniform and usable classification value (Figure 3–6). The principle of the impact test is the same as the hammer test except that the rebound of the impactor is measured. The harder (stronger) the rock substance, the greater the elastic rebound of the impact hammer. This tool must also be calibrated with the results of unconfined compression tests, but once calibrated, the impact hammer is a simple, light-weight engineering classification tool that is more consistent than the "hammer" test.

Point-load testing of irregular-sized rock samples is another useful field method to determine rock properties. In the point-load test (shown schematically in Figure 3–7), the rock sample is placed in a loading device, and the load required to split the sample is determined. The point load index, I_p, is:

$$I_p = \frac{F}{D^2}$$

where

F = load at rupture
D = diameter of the specimen

The diameter (D) affects the results of the test; however, calibration charts have been developed that relate D to a standard diameter (D =

FIGURE 3–6 Schmidt hammer test of a rock core for strength properties.

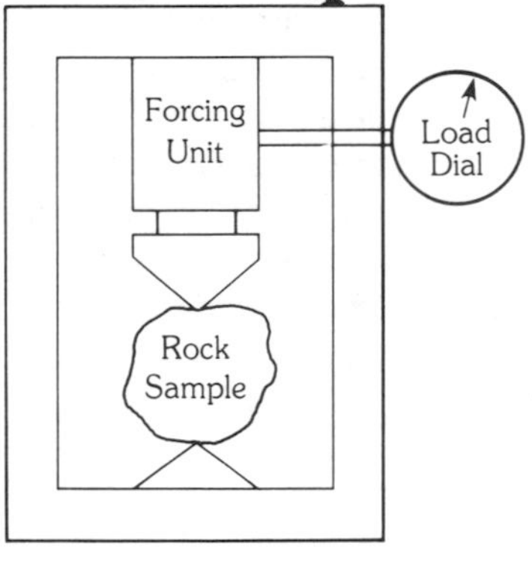

FIGURE 3–7 Diagram of a point loading test.

50 mm), so that I_{50} can be determined. An empirical relationship exists between the unconfined compression strength (σ_a) and I_{50} so that

$$\sigma_a = 24(I_{50})$$

As with any empirical relationship, the engineering geologist should run calibration tests before starting a new project area.

Regardless of the test method, the determination of unconfined compression strength provides valuable strength data for the classification of any rock substance. Table 3–2 presents one classification system for rock substances.

The determination of unconfined compressive strength does not give a complete picture because it does not define the behavior (stress-strain relationship) of the rock. In terms of engineering problems, a rock that behaves as a "true" elastic body with a brittle failure presents a significantly different condition from one that has a ductile phase prior to failure. In the first case, failure comes suddenly without warning, while the other rock fails more gradually and does give warning of impending disaster.

The determination of the stress-strain relationships of a rock is based on the unconfined compression testing program. Field determination of this property is difficult. In the field, the engineering geologist must rely on experience and knowledge of how rocks deform. The stress-strain relationships of any rock are controlled to a large degree by the mineralogy, bedding planes, banding, and other characteristics of rock structure or fabric. Figure 3–8 relates the deformation behavior (stress-strain) of various rock materials to the geologic characteristics of the material.

In the case shown in Figure 3–8A, the rock material is dense, massive, and uniform. During the loading cycle, the rock deforms elastically and does

TABLE 3–2 Classification system for rock substances

Class	Unconfined Compression Strength	Description
A	Over 32,000 psi	Very high strength
B	16,000–32,000	High strength
C	8,000–16,000	Moderate strength
D	4,000–8,000	Low strength
E	Below 4,000	Very low strength

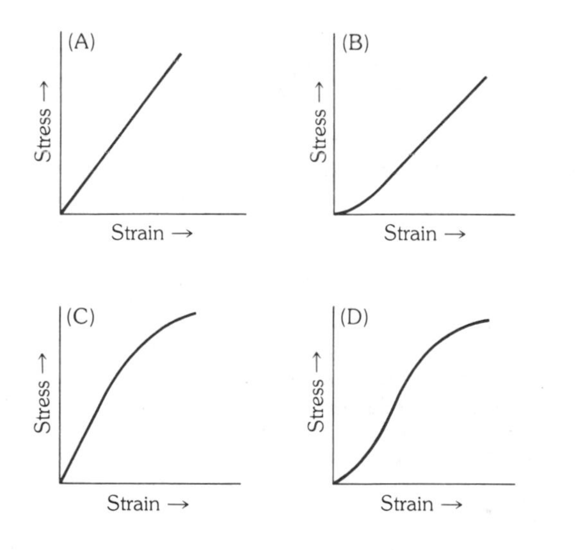

FIGURE 3–8 Stress-strain relationships of rocks. A is massive, very hard rock material; B is hard rock that undergoes some densification during initial loading; C is hard rock with mixed composition where the weaker components gradually fail; D is rock that experiences densification during initial loading and then gradual failure of weaker components (the most common stress-strain relationship of rocks).

not undergo either densification (the closing of pores, micro-fractures, and so on) or progressive failure during which individual minerals fail. The rock sample in Figure 3–8B is uniform and massive with some pore space, as shown by the initial loading, which causes nonelastic (*ductile*) deformation as the rock densifies. The final abrupt (*brittle*) failure results when all mineral components fail as a unit. Rock samples in Figures 3–8C and 3–8D differ from 3–8A and 3–8B in that they have a ductile, or progressive, failure in which the weakest minerals fail first and the failure surface progresses through the rock sample.

Since laboratory-determined stress-strain data are presented in graphic form, a verbal description of the stress-strain curve is needed to classify the data (Table 3–3).

A natural relationship exists between the unconfined compressive strength, the deformation behavior, and the engineering character of any rock substance. A very high-strength rock with elastic deformation behavior presents significantly different problems from a very high-strength rock having an elastic-plastic deformation behavior. In the first case, any failure is sudden and explosive (*rock burst*), while in the second case the plastic or ductile deformational phase warns of impending failure. Rock-burst failure presents a serious health and safety risk, since workers cannot be warned prior to the failure. A classification procedure developed by Deere and Miller (1966) utilizes a modulus ratio (M_R) to relate strength to the slope of the stress-strain curve.

$$M_R = \frac{E_{t50}}{\sigma_a}$$

where

E_{t50} = tangent elastic modulus at 50 percent ultimate strength

σ_a = ultimate compressive strength

The modulus ratio has been divided into three classes (Table 3–4); the differentiation between them is based on the observation that average rocks have a medium modulus ratio. A brittle or rigid rock with lower compressive

TABLE 3–3 Stress-strain relationships

Class	Relationship
E	Elastic
PE	Plastic-elastic (ductile-elastic)
EP	Elastic-plastic (elastic-ductile)
PEP	Plastic-elastic-plastic (ductile-elastic-ductile)

TABLE 3–4 Classes of modulus ratio M_R

Class	Description	Ratio
H	High modulus ratio	over 500:1
M	Medium modulus ratio	200:1–500:1
L	Low modulus ratio	below 200:1

Source: Deere and Miller, 1966.

strength has a high modulus ratio, while a ductile rock with higher strength has a low modulus ratio. Figure 3–9 is a plotting sheet in which the tangential Young's modulus (E_{t50}) is plotted against uniaxial compressive strength to determine the strength and modulus ratio classes.

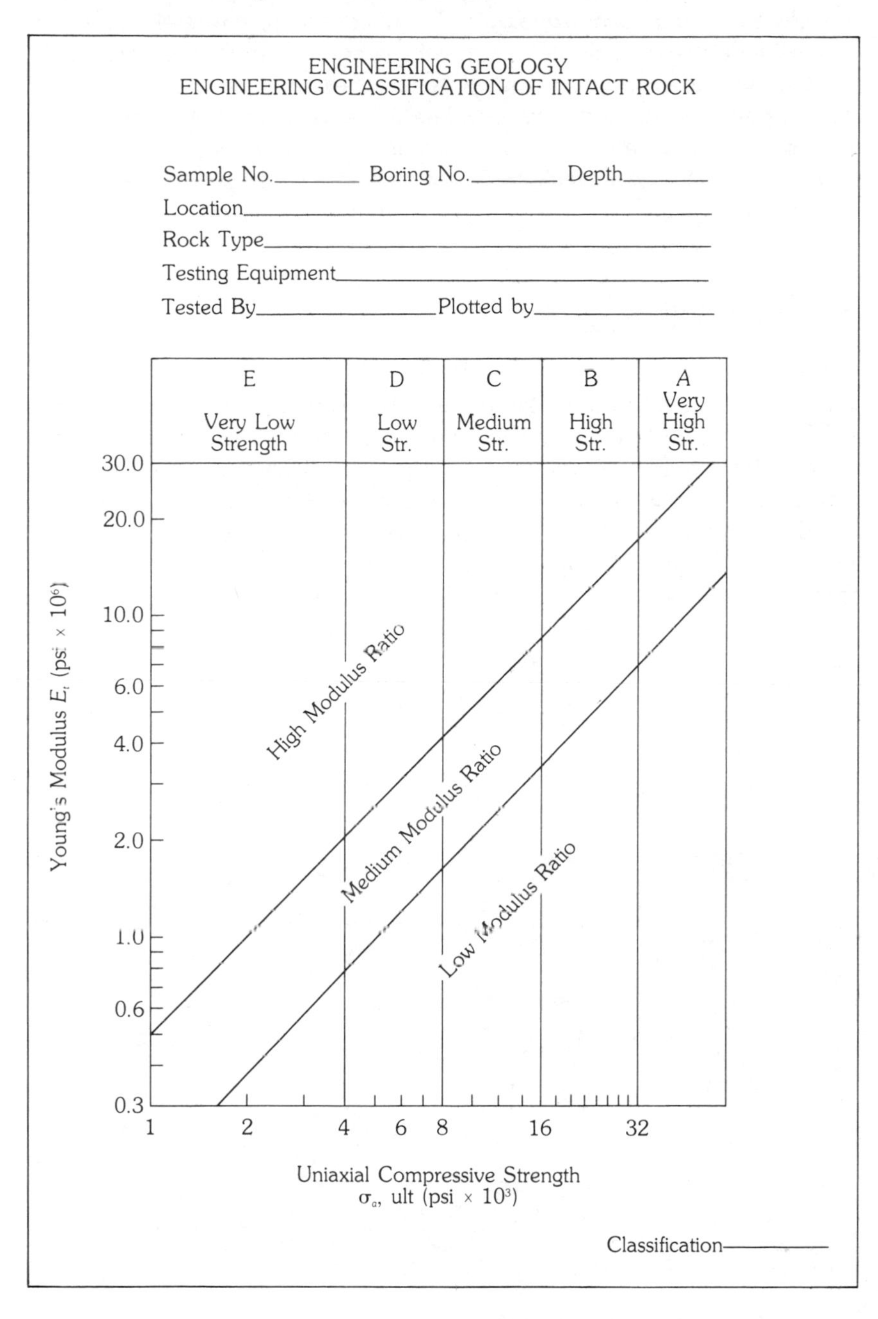

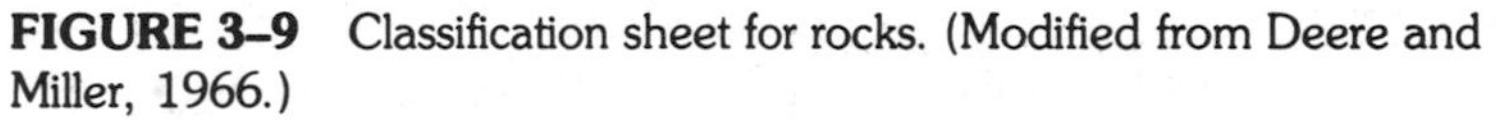
FIGURE 3–9 Classification sheet for rocks. (Modified from Deere and Miller, 1966.)

Because rock materials vary greatly, an inherited characteristic from the rock-forming processes provides a significant amount of information—the geologic name of the rock. The geologic name immediately gives the rock-forming process, mineralogy, geochemistry, and average stress-strain behavior to the engineering geologist. The unconfined compressive strength relates the degree of cementation, massiveness, and other general strength information, and the modulus ratio describes the characteristics of the mode of deformation and failure. Any complete description of a rock substance for purposes of engineering geology must, therefore, carry all three components of the classification: *geologic name, strength,* and *deformation characteristics.*

ROCK-MASS CLASSIFICATION

Just as the intact properties of a rock provide vital information about the strength characteristics and deformation behavior of a rock sample, the rock-mass properties provide vital information about the character of the entire rock in the field. A very high-strength, highly fractured rock with the largest piece only ten centimeters on a side and a medium modulus ratio would behave differently than the same rock would if it were massive and unfractured (Figure 3–10). Rock-mass classification is based on field observation and is generally reported in a descriptive form. Since numerous workers have proposed different terms for rock-mass properties, a table defining the terms used should be included in any report. Table 3–5 gives common rock-mass terms.

Observation of the rock in the field is not commonly possible because it is covered with soil or the area of interest is below ground level, as in a tunnel or deep excavation. Rock properties are, however, important to a

FIGURE 3–10 Rock structure (discontinuities). Notice the massive nature of the rocks on the left compared with those on the right.

design engineer working on a tunnel project, for example. In such cases the site investigation for the project will involve drilling cores to obtain rock samples for classification and testing (Figure 3–11).

To provide a simple, direct means to indicate rock-mass properties, Deere (1968) developed the **rock quality designation (RQD),** which is based on the core recovery. *Core recovery* is the ratio of the length of core recovered to the length drilled and ranges from 0 percent for no core recovery to 100 percent for total recovery. The RQD is a modification of core recovery, in that only the intact pieces of core that are more than ten centimeters long are added together in calculating length recovered. RQD

TABLE 3–5 Rock-mass description

Description	Average Fracture Spacing
Massive bedding	Bedding planes greater than 2 m apart
Layered	Bedding planes less than 2 m apart
Very close fractures	Less than 5 cm apart
Close fractures	5 cm to 30 cm apart
Moderately wide fractures	30 cm to 1 m apart
Wide fractures	1 m to 3 m apart
Very wide fractures	Greater than 3 m apart
Very broken rock	Less than 5 cm between fractures
Broken rock	8 cm to 30 cm between fractures
Blocky rock	30 cm to 2 m between fractures
Solid rock	Greater than 2 m between fractures

FIGURE 3–11 Rock core from the Downey Slide in British Columbia, Canada. Notice the high degree of fracturing of the rock.

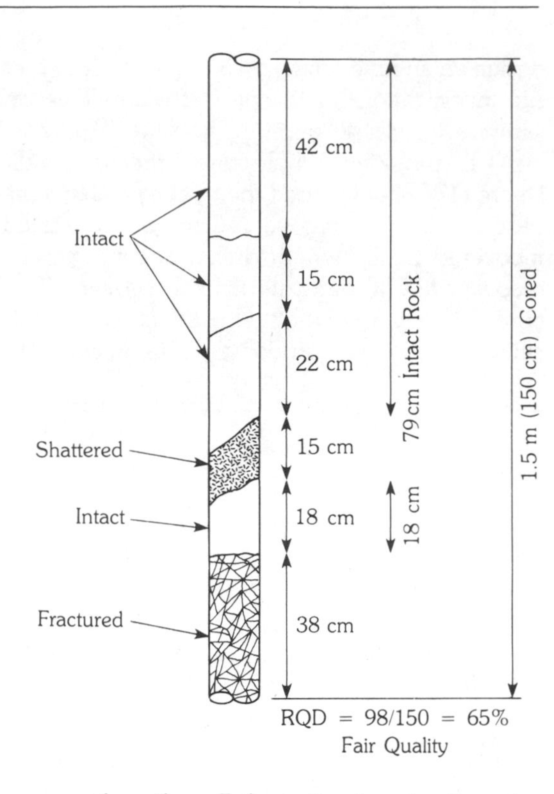

FIGURE 3–12 Rock Quality Designation (RQD).

should not be applied to a core less than 5.4 centimeters in diameter because smaller cores are frequently broken by the coring operation and consequently yield a false RQD. Figure 3–12 shows a sample RQD determination. RQD is related to rock-mass properties in Table 3–6.

Further information, in addition to the spacing of fractures, is also important to the engineering geologist because fractures may represent planes of weakness where undesirable movements can take place. In rocks of medium strength or greater, failures are often related to the shape, continuity, and orientation of fractures. A complete rock-mass description must, therefore, include the characteristics of the fractures as well as their spacing, infilling, smoothness, and width. Figure 3–13 is a descriptive guide to fracture character.

The classification and description of rocks may sound like just so much memorization, but without a complete description or classification you do

TABLE 3–6 RQD descriptions

RQD	Description	Approximate Equivalent Fracture Spacing
0–25%	Very poor	Very close
25–30	Poor	Close
30–75	Fair	Moderately wide
75–90	Good	Wide
90–100	Excellent	Very wide

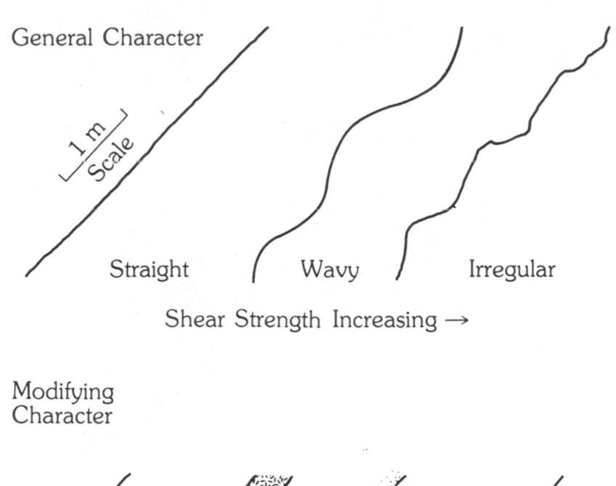

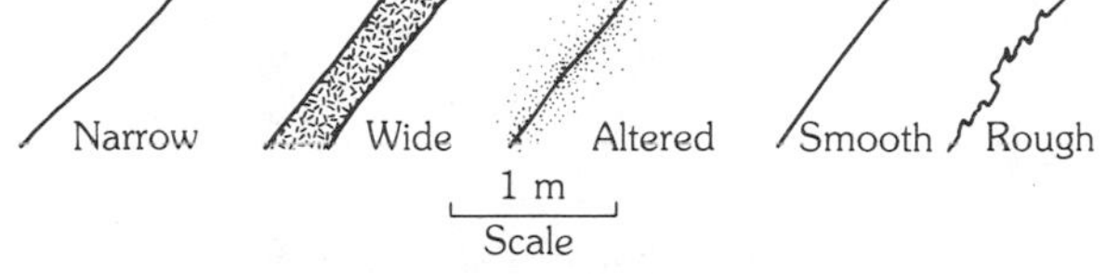

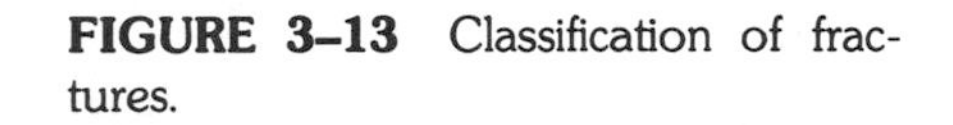
FIGURE 3–13 Classification of fractures.

not have a means of communicating your observations to anyone else. The complete description of a rock material must include the *geologic name* (genetic and mineralogic information), the *rock substance classification* (strength and deformation characteristics), and the *rock-mass description* (continuity of the material). If you omit any of these components from the description, the user of your information is at a serious disadvantage.

Igneous Rocks

four

ORIGIN OF IGNEOUS ROCKS

As the definition states, igneous rocks formed from a molten mass when it cooled. The characteristics of the rock are controlled by two basic factors: the rate of cooling and the chemical composition of the molten mass. The cooling rate is further controlled by the geometry (shape) of the mass and amount of insulation surrounding the melt during the cooling phase.

The cooling rate controls the size of the mineral crystals that form in the melt. Thus, igneous rocks with similar chemical characteristics (similar mineral composition) can be glassy or can have crystal sizes ranging from very fine to very coarse, or a combination. Crystal size is a diagnostic feature and is therefore one component of the classification of igneous rocks. Rapid cooling precludes the growth of crystals, while slow cooling allows their growth. **Extrusive igneous rocks** form on or near the surface of the earth and cool rapidly because they are not well insulated; they are fine crystalline to glassy rocks. **Intrusive igneous rocks** are intruded into deeper parts of the crust, insulated, and cooled slowly; they are medium to coarse crystalline rocks. Figure 4–1, a geological cross section of the crust, shows the environments of formation of igneous rocks. Notice that thin, rapidly cooled bodies containing glasses and fine crystals are characteristic of extrusive igneous rocks, while thick, slowly cooled bodies containing medium to coarse crystals are characteristic of intrusive igneous rocks.

Simply by knowing that the rock is extrusive igneous, the engineering geologist can estimate the characteristics of the rock body: It is thin and possibly discontinuous in nature, is complex in character, and may exist over a limited area (Figure 4–2). An intrusive igneous rock, on the other hand, can be expected to be thick and more uniform in character and to extend over a large area (Figure 4–3). At a construction site, an extrusive rock would probably require more extensive exploration than an intrusive rock in order to define its areal extent and characteristics.

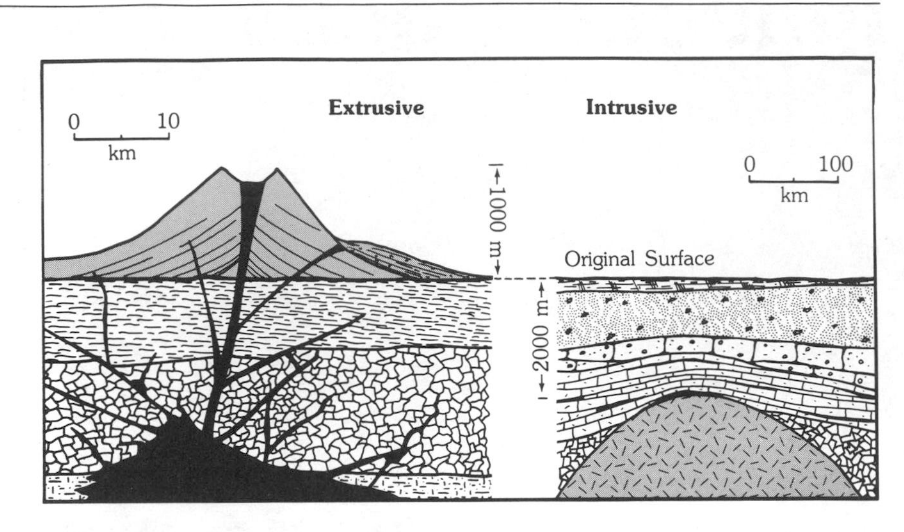

FIGURE 4–1 Environments of formation of extrusive and intrusive igneous rocks.

FIGURE 4–2 Extrusive igneous rock (designated by arrow). A thin flow covers older rock units.

GEOCHEMISTRY OF IGNEOUS ROCKS

The chemical composition of the molten mass is the second factor in the classification of igneous rocks, because the chemistry controls the rock's mineralogy. Since igneous rocks are derived from a molten mass of earth material, the basic elemental composition of the earth establishes the chemistry of the melt. Almost 98 percent of the earth's crust is composed of only eight elements, as Table 4–1 shows. This very limited list of common elements in the earth's crust has a significant chemical impact on the number of different variations that can form, particularly when one realizes that the

FIGURE 4–3 Intrusive rock body. The entire mountain range in this photograph is composed of the same type of rock.

TABLE 4–1 Elemental composition of the earth's crust

Element	Symbol	Ion	% of Crust
Oxygen	O	$O^{=}$	49.52
Silicon	Si	Si^{+4}	25.75
Aluminum	Al	$Al^{+2,+3}$	7.51
Iron	Fe	$Fe^{+3,+4}$	4.70
Calcium	Ca	Ca^{+2}	3.39
Sodium	Na	Na^{+}	2.64
Potassium	K	K^{+}	2.40
Magnesium	Mg	Mg^{+2}	1.94
		Total	97.85%
	All other elements		2.15
			100.00%

only *anion* in this list (negatively charged ion) is oxygen. The next most abundant anion is sulphur (0.048 percent).

A **mineral** is defined as a natural, inorganic substance composed of one or more elements with a unique chemical composition, arrangement of elements (*structure*), and distinctive physical properties. Oxygen can combine with the other seven common elements to form the *oxides;* however, the abundance of the *silicates,* oxygen (49.52 percent) combined with silicon (25.75 percent), exceeds that of the oxides by a significant amount. The silicates are therefore the common igneous rock-forming minerals. There are six of them: feldspars, quartz, amphiboles, pyroxenes, micas, and olivine.

FELDSPARS

The feldspar minerals are potassium, sodium, calcium, and aluminum silicates and are usually light in color. These minerals comprise about 59 percent of the "average" igneous rock. The feldspar family is divided into two basic minerals: *orthoclase,* a potassium-aluminum silicate, and *plagioclase,* a calcium-sodium aluminum silicate. The plagioclase feldspars form a solid solution series between the calcium and sodium end members.

Orthoclase:	$KAlSi_3O_8$
Hardness:	6
Density:	2.56
Crystal:	Monoclinic, usually short prismatic

Plagioclase:	$NaAlSi_3O_8 \longleftrightarrow CaAl_2Si_2O_8$
Hardness:	6
Density:	2.62–2.76
Crystal:	Triclinic, usually tabular, commonly as irregular grains and cleavable masses

QUARTZ

Quartz makes up about 12 percent of the "average" igneous rock and is the second most abundant mineral. Quartz is a silicon oxide that is usually light to clear in color.

Quartz:	SiO_2
Hardness:	7
Density:	2.65
Crystal:	Trigonal, commonly prismatic

AMPHIBOLES

The amphibole family is a large group of closely related minerals that make up about 8 percent of an "average" igenous rock. These minerals are complex calcium, magnesium, iron, and aluminum silicates that have doubled chains of linked SiO_4 tetrahedrons. This double chain results in the formation of a generally simple, six-sided, long, prismatic crystal. Hornblende is the common amphibole mineral.

Hornblende:	$NaCa_2\ (Mg,Fe,Al)_5\ (Si,Al)_8O_{22}(OH)_2$
Hardness:	6
Density:	3.0–3.4
Crystal:	Monoclinic, sometimes in prismatic crystals, also as irregular grains and massive

PYROXENES

The pyroxene family includes a large group of related complex calcium, magnesium, iron, and aluminum silicates that make up about 8 percent of an "average" igneous rock. This mineral family differs from the amphibole family in that it has only a single chain of SiO_4 tetrahedrons, producing a more complex, four- or eight-sided, short, prismatic crystal form. The common pyroxene mineral is augite.

Augite: $Ca(Mg,Fe,Al)(Al,Si)_2O_6$
Hardness: 6
Density: 3.25–3.55
Crystal: Monoclinic, short prismatic crystals, often square in cross section, also granular and massive

Notice the difference in the chemical relationship between the aluminum-silicon-oxygen ratio in hornblende (Si:O = 4:11) and augite (Si:O = 1:3), reflecting the double chain versus the single chain of SiO_4 tetrahedrons.

MICAS

The mica family includes layer silicate minerals that form "sheetlike" hexagonal crystals. Muscovite (white mica) is a light-colored, hydrous, potassium aluminum silicate while biotite (black mica) is a dark-colored, hydrous, iron-magnesium potassium aluminum silicate. The mica family makes up about 8 percent of an "average" igenous rock; both forms of mica are found in igneous rocks.

Muscovite: $KAl_2(AlSi_3O_{10})(OH)_2$
Hardness: 2.5 on cleavage, 4 across cleavage.
Density: 2.8–2.9
Crystal: Monoclinic, usually in layered masses or small flakes, crystals tabular with a hexagonal outline

Biotite: $K(Mg,Fe)_3(AlSi_3O_{10})(OH)_2$
Hardness: 2.5 on cleavage
Density: 2.8–3.4, increasing with iron content
Crystal: Monoclinic, crystals are pseudohexagonal prisms but habit is more commonly layered plates without crystal outline

OLIVINE

Olivine is a common mineral in some igneous rocks (usually extrusive) and constitutes about 4 percent of an "average" igneous rock. Chemically, olivine is a magnesium, iron silicate mineral and is usually green in color.

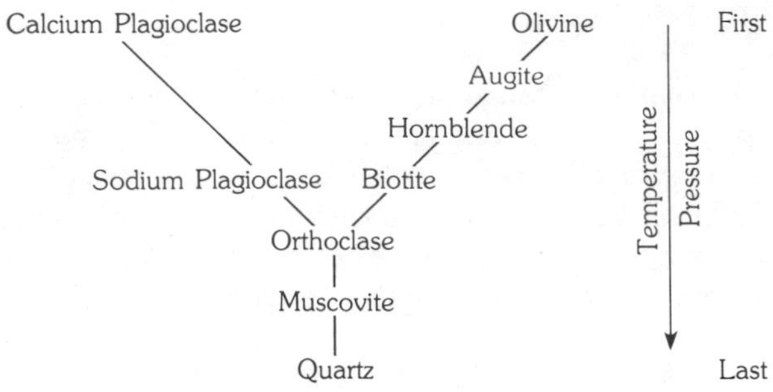

FIGURE 4–4 Bowen reaction series.

Olivine:	$(Mg,Fe)_2SiO_4$
Hardness:	6.5
Density:	3.22–4.39
Crystal:	Orthorhombic, rarely as crystals, usually in granular masses, and as rounded grains in igneous rocks

These six families are the basic mineral components of igneous rocks and have specific temperature and pressure relationships of formation. Some of them form at higher temperatures than others and may be redissolved in the melt as it slowly cools. These early-formed minerals may also settle to the bottom of the molten mass and therefore increase the potassium-silicon-oxygen ratio of the melt. The temperature relationships were worked out by Bowen and are known as the Bowen reaction series (Figure 4–4). High-temperature minerals, formed early, are related to each other as are low-temperature minerals, formed later. This series explains, for those interested in the science of igneous rock formation, why specific minerals appear together to form one rock type while others form another rock. The Bowen reaction series is important to the engineering geologist because it relates the geochemical environment that existed when the mineral formed and therefore is helpful in understanding mineral weathering and decay.

ENGINEERING GEOLOGIC CLASSIFICATION OF IGNEOUS ROCKS

Igneous rocks have two easily identified diagnostic features: *crystal size,* related to the origin of formation (intrusive or extrusive), and *mineralogy,* related to the geochemistry of the melt. As the Bowen reaction series shows, as temperature decreased during slow cooling, the minerals formed were lighter in color. Thus color is a reasonable representation of mineral composition. Color, however, may not always be a diagnostic feature, especially for rarer forms of igneous rocks.

Figure 4–5 shows a classification system of igneous rocks for most engineering geologic purposes. (See Figure 4–6 for a geologic classification.) Two igneous rocks constitute by far the majority: **granite** and **basalt.** Granite is an intrusive, light-colored, coarse crystalline rock resulting from a

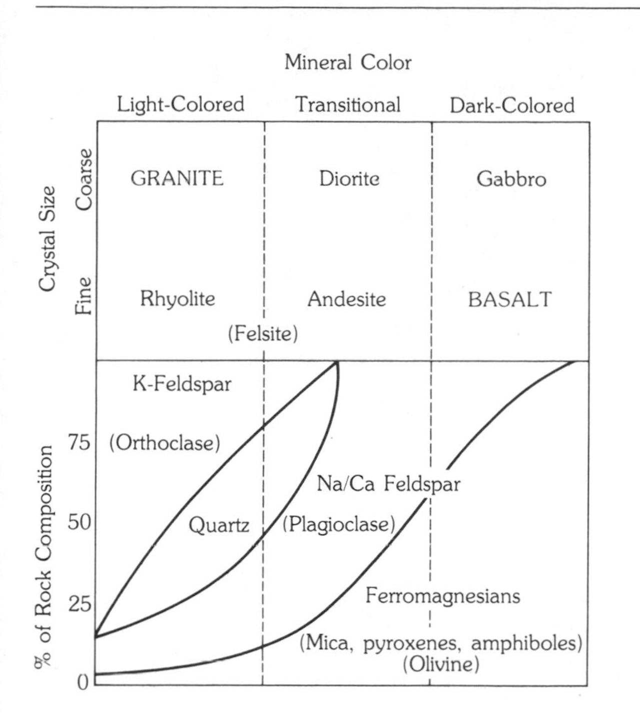

FIGURE 4–5 Classification of igneous rocks for engineering geology.

very slow cooling process, while basalt (lava) is an extrusive, dark-colored, fine crystalline to noncrystalline rock resulting from a rapid cooling process. In cases where the engineering geologist cannot identify the exact classification of an igneous rock in the field, the general terms **granitic** or **basaltic** are used. These general determinations are still valuable because the origin and chemical (mineralogic) composition are indicated, and the engineering geologist can keep in mind a general picture of the rock.

Igneous rocks, having formed during the cooling of a molten mass, are characteristically composed of interlocking crystals. Figure 4–7A is a microscopic view of the crystal interlocking in a granite. Notice that there are no pore spaces between the crystals and that the pattern of crystal orientation is random. Figure 4–7B, by contrast, is a microscopic view of a basalt that cooled rapidly, so that the crystals are very small and of nearly uniform mineralogy. In some basalts, spherical voids form in the rock as a result of gas trapped in the molten mass by the rapid cooling. These two figures suggest that intrusive igneous rocks should be more uniform in their engineering properties (strength and modulus ratio) than extrusive igneous rocks. Figure 4–8 is a generalized plot of the distribution of the engineering properties of igneous rocks. Notice that intrusive rocks are characteristically high- to very high-strength rocks while the extrusive rocks range from very low to very high strength.

The engineering geologic characteristics of igneous rocks summarized in Table 4–2 can be related to the environment of formation of the rock. Intrusive igneous rocks are high to very high strength materials that usually have elastic-plastic, stress-strain characteristics, regardless of their geologic

Megascopic Classification of the Igneous Rocks

General Color	Light-Colored (except certain glasses)			Intermediate		Dark-Colored (except Anorthosite and Dunite)								
								Chiefly			Practically All			
Dominant Minerals	Alkali Feldspars (Orthoclase and/or Microcline)			Na/Ca Plagioclase, Hornblende		Pyroxene Ca/Na Plagioclase		Pyrox. or Hornbld.	Olivine	Ca/Na Plag.	Pyrox-ene	Horn-blende	Oliv.	Ca/Na Plag.
Additional Minerals	± Biotite ± Hornblende			± Biotite ± Pyroxene		± Hornblende ± Mgt. or Ilmenite		Ca/Na Plagiocl. ± Oliv. ± Mgt. or Il.	± Pyroxene ± Mica ± Mct. ± Crm.	± Pyrox. ± Mgt. or ± Il.				
	+ Quartz − Fspds.	− Quartz − Fspds.	− Quartz + Fspds.	+ Quartz	− Quartz	− Oliv.	+ Oliv.							
Texture Pegmatitic	Granite Pegmt.	Syenite Pegmt.	Fspdl. syen. Pegmt.	Qtz. dior. Pegmt.	Diorite Pegmt.	Gabbro Pegmt.	Oliv. Gabbro Pegmt.	Perknite Pegmt.	Peridotite Pegmt.	Anorthosite Pegmt.				
Granitoid (equigranular)	Granite	Syenite	Fspdl. Syenite	Quartz diorite	Diorite	Gabbro	Olivine Gabbro	Perknite	Peridotite	Anorthosite	Pyroxenite	Horn-blendite	Dunite	Anorthosite
Porphyritic: Gm. Phaneritic	Granite Porphyry	Syenite Porphyry	Fspdl. syen. Porphyry	Qtz. dior. Porphyry	Diorite Porphyry	Gabbro Porphyry	Oliv. Gab. Porphyry	Perknite Porphyry	Perid. Porph (Kimberlite)	Porphyritic rocks seldom if ever occur in these clans.				
Porphyritic: Gm. Aphanitic	Rhyolite Porphyry	Trachyte Porphyry	Phonolite Porphyry	Dacite Porphyry	Andesite Porphyry	Basalt Porphyry	Oliv. Basalt Porphyry	Augitite Porphyry	Limburgite Porphyry					
Porphyritic: Gm. Aphanitic Very few Phenocrysts	Rhyolite	Trachyte	Phonolite	Dacite	Andesite	Basalt	Oliv. Basalt	Augitite	Limburgite					
Aphanitic	Felsite					Basalt				Aphanitic rocks do not occur in these clans.				
Porphy.-glassy	Vitrophyre							Glassy rocks do not occur in these clans.						
Glassy	Obsidian Pitchstone Perlite Pumice					Tachylite — — — — Scoria		Massive, smooth, glassy luster, prominent conchoidal fracture. Luster resembles tree pitch or gum. Concentric cracks make pearllike nodules. Very vesicular.						
Frag.: Unconsolidated	Blocks, bombs, cinders (lapilli), ash, dust							Named according to grain size, regardless of composition.						
Frag.: Consolidated	Volcanic breccia (angular) or agglomerate (rounded), and tuff													

Fspdl. Feldspathoidal; Fspds. Feldspathoids; Mgt. Magnetic; Il. Ilmenite; Crmt. Chromite; Oliv. Olivine;
Gm. Groundmass; Frag. Fragmental; Pegmt. Pegmatite; Syen. Syenite; Dior. Diorite; Gab. Gabbro; Porph. Porphyry.

FIGURE 4–6 Complete classification of igneous rocks. (Reprinted by permission, H. Blank, Texas A&M University.)

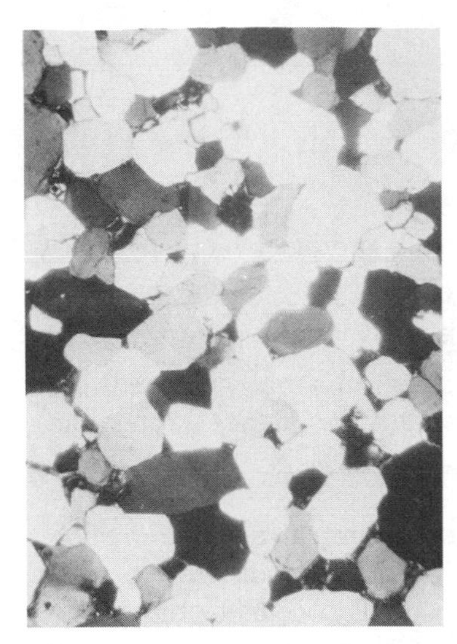

(A)

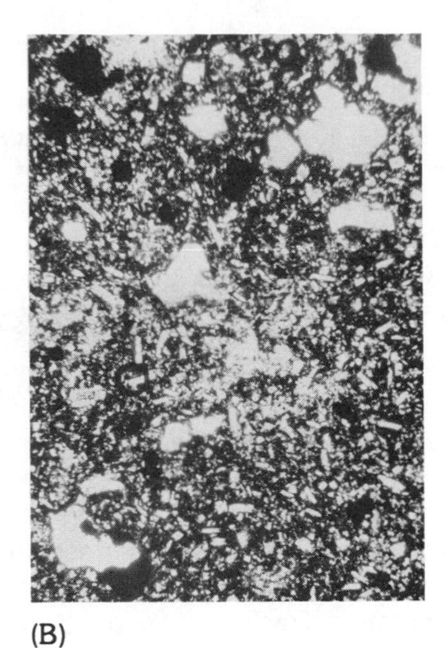

(B)

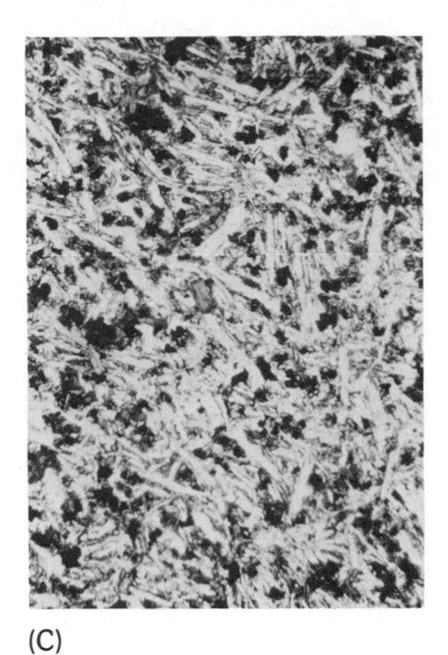

(C)

FIGURE 4–7 Photomicrographs of (A) an intrusive igneous rock and (C) an extrusive igneous rock showing the significant difference in crystal size. Photomicrograph (B) shows an igneous rock with two crystal sizes (porphyritic texture), indicating two cooling rates. Photographs are the same scale.

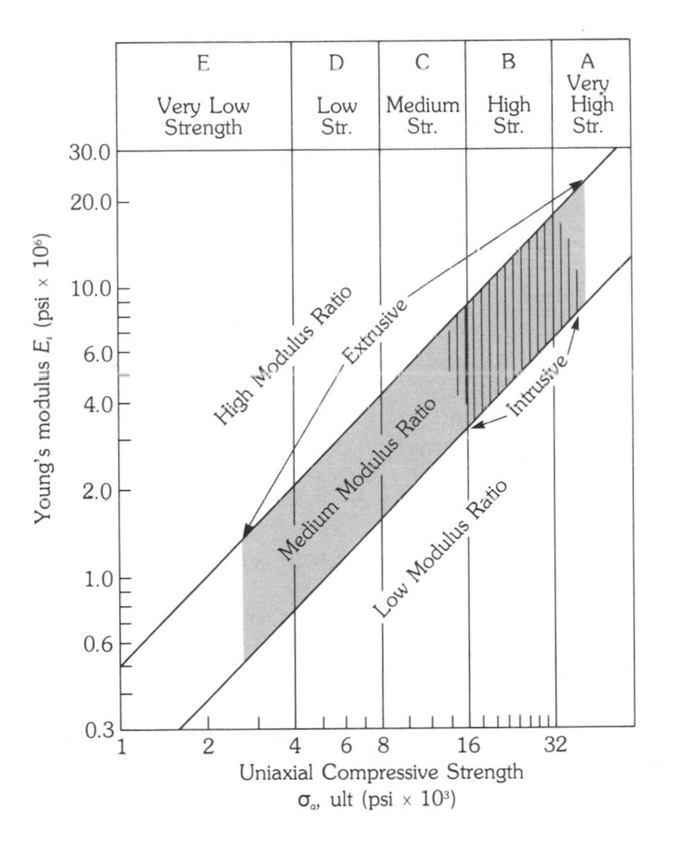

FIGURE 4–8 Engineering classification of igneous rocks. (After Deere and Miller, 1966.)

name. These engineering geologic characteristics are the result of the following factors:

1. Interlocking of high strength crystals produces high strength.
2. Crystal interlocking does not allow pore space to develop; thus the stress-strain relationship is elastic when the rock is loaded.
3. Plastic deformation at failure occurs because the weakest crystals in the rock fail first and a failure surface progresses through the rock (*progressive failure*).

Extrusive igneous rocks have highly variable strength and deformation characteristics because they are formed under variable conditions at or near the earth's surface. Extrusive igneous rocks generally occur as porous, low density materials (ash, tuff, and pumice) or as massive, high density materials (flows, sills, and dikes). Porous units have very low strength and either plastic-elastic or elastic deformation characteristics depending upon the character of the rock. These engineering geologic characteristics result from the following conditions:

1. Loading causes the thin walls of the pores to fail, thus producing a very low strength rock.

TABLE 4–2 Engineering geologic characteristics of igneous rocks

Characteristic	Description		
Environment of formation	Intrusive igneous rock		
Regional extent	Large area, occasionally can occur as small bodies		
Geologic rock name	Granite	Diorite	Gabbro
General rock color	Light to pink	Gray	Black
Crystal size	Large; visible to the unaided eye		
Rock substance properties			
(a) Rock strength	High to very high strength (16,000 psi or greater)		
(b) Deformation characteristics	Usually elastic-plastic		
Environment of formation	Extrusive igneous rock		
Regional extent	Highly variable, occurring as flows or shallow intrusions		
Geologic name	Rhyolite	Andesite	Basalt
General rock color	Light to pink	Gray	Black
Crystal size	Fine grained; frequently not visible to unaided eye		
Rock substance properties			
(a) Rock strength	Variable from very low to very high strength		
(b) Deformation characteristics	Variable from plastic-elastic to elastic		

2. If the pores in the rock are not fractured and well formed, the rock will behave as an elastic body because no densification can occur.
3. If the rock is fractured or the pores are poorly formed, the rock will initially densify, producing plastic deformation upon loading.
4. Elastic behavior at failure is common because the rock does not have a mixture of weak and strong crystals.

In contrast, massive extrusive igneous rocks are very high strength materials with an elastic deformation behavior because

1. Interlocking high strength microcrystals provide high strength.
2. Interlocking of crystals does not allow pore spaces to develop; thus elastic behavior is expected upon loading.
3. A lack of mixture of weak and strong crystals does not allow a gradual failure; thus the entire rock fails as a single unit (*general failure*).

Sedimentary Rocks

five

ORIGIN OF SEDIMENTARY ROCKS

As their definition implies, sedimentary rocks were derived from the process of settling, deposition, or precipitation of earth materials in a fluid medium. Sedimentary rocks have been divided into three general classes. **Clastic** sedimentary rocks are composed of particles eroded from other rocks, transported to a site, and deposited. **Organic** sedimentary rocks are formed from the accumulation of organic matter from either plants or animals. **Chemical** sedimentary rocks are derived from chemical processes such as evaporation or replacement.

ENVIRONMENTS OF DEPOSITION OF CLASTIC SEDIMENTARY ROCKS

Clastic rocks are the most common sedimentary type and are generally classified by the size of the grains (particles) that make up the rock. Since clastic sedimentary rocks are composed of particles that have been transported and deposited, the grain size of the particles is controlled by the kinetic energy of the transport media. All transport mechanisms, including sliding, rolling, ice movement, flowing water, and wind, are basically driven by gravity. Each has unique features and characteristics that are imprinted in the geologic record, so that frequently identification of the environment of deposition is possible. From this information, the engineering geologist can predict the geologic aspects of a site.

The degree of sorting (grain-size distribution) and the shape of a sedimentary rock particle is an indication of transport distance. The relationship between different sedimentary rock units, both vertically and horizontally, is an important indicator of the environment of deposition. Grain size is directly related to the kinetic energy of the depositional environment. Sedimentary structures, orientation, and shape of the beds reflect the active processes during the time of deposition. Table 5–1 shows the general relationship between the characteristics of a clastic sedimentary rock and its transport mechanism.

Environments of deposition of clastic sedimentary rocks range from talus deposits at the base of mountains to deep oceanic deposits far from shore. **Talus** deposits are characterized by jumbled bedding, and, depending upon the materials making them up, range from well sorted to poorly sorted. As shown schematically in Figure 5–1, they are the result of the accumulation of a wide variety of materials deposited at the base of a

TABLE 5–1 Character of clastic sedimentary rocks

General Characteristic	
Sorting: Well sorted	Long travel distance
Poorly sorted	Short travel distance
Grain Size: Fine to very fine	Low kinetic energy
Medium to large	High kinetic energy
Grain Shape: Well rounded	Long travel distance
Angular	Short travel distance

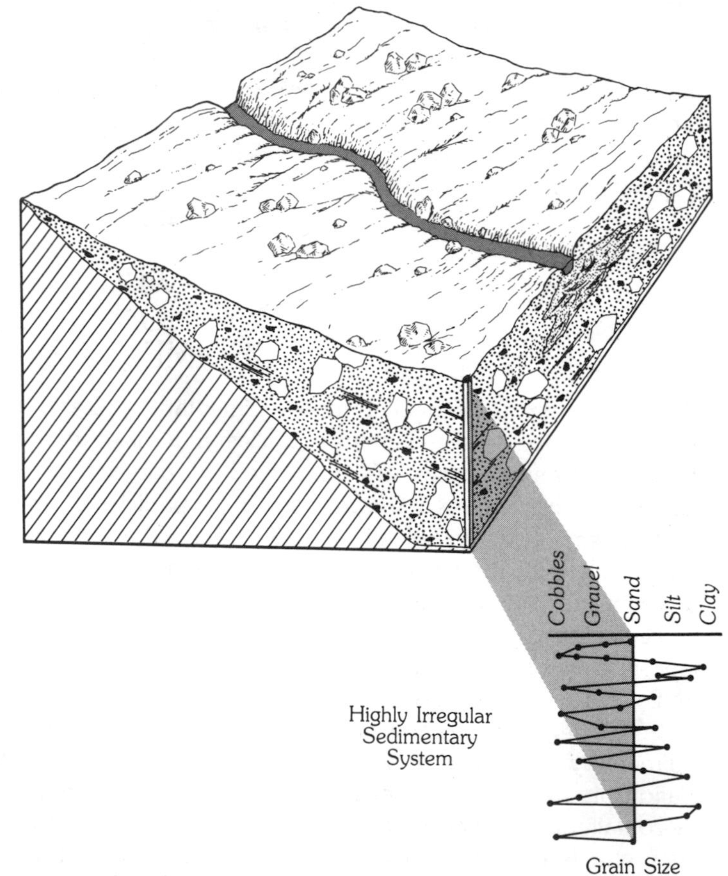

FIGURE 5–1 Characteristics of clastic sediments deposited at the base of a mountain as talus.

mountain slope. Notice the fine-grained sediments associated with the stream and the poorly sorted slide deposits throughout the area.

Sediments can be eroded and transported away from these mountain-based deposits by wind or running water. Wind transportation **(aeolian processes)** is most active in areas of low rainfall where vegetation is sparse. A desert valley is shown schematically in Figure 5–2. Notice the underlying talus deposits covered by finer-grained wind-blown sands and silts. The vertical variation in grain size is the result of variation in wind energy.

Stream sedimentation **(fluvial processes)** is controlled by the stream gradient, quantity of flowing water, and amount of sediment being transported. These three basic factors lead to two general types of stream channels. A **braided stream**, a series of channels and islands (Figure 5–3), generally forms in areas of steep gradient and high sediment load. Deposits left by braided channels are highly irregular, interfingering lenses of gravels, sands, and silts. **Meandering streams** develop where the stream gradient is low and the quantity of water is large compared to the sediment load. Sediments deposited in valleys of meandering streams are also irregular deposits of gravel and sand surrounded by finer-grained floodplain **(overbank)** deposits (Figure 5–4).

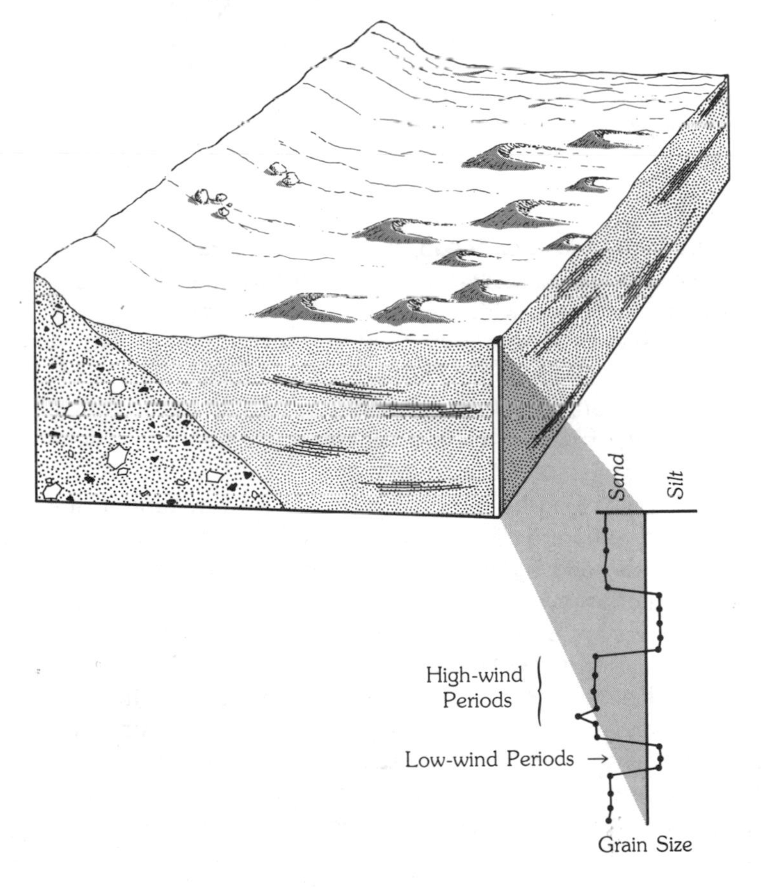

FIGURE 5–2 Environment of deposition of windblown (aeolian) clastic sediments.

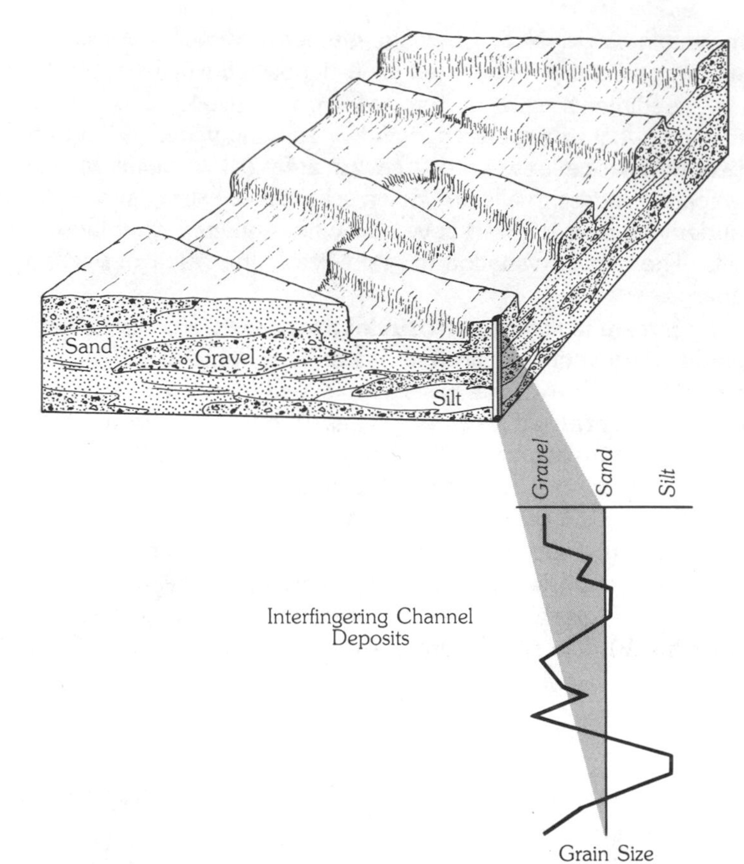

FIGURE 5–3 Environments of deposition of clastic sediments by a braided channel.

As a stream enters a lake or ocean, the gradient decreases, and the sediments that it carries are deposited, forming a **delta.** Since the kinetic energy of a system decreases with increasing distance from the shore, sediments are deposited in a sequence from coarse- to fine-grained. A vertical section downward through a delta starts with floodplain deposits of fine sediments and organics, overlying channel sands, then pro-delta, fine-grained sediments. Figure 5–5 is a schematic diagram of a delta showing the general sequence of sediments.

Wave forces and longshore currents erode and redeposit the sediments in the delta and form coastal sedimentary features (Figure 5–6). **Bars, barrier islands,** and **spits** are accumulations of coarse-grained sediments that have been moved along the shoreline. These features reduce the incoming wave energy and allow for the deposition of fine-grained sediments in **lagoons** and **tidal flats.**

Finally, as the kinetic energy of the system approaches zero, the very fine-grained materials are deposited. Very low-energy environments include **lakes** and deep ocean basins **(marine).** The sediments in these environments are characterized by continuous layers of clays and silt-sized particles that are limited only by the size of the lake or ocean basin.

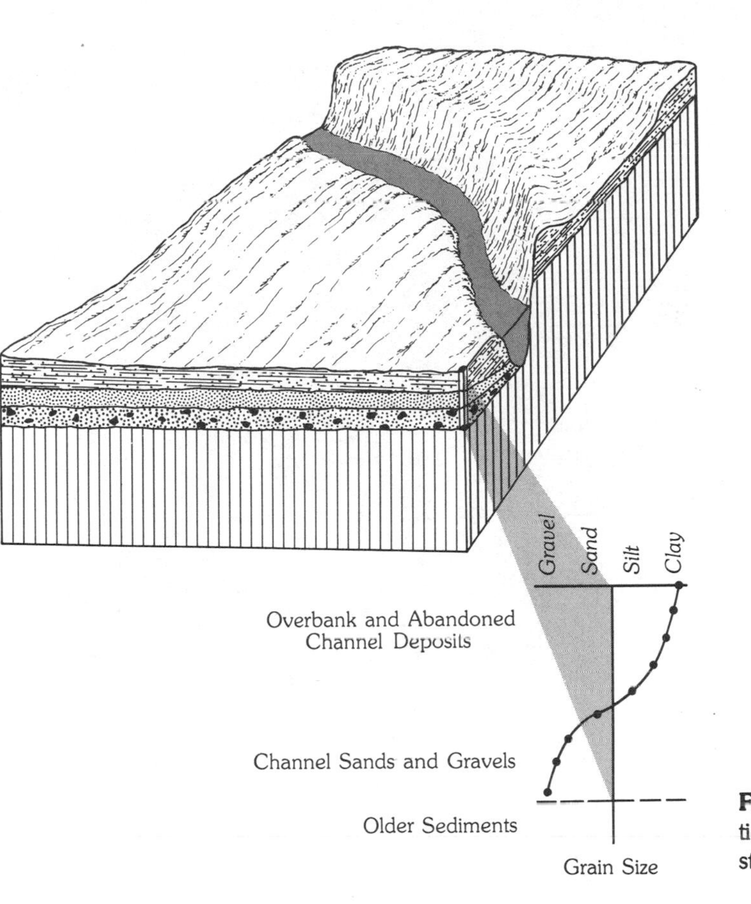

FIGURE 5–4 Environments of deposition of clastic sediments by a meandering stream channel.

ENVIRONMENTS OF DEPOSITION OF ORGANIC SEDIMENTARY ROCKS

Organic sedimentary rocks form in two basic environments: in marine environments and as accumulations of plant debris in swamps. In the marine environment, organisms ranging from microscopic plankton to massive corals secrete hard parts composed of calcium carbonate ($CaCO_3$) or silica (SiO_2). Planktonic organisms float freely and produce new shells as they reproduce by cell division. Thus, a growing and thriving population contributes a continuous rain of calcium carbonate or silica particles to the sea floor. Accumulations of shells from bottom-living organisms can form shell deposits or even reefs. Corals build reef structures that can produce significant accumulations of massive calcium carbonate.

Calcium carbonate rocks derived from organic processes range from continuous, fine-grained beds occasionally mixed with fine-grained clastic sediments to isolated mounds **(reefs)** of carbonate rocks surrounded by clastic rocks (Figure 5–7). Clastic carbonate rocks can form when a reef or shell deposit (primary source) is eroded, transported, and redeposited by waves and currents.

Organic deposits that form in swamps and marshes are associated with clastic deposits (Figure 5–8); they occur when plant debris accumulates in an anaerobic environment where it does not decay. If the depositional site

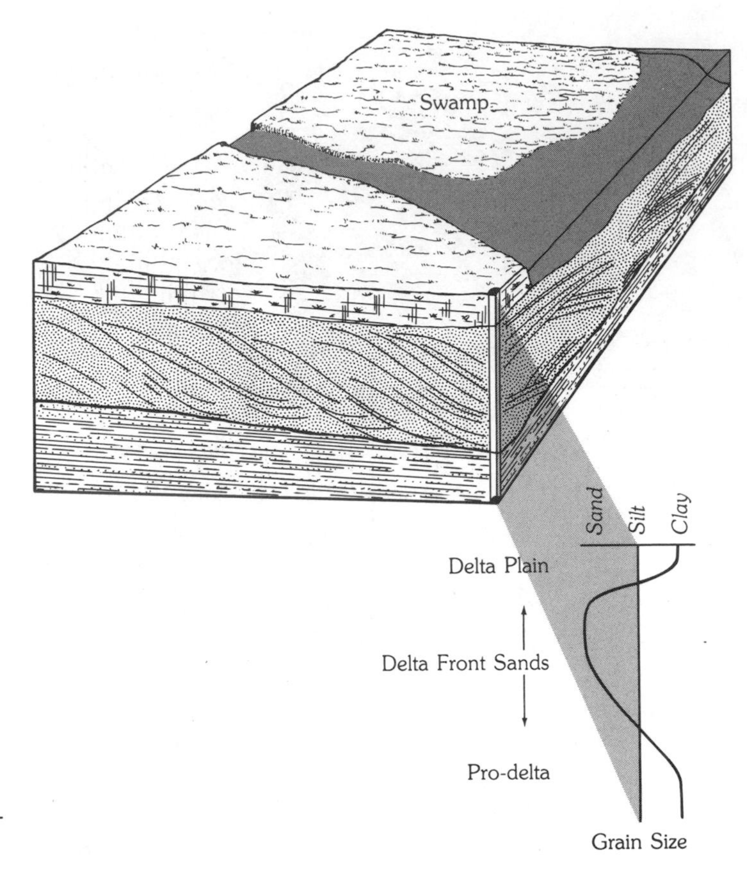

FIGURE 5–5 Environments of deposition of clastic sediments in a delta.

subsides at a rate that keeps up with the accumulation of organic matter, thick organic deposits can form. These deposits are influenced by clastic depositional processes and can form a complete range of organic materials: organic "clays" to "clayey" organics to pure organic.

ENVIRONMENTS OF DEPOSITION OF CHEMICAL SEDIMENTARY ROCKS

Chemical sedimentary rocks form in environments where chemical processes cause the precipitation and concentration of compounds from fluids. The primary type of chemical sedimentary rock is the **evaporite,** formed by the evaporation of seawater in a restricted bay or lake (Figure 5–9). Most commercially available salts are mined from evaporite deposits.

This general discussion of environments of deposition of sedimentary rocks points out the complexity of these rocks and their potential variability over short distances, both vertically and horizontally. It is important for the engineering geologist to understand this variability when attempting to describe the characteristics of a site. The complexity of sedimentary environments means that an engineering structure may be founded on two or more kinds of material, each with significantly different strength and compressibility characteristics.

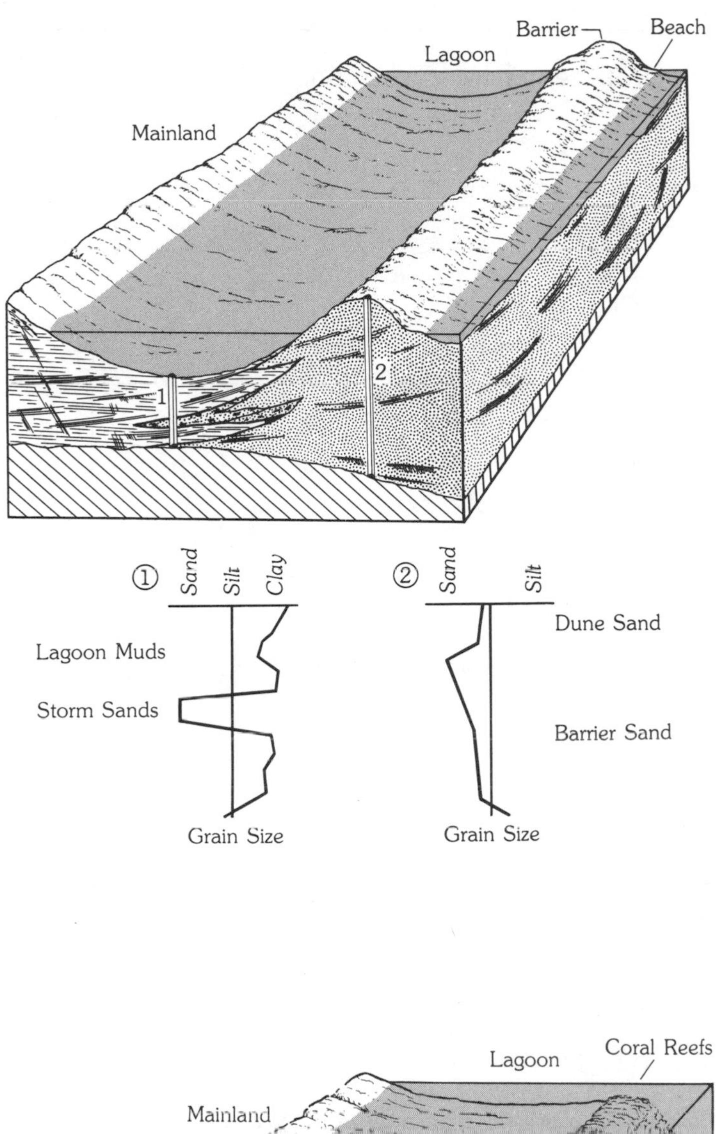

FIGURE 5–6 Environments of deposition of clastic sediments in the coastal zone.

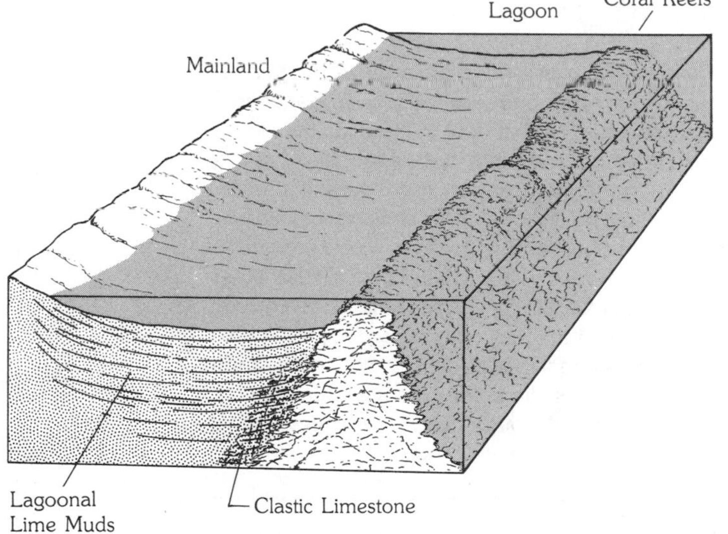

FIGURE 5–7 Environments of deposition of organic carbonate sedimentary rocks.

CASE IN POINT

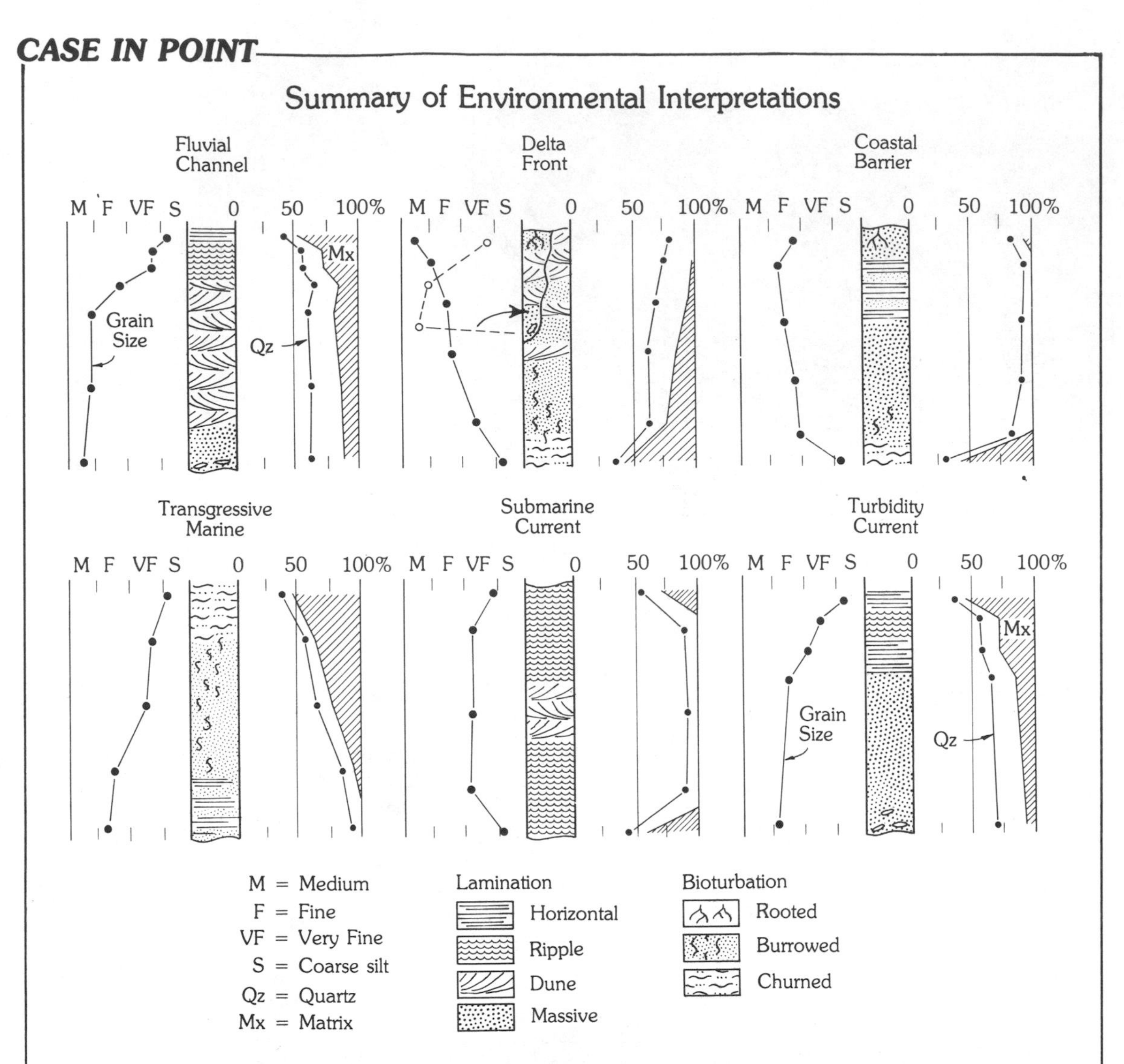

Sandstones commonly occur in regular successions characterized by vertical changes in sedimentary structures, textures, and composition. These ordered sequences reflect the processes that transported the sand and the conditions under which it was deposited. When observed in a vertical sequence, as in a core, the succession of rock properties permits the interpretation of sedimentary processes and may allow predictions for the size and shape of the sandstone body. The most common, and therefore most important, sandstone successions are probably those transported and deposited in either fresh or marine water. The characteristics of these sequences can be summarized in a general manner. Individual sections may vary considerably from the ideal, depending upon local conditions of deposition, nature of the source sediment, and other variables of sedimentary processes. In general, however, the idealized sections will form a guide for the interpretation of depositional environments of most sandstone.

Adapted from R. R. Berg, Exploration for sandstone stratigraphic traps. American Association of Petroleum Geologists Short Course Series 3, 1980.

FIGURE 5–8 Environment of deposition of organic (peat) sediments: a stagnant swamp.

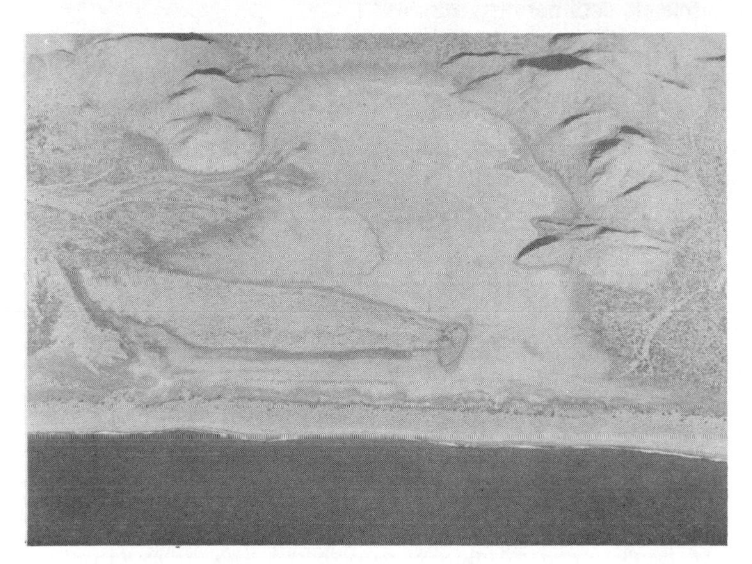

FIGURE 5–9 Environment of deposition of evaporites. Marine waters are washed over the barrier and into the restricted bay where the water evaporates, leaving behind the chemicals that were dissolved.

CLASTIC MINERALS

Clastic sedimentary particles are derived from the **weathering** of other rocks by either chemical or physical processes. Once discrete particles have formed, they can be transported and redeposited to form clastic sedimentary rocks. Since igneous rocks make up more than 95 percent of the volume of the earth's crust, weathering of igneous rocks produces most sedimentary particles. Reviewing the Bowen reaction series, you can see that some minerals are formed at higher temperatures and pressures than others. The conditions of formation indicate the degree of disequilibrium between the mineral and its new environment when exposed at the surface. High-temperature, high-pressure minerals are further removed from equilibrium conditions relative to the surface environment and therefore are more susceptible to chemical weathering processes.

Chemical weathering is the response of materials within the **lithosphere** to conditions at or near its contact with the **atmosphere,** the

hydrosphere, and perhaps more importantly the **biosphere** (Keller, 1957). The most important chemical weathering process is the reaction of a mineral with water—**hydrolysis.** This process is the reaction H^+ and OH^- ions in the water with elements or ions in the mineral. A generalized hydrolysis reaction for a silicate mineral takes the form of

$$MSiAlO_n + H^+ + OH^- \rightleftharpoons Al(OH)_3 + (M_1H)Al\ SiAlO_n$$

$$(M_1H)Al\ SiAlO_n \rightleftharpoons M^+ + OH^- + [Si(OH)_n]_n + [Al(OH)_6]_n$$

where M = any metal cation.

In this reaction the metal cation (K^+, Na^+, Ca^+, or another) is removed from the mineral and carried out of the system in the water, leaving behind H^+ cations and clay minerals. As a result of the hydrolysis reaction, the clay mineral content of the land surface is increased, making clay minerals the most common clastic sedimentary minerals.

Other process of chemical weathering are

1. *Chelation,* the reaction of cations with organic compounds.
2. *Cation exchange,* the substitution of H^+ or metal cations for other cations in a mineral.
3. *Dialysis,* the removal of an ion from the crystal lattice.
4. *Oxidation,* the reaction of an element with oxygen (iron oxide = rust).
5. *Hydration,* the process of taking water into the mineral lattice.

TABLE 5–2 Mineral resistance to chemical weathering

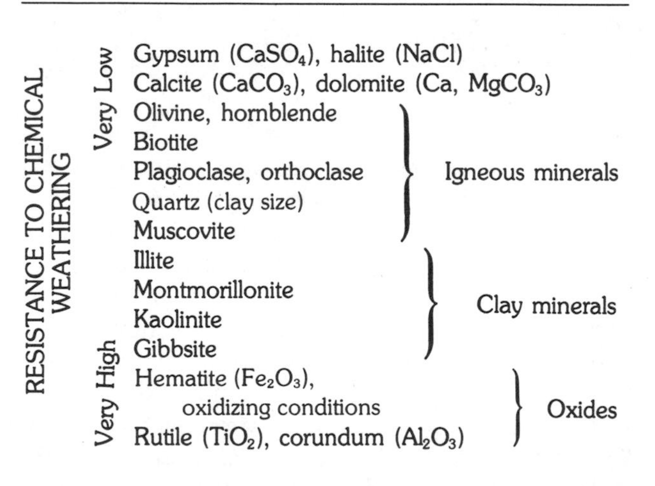

RESISTANCE TO CHEMICAL WEATHERING	Mineral	Group
Very Low	Gypsum ($CaSO_4$), halite (NaCl)	
	Calcite ($CaCO_3$), dolomite (Ca, $MgCO_3$)	
	Olivine, hornblende	Igneous minerals
	Biotite	Igneous minerals
	Plagioclase, orthoclase	Igneous minerals
	Quartz (clay size)	Igneous minerals
	Muscovite	Igneous minerals
	Illite	Clay minerals
	Montmorillonite	Clay minerals
	Kaolinite	Clay minerals
	Gibbsite	Clay minerals
	Hematite (Fe_2O_3), oxidizing conditions	Oxides
Very High	Rutile (TiO_2), corundum (Al_2O_3)	Oxides

Source: Modified from M.L. Jackson et al., Weathering sequence of clay-size minerals in soils and sediments, *Journal of Physical Colloidal Chemistry* 52: 1237 - 60. Copyright © 1948 by Williams and Wilkins Co., Baltimore.

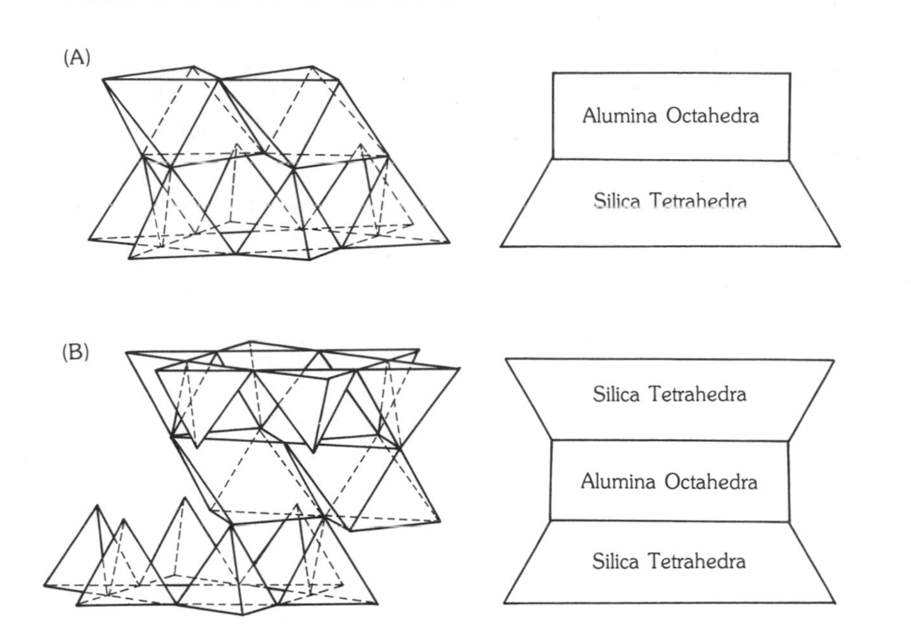

FIGURE 5–10 Structure of (A) two-layer and (B) three-layer clay minerals.

Of the common igneous minerals, quartz (SiO_2) is the most stable as the Bowen reaction series suggests. Table 5–2 lists a series of common minerals in order of increasing resistance to chemical weathering.

Minerals found in the common clastic sedimentary rocks are those that either resist chemical weathering or are broken fragments of the source rock that have not yet been altered by chemical weathering. In areas close to the source material, such as a mountain stream bed, the mineralogy of the clastic sedimentary rocks will be similar to the mineralogy of the source rocks. Far from the source, however, weathering will have altered the unstable minerals to clay, quartz, and the very stable oxide minerals.

Clay minerals are complex, layered, hydrous aluminum silicates that are related by their chemical composition and structure. There are two basic structural types of clays. A 1:1 or two-layer mineral consists of one sheet of alumina octahedra and one sheet of silica tetrahedra bonded together to form a layer (Figure 5–10A). Mineral layers are stacked on top of each other to form crystals or clay platelets with a strong ionic bonding between the sheets and no net charge imbalance. A 2:1 or three-layer mineral consists of an octahedral sheet sandwiched in between two tetrahedral sheets (Figure 5–10B). Unlike the 1:1 minerals, ionic substitution often occurs within the octahedral and tetrahedral sheets of 2:1 layer clays, and produces a charge imbalance.

KAOLINITE

Kaolinite is a two-layer, 1:1 mineral with one octahedral sheet bonded to a tetrahedral sheet. This mineral often occurs in well-crystallized layers that are strongly bonded together.

Kaolinite: $Al_4Si_4O_{10}(OH)_8$
Hardness: 2
Density: 2.6
Crystal: Triclinic, usually in earthy aggregates, pseudohexagonal platy crystals sometimes distinguishable under the microscope

ILLITE

Illite is a three-layer, 2:1 mineral that is very similar to muscovite.

Illite: $(K,Ca,Na,H_2O)_x(Al,Mg,Fe)_2(Si_{4-x},Al_x)O_{10}(OH)_2$
Hardness: 2
Density: 2
Crystal: Monoclinic, found in irregular matted flakes

SMECTITE

Smectite is a three-layer, 2:1 mineral that has a significant amount of ionic substitution in the octahedral sheet. Smectite minerals form a class of clay minerals that are characteristically very fine-grained and highly reactive with water. Montmorillonite is a specific smectite mineral, but the name is frequently used to identify the entire smectite class of minerals.

Montmorillonite: $Al_2Si_4O_{10}(OH)_2{\cdot}xH_2O$
Hardness: 2–2.5
Density: 2.0–2.7, decreasing with increasing water content.
Crystal: Monoclinic, always in earthy masses, crystals not distinguishable even in the electron microscope

Other common minerals found in clastic sedimentary rocks are the weather-resistant oxides hematite, rutile, and corundum.

HEMATITE

Hematite is a steel gray to earthy red mineral that contains about 69 percent iron.

Hematite: Fe_2O_3
Hardness: 5–6
Density: 5.26

Crystal: Trigonal, micaceous to platy, also granular friable to compact

RUTILE

Rutile is a titanium oxide mineral that is white when pure, but more frequently it is reddish brown to black in color.

Rutile: TiO_2
Hardness: 6–6.5
Density: 4.25
Crystal: Tetragonal, crystals commonly prismatic so slender or acicular

CORUNDUM

Corundum is extremely hard and composed of aluminum oxide that is white when pure. Blue corundum is the gemstone sapphire, while red to blood red corundum is the gemstone ruby.

Corundum: Al_2O_3
Hardness: 9
Density: 4.0–4.1
Crystal: Trigonal, commonly rough, rounded, barrel-shaped crystals, rarely rhombehedral

Industrial-grade corundum is mined and used for abrasive purposes. Rutile and corundum are formed at high temperatures and pressures either as accessory minerals in igneous rocks or as metamorphic minerals. Their weather-resistant quality, however, leads to their association with clastic sedimentary rocks. Rutile, for example, is mined from beach sands as a source of titanium metal.

ORGANIC SEDIMENTARY MINERALS

Calcite and dolomite are the most common carbonate minerals. Calcite is pure calcium carbonate ($CaCO_3$), and dolomite is a calcium-magnesium carbonate ($CaMg(CO_3)_2$). Calcite and dolomite are colorless and transparent to white when pure but vary widely in color due to impurities.

Calcite: $CaCO_3$
Hardness: 3
Density: 2.71
Crystal: Common and extremely varied in development, massive, usually concentrically banded and internally radiating

Dolomite: $CaMg(CO_3)_2$
Hardness: 3.5–4
Density: 2.85
Crystal: Crystal commonly rhombohedral, massive, coarse to fine grained, columnar

CHEMICAL SEDIMENTARY MINERALS

GYPSUM

Gypsum ($CaSO_4 \cdot 2H_2O$) is normally the first salt deposited in the evaporation of seawater and is therefore very soluble. Gypsum is usually colorless to transparent in crystal form and white, gray, or yellowish in color in the massive form. The mineral is very soft and can be scratched with a fingernail.

Gypsum: $CaSO_4 \cdot 2H_2O$
Hardness: 2
Density: 2.32
Crystal: Monoclinic, frequently with warped or curved surfaces, and long prismatic crystals are often bent or curled

HALITE

Halite (NaCl) is the most common water-soluble mineral and is the usual cause of saline groundwater. Halite crystallizes in a cubic form or occurs as massive units and is usually light-colored. A salty taste is characteristic of halite.

Halite: NaCl
Hardness: 2.5
Density: 2.16
Crystal: Isometric, usually as cubes, often with cavernous and stepped faces, hopper crystals

ENGINEERING GEOLOGIC CLASSIFICATION OF SEDIMENTARY ROCKS

Clastic sedimentary rocks are classified by the grain size of the mineral particles (Table 5–3). As has been discussed, the grain size of a clastic sedimentary rock is related to the transport history and the environment of depositions of the sediments. Therefore, a classification system related to origin of the rock provides important information about the character of the rock in three-dimensional space. Grain size is also a constituent property that can be recognized in any earth material.

Organic and chemical sedimentary rocks are classified by the mineralogy of the rock (Table 5–3). Organic sedimentary rocks formed from the accumulation of plant debris do not contain true minerals, since they are composed of organic matter. However, since the organic matter occurs in association with sedimentary rocks that are composed of mineral grains, these sediments are considered to be sedimentary rocks. Figure 5–11 gives a classification system of sedimentary rocks that includes variations between the classifications given in Table 5–3.

Sedimentary rocks form as individual mineral or rock particles are deposited, as organic debris accumulates or as chemicals precipitate out of a water body. The environment of deposition of each sedimentary rock can be highly variable both horizontally and vertically. Secondary, postdepositional processes such as **cementation, compaction, void filling,** or **dissolution** can further modify the characteristics of the rock. As a result, the engineering properties of sedimentary rocks are highly variable, ranging from loose, soillike deposits to high-strength, well-cemented materials.

TABLE 5–3 Classification of sedimentary rocks

CLASTIC	
Grain size	*Rock Name*
Coarse, pebbles to clay	Conglomerate
Angular broken rock	Breccia
Sand	Sandstone
Silt and clay	Shale
ORGANIC AND CHEMICAL	
Materials	*Rock Name*
Calcium/dolomite	Limestone
Plant	Coal
Salts, gypsum	Evaporite

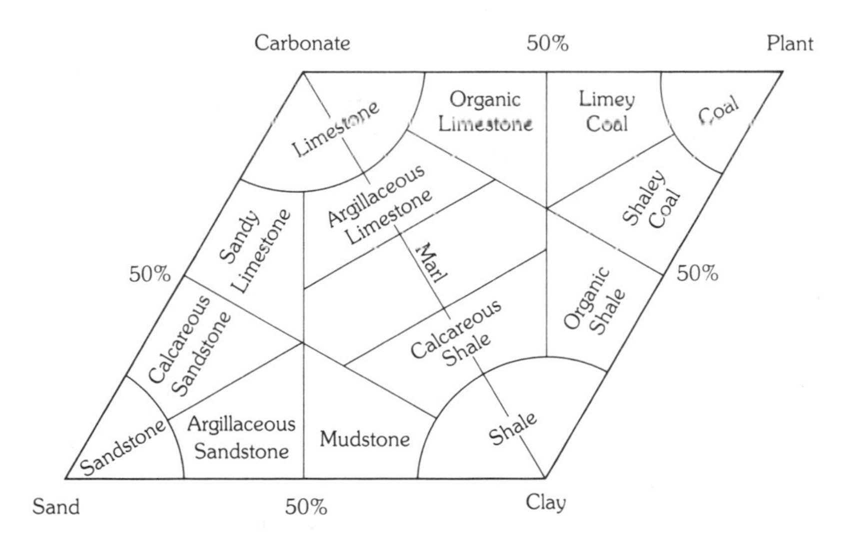

FIGURE 5–11 Classification of sedimentary rocks.

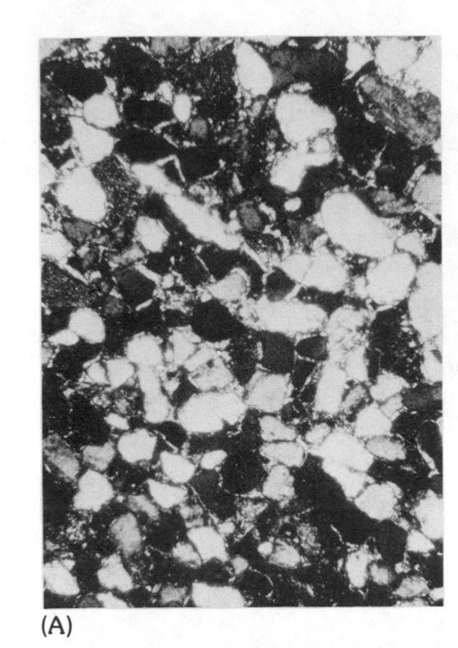
(A)

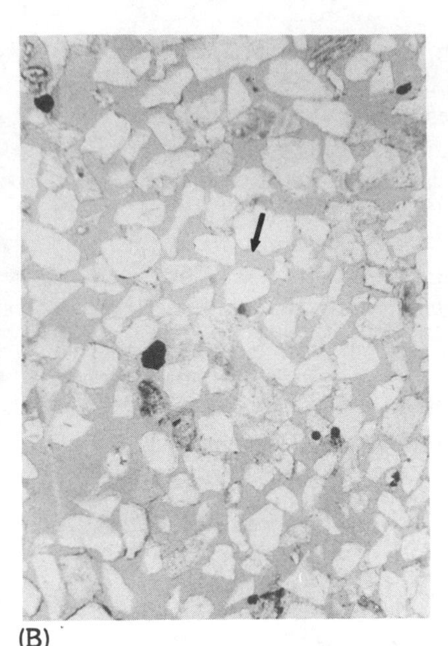
(B)

FIGURE 5–12 Photomicrographs of (A) a cemented sandstone and (B) an uncemented sandstone. Notice the difference in the amount of voids in these two samples (designated by arrow). Photographs are the same scale.

Figure 5–12A is a photomicrograph of a weakly cemented sandstone, while the sandstone in Figure 5–12B is well cemented and has high strength. Similar ranges occur in the engineering properties of the other clastic rocks (Figure 5–13). The organic and chemical sedimentary rocks are more uniform in their composition and therefore tend to vary less in their engineering properties as Figure 5–14 shows.

The engineering geologic characteristics of clastic sedimentary rocks are highly variable, depending upon the environment of deposition; mineralogy, sorting, and shape of the sedimentary particles; the amount of post-depositional cementation and void filling; and the chemistry of the cement or void-filling material. In contrast, organic and chemical sedimentary rocks have more uniform engineering geologic characteristics that depend upon the mineralogy or chemistry of the rock material. Table 5–4 gives a summary of the engineering geologic characteristics of sedimentary rocks.

In clastic sedimentary rocks the engineering geologic characteristics result from the following factors:

1. High-strength mineral grains produce a higher-strength rock.
2. Improved sorting descreases rock strength.
3. Improved particle rounding decreases rock strength.
4. Decreasing grain size usually decreases strength.
5. Dense grain packing increases strength and decreases total strain to failure.

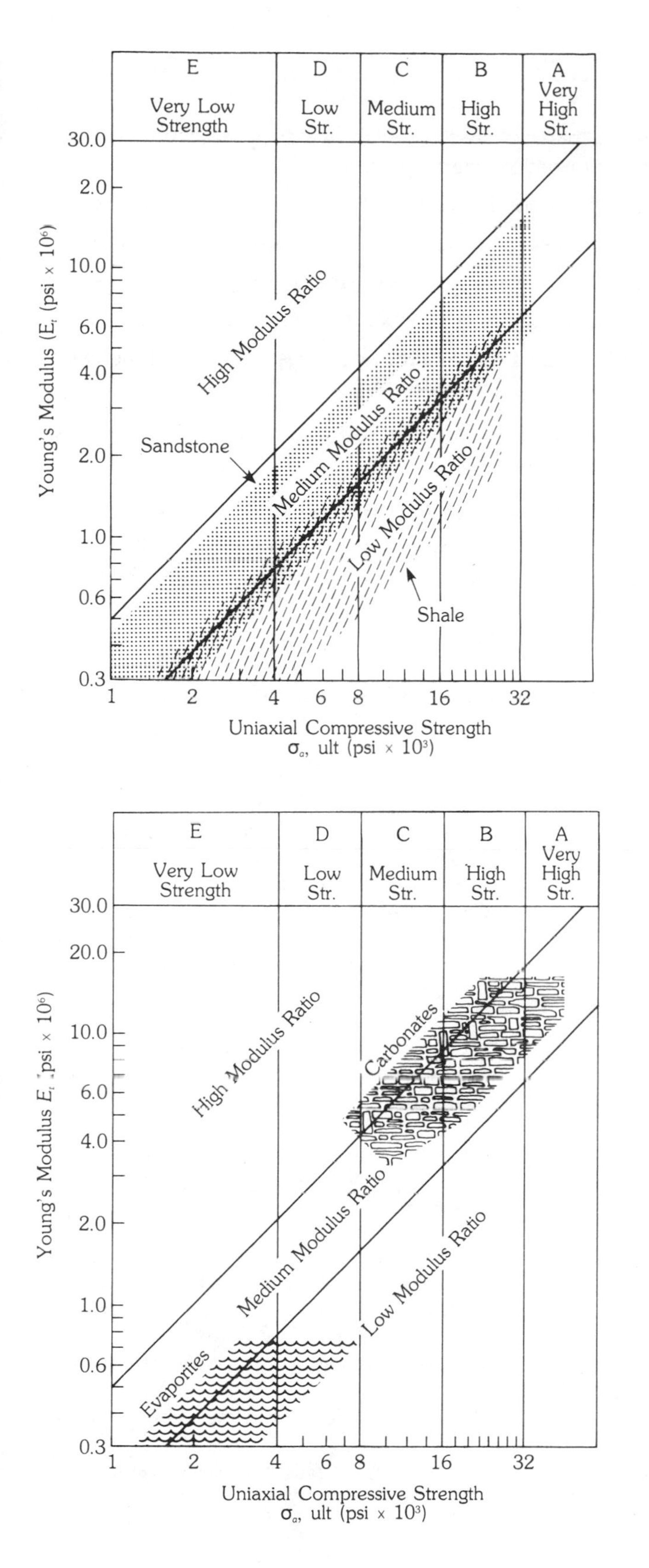

FIGURE 5–13 Engineering classification of clastic sedimentary rocks. The variations in the engineering properties are caused by the influence of bedding and cementation. (After Deere and Miller, 1966.)

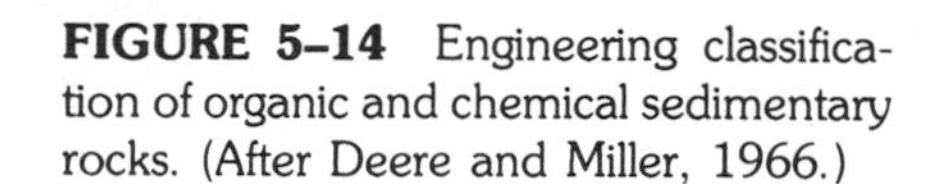

FIGURE 5–14 Engineering classification of organic and chemical sedimentary rocks. (After Deere and Miller, 1966.)

6. Increased particle reaction with water (plasticity) increases plastic deformation characteristics.
7. Densification upon loading produces early plastic deformation.
8. Grain interlocking produces elastic deformation at failure.
9. Increased cementation increases rock strength.
10. High-strength cements produce higher-strength rocks.
11. Void filling with fine particles decreases sorting and increases strength.

The engineering geologic characteristics of limestones (carbonates) result from the following factors:

1. Rock strength increases as crystals or grain size decreases.
2. In clastic limestones the strength of the rock is controlled by the strength of the limestone particles (fossils or broken aggregates of other limestone bodies).

TABLE 5–4 Engineering geologic characteristics of sedimentary rocks

Characteristic	Description		
Environment of formation	Clastic sedimentary rock		
Regional extent	Highly variable from regional to local		
Geologic rock name	Conglomerate	Sandstone	Shale
Kinetic energy	Very high	High Medium	Low
Grain size	Very coarse	Coarse Fine	Very fine
Grain sorting	Poor	Poor to well	Well
Grain shape	Angular to rounded, depending upon transport distance		
Rock substance properties			
(a) Rock strength	Variable, very low to very high depending upon post-depositional cementation and grain interlocking		
(b) Deformation characteristics	Variable plastic, plastic-elastic-plastic, to plastic elastic, depending upon post-depositional cementation		
Environment of formation	Organic and chemical sedimentary rock		
Regional extent	Usually large area; two-dimensional body		
Geologic rock name	Limestone	Coal	Evaporite
Kinetic energy	Usually low energy		
Grain size	Variable from very fine to coarse crystalline		
Rock substance properties			
(a) Rock strength	Medium to high	Very low to low	Very low to low
(b) Deformation characteristics	Plastic-elastic-plastic to elastic-plastic		

3. Coral or algal reefs and masses have medium to high strength, depending upon the properties of the mass.
4. Very fine-grained limestones, produced by chemical processes, have high to very high strengths.
5. Dissolution of the rock mass produces cavities and zones of weakness that decrease rock strength.
6. Void filling and cementation increase rock strength.
7. Densification upon initial loading produces a plastic phase.
8. Preferential orientation of the crystal lattice produces a plastic phase prior to failure.

Evaporites have engineering geologic characteristics resulting from:

1. The chemistry and mineralogy of evaporites indicate that they are low-strength materials.
2. Coarse crystals decrease rock strength.
3. Dissolution produces zones of weakness which decrease rock strength.
4. Evaporites usually lack voids and have an initial elastic phase of very limited range.
5. Low strength and creep properties of evaporite minerals produce an extended plastic phase prior to failure.

TABLE 5–5 Common characteristics of clastic sedimentary rock units

Environment of Deposition	Common Characteristics of the Clastic Unit
Talus	Conglomerate; jumbled bedding
Aeolian	Sandstone to shale; well sorted
Fluvial	
Braided channel	Conglomerate to shale; grain size decreases upward
Meadering channel	Sandstone to shale; grain size decreases upward
Overbank	Shales, laminated beds; grain size decreases upward
Delta	Sandstone to shale; grain size increases upward; sand bodies irregular in shape
Coastal	
Beach	Gravel to sandstone; well sorted; may contain shells or shell fragments
Barrier island	Sandstone, aeolian dune deposits; well sorted
Tidal flats	Shale, laminated beds with occasional sandstone layers from storm washover; may contain shells or shell fragments
Marine or lake	Shale; very fine grained; laminated beds

Sedimentary rocks are the most common rock exposed on the earth's surface. As a result, most of the engineering geologist's work will be concerned with these rocks. In sedimentary rock areas the engineering geologist should remember that, of all rock types, sedimentary rocks have the widest range of uncertainty in their engineering geologic characteristics. In addition to the inherent uncertainty in the rock substance properties, the irregular shapes of clastic sedimentary rocks affect their rock mass properties. Table 5–5 presents a summary of the common rock mass characteristics of clastic sedimentary rock units and their relationship to their environment of deposition.

Metamorphic Rocks

six

ORIGIN OF METAMORPHIC ROCKS

Metamorphic rocks are derived from the modification and alteration of other rocks through increased heating or pressure or both. The degree, grade, or level of metamorphism is directly controlled by the heating and pressure conditions. Since the earth system is a continuum—that is, there are both thermal and pressure gradients acting upon any material—metamorphic rocks can form under widely varying conditions.

There are two basic classes of metamorphism. **Contact** or **thermal metamorphism** occurs when a rock is metamorphosed by direct heating from an igneous body (Figure 6–1). **Regional metamorphism** is the result of large-scale, widespread high pressures, which are caused by excessive overburden or tectonic stresses; there is also a gradual increase in temperature (Figure 6–2). Levels of metamorphism and the relationship between the metamorphic rock and the surrounding unmetamorphosed rocks are different in these two classes. In contact metamorphism, the level of metamorphism decreases as one moves away from the heat source; therefore, contact metamorphism is generally local. Regional metamorphism covers large areas, and significant changes in the level of metamorphism may not be visible in the rocks exposed in a project area. Complex metamorphic provinces, reflecting more than one stage of metamorphism, are common in many regions, thus posing complicated geologic problems.

GEOCHEMISTRY OF METAMORPHIC ROCKS

The same chemical elements of the earth's crust that form igneous rock minerals also form metamorphic minerals. However, high-temperature, or **hydrothermal,** fluids can concentrate some of the less common elements

FIGURE 6–1 Contact metamorphic zones (designated by arrows) below each of the flow basalts.

and provide the geochemical conditions necessary to produce new minerals. In addition, new and highly variable temperature-pressure conditions are involved, leading to the formation of a new suite of minerals. The geologist utilizes the mineral content of the rock to determine the *grade* or *level* of metamorphism.

Quartz, feldspar, and the other igneous minerals are, of course, common to metamorphic rocks; therefore, the metamorphic minerals are important indicators of the history of the rock. Geologic research has determined that certain suites of minerals form at different temperature and pressure conditions, and it has established three grades of metamorphism (Figure 6–3). Table 6–1 gives the common metamorphic minerals, grouped by grade (level) of metamorphism.

FIGURE 6–2 Regional metamorphic terrain. The entire mountain range is composed of the same type of metamorphic rocks.

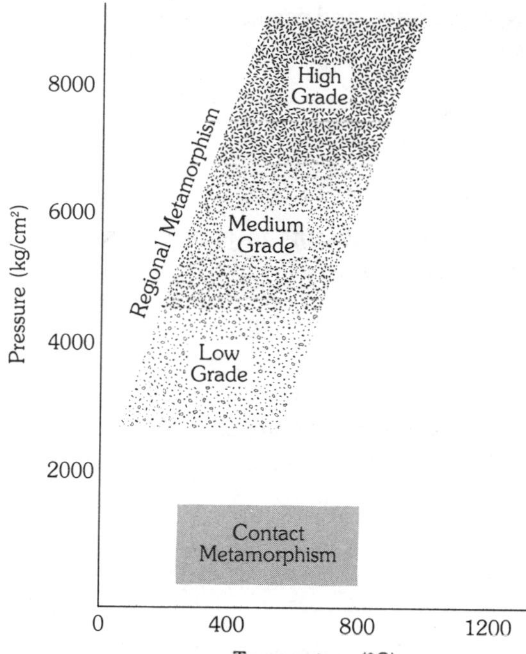

FIGURE 6–3 Temperature-pressure conditions for various levels of metamorphism.

TABLE 6–1 Common metamorphic minerals

Name	Formula
Low Grade	
Serpentine	$Mg_6Si_4O_{10}(OH)_8$
Talc	$Mg_3Si_4O_{10}(OH)_2$
Chlorite	$(Mg,Fe,Al)_6(Al,Si)_4O_{10}(OH)_8$
Epidote	$Ca_2(Al,Fe)_3Si_3O_{12}(OH)$
Medium Grade	
Kyanite	Al_2SiO_5
Andalusite	Al_2SiO_5
Staurolite	$Fe_2Al_9O_6(SiO_4)_4(O,OH)_2$
Biotite	$K(Mg,Fe)_3(AlSi_3O_{10})(OH)_2$
Hornblende	$Ca_2(Mg,Fe)_5(Al,Si)_8O_{22}(OH)_2$
High Grade	
Sillimanite	Al_2SiO_5
Fosterite	Mg_2,SiO_4
Garnet	$(Ca,Mg,Fe,Mn)_3(Al,Fe,Cr)_2(SiO_4)_3$
Wollastonite	$CaSiO_3$

LOW-GRADE METAMORPHIC MINERALS

SERPENTINE

Serpentine is usually green but may be yellow, brown, reddish brown, or gray in color. This mineral may form crystals that have layered structures, like kaolinite, or fibrous forms, as asbestos, and occurs either structureless or occasionally as fibrous masses.

Serpentine:	$Mg_6Si_4O_{10}(OH)_8$
Hardness:	4–6
Density:	2.5–2.6
Crystal:	Monoclinic, usually occurs in structureless masses, occasionally fibrous

TALC

Talc is chemically similar to serpentine except that it is very soft. Talc usually occurs as a compact mass of fine-grained aggregates or as a banded (foliated) mass that is characteristically pale green.

Talc:	$Mg_3Si_4O_{10}(OH)_2$
Hardness:	1
Density:	2.82
Crystal:	Monoclinic, foliated masses or compact fine-grained aggregates

CHLORITE

Chlorite is a series of minerals ranging from $(Mg_4Al_2)Al_2Si_2O_{10}(OH)_8$ to $(Fe_4Al_2)Al_2Si_2O_{10}(OH)_8$ in chemical composition. In crystalline form this mineral looks like a green mica.

Chlorite:	$(Mg,Fe,Al)_6(Al,Si)_4O_{10}(OH)$
Hardness:	2.5, on cleavage
Density:	2.6–3.3, increasing with iron content
Crystal:	Monoclinic, pseudohexagonal crystals, usually as scaly aggregates and as fine-grained and earthy masses

EPIDOTE

Epidote is another mineral group that contains Ca,Fe,Al,Mg, and rare earth minerals. The epidote minerals occur as prismatic, massive, fibrous, or tabular crystals, and are usually a light pistachio green to black in color, depending upon the chemistry of the mineral.

Epidote:	$Ca_2(Al,Fe)_3Si_3O_{12}OH$
Hardness:	7

Density:	3.3–3.6, increasing with iron content
Crystal:	Monoclinic, massive, fibrous, or granular

MEDIUM-GRADE METAMORPHIC MINERALS

KYANITE

Kyanite is a patchy, light blue, green, white, or gray mineral that in crystalline form occurs as tablets or as bladed masses. Kyanite is associated with low- to medium-grade regional metamorphism.

Kyanite:	Al_2SiO_5
Hardness:	Variable
Density:	3.63
Crystal:	Triclinic, crystals are distinctly flexible and often bent or twisted

ANDALUSITE

Andalusite has a composition similar to kyanite but forms at higher temperatures. The color varies; it may be white, gray, rose red, brown, and sometimes green. Crystals of andalusite are coarse prisms.

Andalusite:	Al_2SiO_5
Hardness:	7.5
Density:	3.15
Crystal:	Orthorhombic, coarse prismatic crystals with a nearly square cross section

STAUROLITE

Staurolite is a brown mineral that often occurs in a cross-shaped crystal form. It has about the same hardness as quartz.

Staurolite:	$FeAl_4Si_2O_{10}(OH)_2$
Hardness:	7
Density:	3.7–3.8
Crystal:	Monoclinic

Biotite, muscovite, and hornblende are common igneous minerals that also can be formed at medium grades of metamorphism. Concentrations of these minerals in excess of igneous concentrations are indicators of metamorphism.

HIGH-GRADE METAMORPHIC MINERALS

SILLIMANITE

Sillimanite is the high-grade metamorphic equivalent of kyanite and andalusite. The mineral is nearly as hard as quartz, is white or occasionally brownish or greenish, and occurs as finely fibrous to coarse prismatic masses.

Sillimanite:	Al_2SiO_5
Hardness:	7
Density:	3.24
Crystal:	Orthorhombic, usually in finely fibrous or coarse prismatic masses

FORSTERITE

Forsterite is a nearly pure form of magnesium olivine that can be concentrated at high levels of metamorphism.

Forsterite:	Mg_2SiO_4
Hardness:	6.5
Density:	3.22
Crystal:	Orthorhombic, rarely as crystals, usually in granular masses

GARNET

Garnet is a chemically complex mineral that has the general chemical formula $X_3Y_2(SiO_4)_3$, where *X* can be Ca, Mg, Fe, or Mn, and *Y* can be Al, Fe, or Cr. The crystal forms are usually dodecahedrons or trapezohedrons or a combination of the two. Garnets are commonly dark red to reddish brown crystals; however, they can also be white, yellow, pink, black, or green.

Garnet:	$X_3Y_2(SiO_4)_3$
Hardness:	7–7.5
Density:	3.6–4.3
Crystal:	Isometric, commonly in crystals, the crystal forms being characteristic, usually dodecahedrons or trapezohedrons or combinations

WOLLASTONITE

Wollastonite is a white to grayish, fibrous to platy mass of crystals. Wollastonite forms from the high-level metamorphism of siliceous limestones.

Wollastonite:	$CaSiO_3$
Hardness:	5
Density:	2.9
Crystal:	Triclinic, usually in cleavable or fibrous masses, sometimes granular and compact

ENGINEERING GEOLOGIC CLASSIFICATION OF METAMORPHIC ROCKS

Metamorphic rocks are initially classified by the physical characteristics of the rock, with the primary metamorphic mineral as a modifier. The physical characteristics include the grain size and foliation, if visible, or crystal character and banding.

Low- to moderate-grade metamorphism of sedimentary rocks increases their cementation, hardness, and brittleness with some destruction of their sedimentary appearance. At low levels of metamorphism, temperature-pressure conditions usually cause limited rather than complete recrystallization.

Very low-level metamorphic rocks may appear very similar to the sedimentary rocks that were metamorphosed, except that the metamorphic rocks are harder and more brittle. In these cases, it is frequently difficult to determine from an isolated sample if the rock is sedimentary or metamorphic (Figure 6–4). As the level of metamorphism increases, fine sedimentary structures, such as some bedding, may be destroyed, and the rock will be massive in appearance. Recrystallization, in the form of cementation and void filling, will be noticeable with a hand lens.

Moderate- to high-grade metamorphism causes recrystallization of the original minerals and forms a rock containing a new or altered suite of minerals (Figure 6–5). Changed mineralogy, structure, foliation, and

FIGURE 6–4 Metasediments exposed in a roadcut in Pennsylvania.

FIGURE 6–5 High-grade metamorphic rocks showing the effect of partial solution that produced the bands.

banding are characteristic of high-level metamorphic rocks. The primary classification tool for these is based on the character of foliation or banding.

Foliation describes the appearance of a rock in which layered minerals, such as the micas or chlorites, are oriented parallel to each other.

FIGURE 6–6 Outcrop of schist. Notice the fine banding or layering.

TABLE 6–2 Classification of metamorphic rocks

	Source Rock	Grade of Metamorphism	
		Low	High
Sedimentary Rocks	Gravel, sandstone	Quartzite	Quartzite
	Clayey, silty sandstone	Quartzite	Quartzite schist
	Clay, shale	Slate	Mica schist
	Limestone (pure)	Marble	Marble
	Dolomitic limestone	Marble	Serpentine marble
	Clayey limestone	Marble	Serpentine marble
	Interbedded shale-limestone	Slate-marble	Amphibolite*
	Coal	Anthracite	Graphite
Igenous Rocks	Granite (felsitic)	—	Gneiss
	Basalt	—	Hornblende schist
			Serpentine schist

*Amphibolite is a crystalline rock composed of amphibole and plagioclase.

This orientation is a reflection of the pressure conditions under which the new suite of layered minerals formed. **Banding** describes the alternation of bands of different mineral suites. These bands also reflect pressure and heating conditions.

There are two general classes of high-level metamorphic rocks. **Schist** has a medium- or coarse-grained crystal character with a noticeable parallel to subparallel orientation of the platy minerals (mica, chlorite). **Gneiss** has alternating bands of coarse granular minerals (quartz and feldspars) with finer schistlike **schistose bands** (Figure 6–6). Table 6–2 is a classification chart of the metamorphic rocks showing their source rocks.

The engineering properties of metamorphic rocks are highly variable, ranging from very high strength, nearly pure elasticity and homogenity for quartzite to low strength, variable plasticity to elasticity, and heterogenity for a schist. Figure 6–7 is a photomicrograph of a quartzite. Notice the uniformity of the rock and the lack of banding. When compared to Figure 6–8, a photomicrograph of a schist, notice the distinctive banding in the rock when viewed parallel to the schistosity.

Many engineering problems, including total failure of the structure, have been related to failure to recognize and design for the anisotropic engineering properties of schistosic and gneissic rocks. Even when these variable properities are recognized, strength may vary with changes in orientation (Figure 6–9). Figure 6–10 is a diagram of the engineering properties of low-level metamorphic rocks, and Figure 6–11 shows the engineering properties of high-level metamorphic rocks.

The engineering geologic characteristics of metamorphic rocks are controlled by the composition of the rock that was metamorphosed and the level of metamorphism. The general effect of low levels of metamorphism is the same as cementation in sedimentary rocks—voids are filled and the rock strength is increased. At high levels of metamorphism geochemical reactions

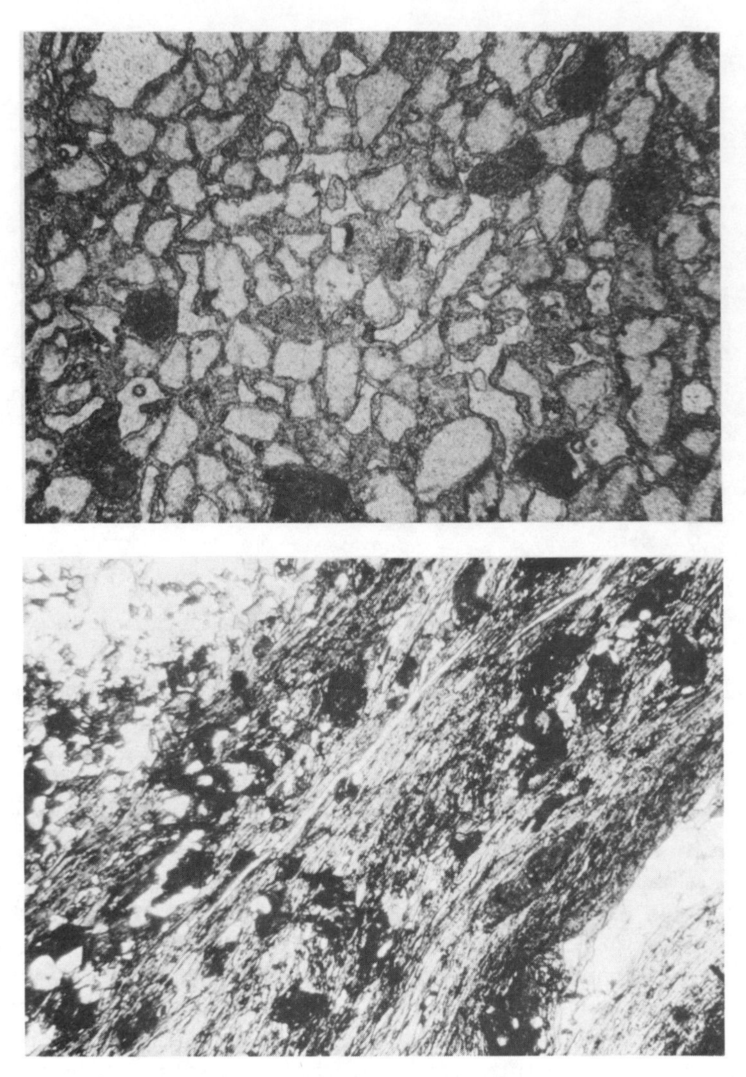

FIGURE 6–7 Photomicrograph of quartzite. Notice its similarity to a well-cemented sandstone shown in Figure 5-12A.

FIGURE 6–8 Photomicrograph of schist. Notice the distinctive banding of the minerals.

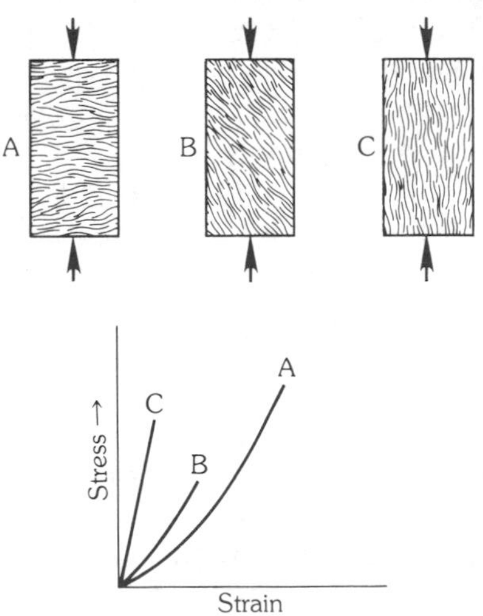

FIGURE 6–9 Variation in the stress-strain behavior of a rock having a distinctive plane of weakness as a function of the relationship between the orientation of the plane and the direction of loading.

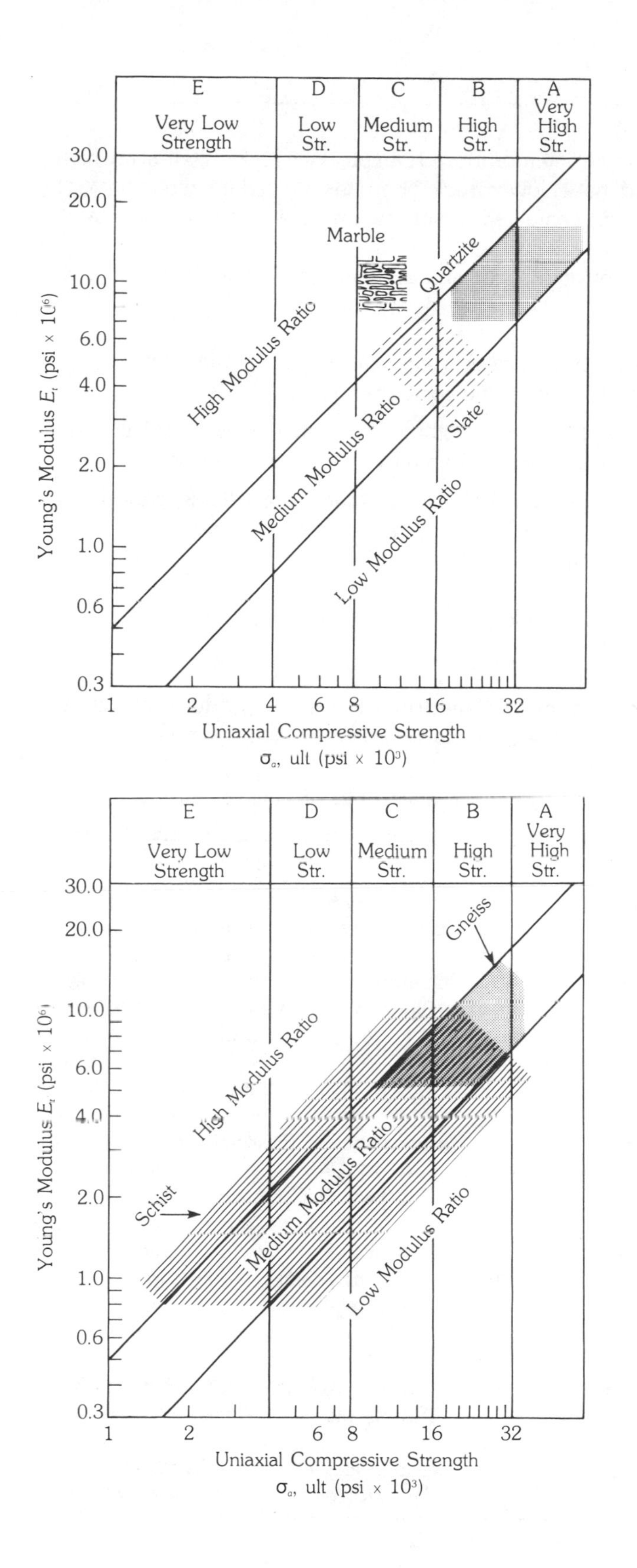

FIGURE 6–10 Engineering classification of low-level metamorphic rocks. (After Deere and Miller, 1966.)

FIGURE 6–11 Engineering classification of high-level metamorphic rocks. The broad range in the engineering properties of schist is caused by the influence of planes of weakness along the bands of schistosity. (After Deere and Miller, 1966.)

within the rock being metamorphosed can produce rock units that are highly anisotrophic and have lower rock strengths in certain directions. The engineering geologic characteristics of metamorphic rocks are summarized in Table 6-3.

The properties of quartzite and slate are determined by the following factors:

1. Metamorphism closes voids and cements the rock particles, thereby increasing the rock strength.
2. Metamorphism hardens the rock, making it elastic at failure, and produces a rock that behaves as a brittle material.
3. The uncertainty in the rock substance properties is reduced since low-strength shales and sandstones are cemented during metamorphism.

The engineering geologic characteristics of marbles result from the following:

1. Metamorphism of limestone tends to produce calcite and carbonate mineral crystals, which weaken the rock.

TABLE 6–3 Engineering geologic characteristics of metamorphic rocks

Characteristic	Description		
Environment of formation	Low-level metamorphism		
Regional extent	Usually large area, except where contact metamorphism occurs		
Geologic rock name	Quartzite	Slate	Marble
Grain or crystal size	Usually visible	Very small	Variable
General rock color	Light to dark	Dark	White
Rock substance properties			
(a) Rock strength	High to very high	Medium to high	Medium
(b) Deformation characteristics	Plastic-elastic	Plastic-elastic	Elastic-plastic

Characteristic	Description	
Environment of formation	High-level metamorphism	
Regional extent	Large area	
Geologic rock name	Schist	Gneiss
Crystal size	Fine, with isolated coarse	Fine to coarse
Crystal shape	Platey	Angular
General rock color	Dark	Light, red to pink
Rock substance properties		
(a) Rock strength	Very low to very high Anisotropic	Medium to very high Maybe slightly anisotropic
(b) Deformation characteristics	Plastic to elastic Anisotropic	Elastic-plastic Maybe anisotropic

2. The chemistry of the rock, however, is not changed; thus a weaker rock still has the same stress-strain relationship, and the modulus ratio is increased.
3. Crystal formation during metamorphism decreases the range in strength values of limestones and reduces the uncertainty in the rock substance properties of marbles when compared with limestones.

At high levels of metamorphism the geochemistry of the original sedimentary rock affects the engineering geologic characteristics of schists and gneisses in the following ways:

1. Layered (platey) minerals (micas and chlorites) are anisotrophic minerals that have a distinct plane of weakness parallel to the plates.
2. The layered minerals form with their plates aligned perpendicular to the direction of maximum pressures during metamorphism.
3. Loading a schist perpendicular to the orientation of the platey minerals results in (a) an increase in plastic deformation due to repressurizing the minerals and (b) a higher-strength rock because there are no planes of weakness.
4. Loading a schist parallel to the orientation of the platey minerals (a) reduces the plastic deformation because of the strength of the plates and (b) reduces the rock strength because the mineral plates may buckle or bend.
5. Loading a schist at an angle to the orientation of the mineral plates produces a rock that falls between the other loading conditions with respect to plasticity, but decreases the rock strength because there is a distinct plane of weakness oriented sub-parallel to the direction of maximum shear.
6. Gneiss tends to be very similar to intrusive igneous rocks with respect to their rock substance properties, except that banded gneiss will have some of the same anisotrophic properties as schists.

Soil Materials

seven

Soil implies different meanings to different people. It is the verb "to soil;" it is a medium for plant growth; it is a low-strength earth material, the result of in situ weathering—among many others. The term *soil* unfortunately has at least three different definitions to the three professions that deal with the earth.

To the agronomist, soils are the weathered, uppermost layers of organic and inorganic earth materials, formed through physical and biochemical processes, that are capable of supporting plant life. A "good soil" is an agriculturally productive soil, while a "poor soil" is one in which plants cannot survive.

To a geologist, soils are simply the products of in situ weathering processes. In short, soils are only the debris that cover the interesting things—the rocks. In glaciated terrain, for example, the geologist may identify a series of "soils"; one is the soil presently forming on the surface, while the others formed prior to the deposition of the overlying sedimentary rocks and are now buried.

To the engineer, soil is an organic or inorganic aggregate of earth materials that will disaggregate when subjected to gentle mechanical force or when submerged in water. To be blunt, it is a soil if you can excavate it without blasting, regardless of its origin. A friable sandstone, deposited in an Eocene river channel and buried 150 m, is soil to the engineer, because any design considerations for this material must recognize its low (soillike) strength.

Occasionally, the term *surface deposits* or *surficial deposits* is used. These terms are as confusing as *soil* because each profession defines them differently. Surface deposits are the loose, unconsolidated sediments that are the parent material of what an agronomist regards as soil. They are unconsolidated, modern sediments (alluvium) to the geologist. To an engineer, they are the upper layers of organic and mineral matter (topsoil) that are removed from the surface when preparing a construction site.

The multiple definition of soil means that the engineering geologist needs to understand the basics of soil science, rock weathering, and soil mechanics. In any discussion about soil materials the engineering geologist must also know with whom he or she is speaking.

AGRICULTURAL SOILS

The engineering geologist can gain a great deal of information and assistance from soil science, particularly when dealing with such problems as erosion, waste disposal, landslides, and flooding. The characteristics of the upper few meters of the earth's surface are established by the active surface processes and therefore record recent earth history. Many building projects, septic tank installations, and landscaping projects fail simply because this record was not considered. In California and other tectonically active regions, displacements of soil horizons and other subtle soil features identify active faults.

Soil science has developed into a recognized scientific study of the physical, mineralogical, biological, and chemical properties of the very thin soil layer. You can easily appreciate the complexity of soil science if you read the college curriculum for the field, and this text cannot do it justice. You should consider a formal course in soil science if one is available or do some outside reading on the subject.

Soils are the results of interactions between the five soil-forming factors:

1. Parent material
2. Topography
3. Climate
4. Organisms
5. Time

Parent material is the "original" material, exposed at the earth's surface, from which soils form. Parent materials include all rock types and

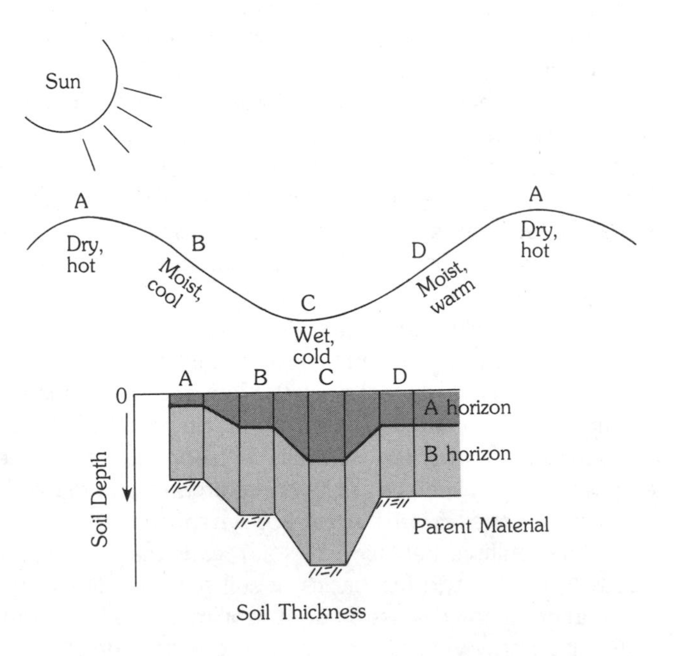

FIGURE 7–1 Effect of microtopography on soil thickness, temperature, and moisture conditions.

establish the basic character of the soil. A quartz-rich, sandy parent material imprints a sandy texture on the soil, while clay-rich parent material yields clayey soils. A permeable parent material leads to a deep soil, while an impermeable parent material has a thin soil layer.

Topography, both macro- and microtopography, acts to modify the environment in which the soil forms. Macrotopographic features influence the temperature and moisture conditions (microclimate) and the plants and animals (organisms) in the area. Steep mountain slopes have little or no soil because of high erosion rates. Microtopography has a far more subtle influence, which is often more important to the engineering geologist than is the influence of macrotopography. Microtopographic changes often occur within the boundaries of an engineering project, resulting in variable soil characteristics on the site (Figure 7–1).

Climate affects the availability of water, freeze-thaw cycles, wetting and drying history, and the types of organisms acting on the soil. Macroclimate leads to different soils in different regions, while microclimate leads to variations in small areas. Mountains have a wide variety of different types of vegetation depending upon the microclimate of the slope (Figure 7–2). At an engineering project site, microclimate, controlled by microtopography or by structures built by people, can create significant variations in moisture and other characteristics of the soil. (This is important in expansive soil areas.)

Both flora and fauna *organisms* play a significant role in soil formation. Black, organic-rich soils differ greatly from red, organic-poor soils as agricultural producers. Humic acid, derived from organic matter lying as litter on a forest floor, is a significant leaching fluid in soil formation. Organic matter also improves the water-holding capacity of a soil, thereby influencing the microclimate. Burrowing organisms act to loosen and mix the soil, and in some cases, they can create a completely new soil profile.

Time is the most subjective of the soil-forming factors because the influence of time is difficult to understand. In a modern sense, the effect of time is easily recognized when the first planting of vegetation on an untreated, reclaimed surface mine comes up bright green and healthy, only to wither and die a few months later. During this short period, chemical

FIGURE 7–2 The effect of microclimate as expressed by the different vegetative zones on the mountain.

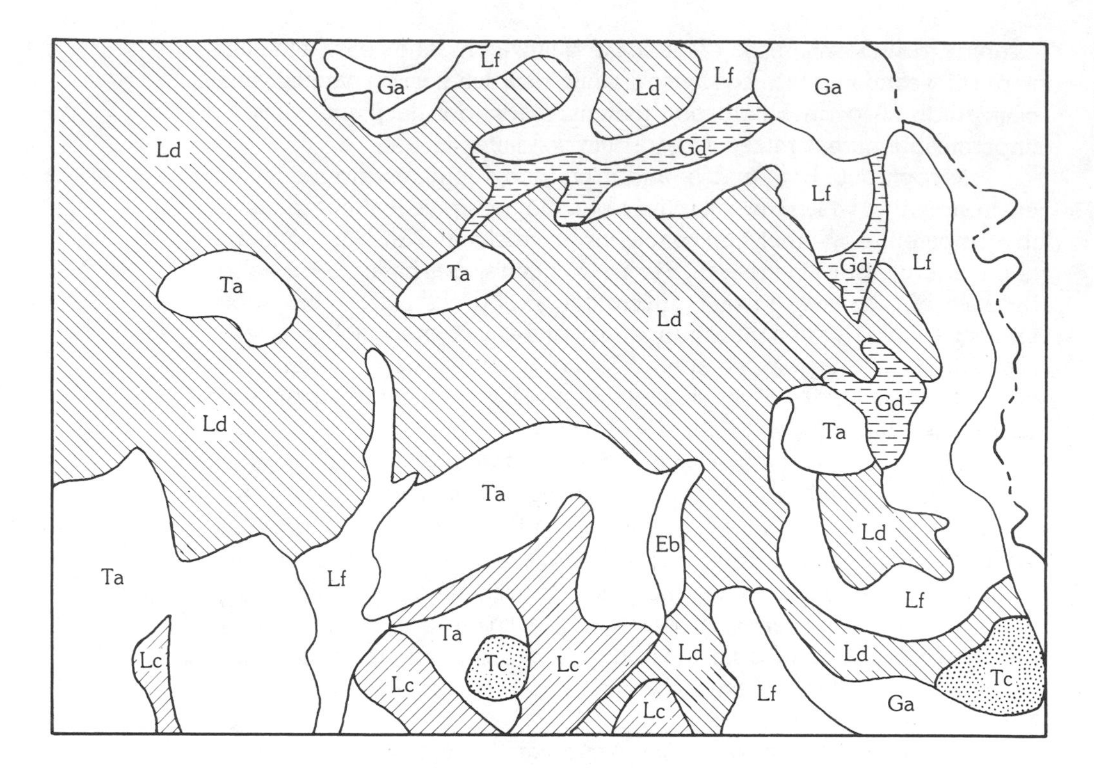

FIGURE 7–3 Generalized soil map of a subdivision showing the variability of agricultural soils over a small area. This tract of land is less than 300 acres.

processes oxidize soil minerals and form a highly acidic and toxic soil. The concept of time as a soil-forming factor assumes that the active soil-forming factors (climate and organisms) will eventually lead to the formation of a common soil, regardless of parent material or topographic conditions. This assumption is probably true, but it is impossible to prove because other active processes, such as erosion, landslides, and tectonics, destroy the evidence.

The interaction of these five soil-forming factors form a wide variety of mappable soil types. Figure 7–3 is a portion of a U.S. Department of Agriculture Soil Conservation Service soil survey of Texas. Notice that even in a generally flat area with a uniform parent material the map shows numerous soil types. This type of map is available for most of the United States and for most states on a county basis. (Reports are usually available from the county agricultural agent, Soil Conservation Service, or a land grant or agricultural college.)

Agricultural soil descriptions, to be complete, must include soil color, dry and moist; soil texture; soil structure; and soil consistency, dry, moist, and wet. These descriptions are used as the basis for soil classification. Soil classification systems are complex, and one system has not yet been universally recognized.

GEOLOGICAL SOILS

To a geologist, soil is the product of rock weathering; it is one source material for clastic sediments. An understanding of rock weathering processes is important to the engineering geologist, not only to warn of possible rock decay, but to predict the general character of a soil, based only on a geologic map or sample. Estimates of the permeability, mineralogy, thickness, and other physical properties of soils can be made based on the geology and location of the site.

ENGINEERING SOILS

The engineer is concerned with the mechanical and material properties of the soil, regardless of its origin, ability to support plant life, or depth of burial. An understanding of the soil-forming factors and geology of soils or sedimentary rocks is an important aspect in the selection of samples for laboratory testing, the interpretation of the test results, and the ultimate design of a foundation or selection of a material to be buried in the soil. Soil mineralogy is important in evaluating how a soil will perform as an engineering material. Some soils, for example, contain high concentrations of sulfuric minerals that, when exposed in an excavation, will oxidize to form sulfuric acid. Highly acid soils can be disastrous to concrete or iron that comes in contact with them.

SOIL DESCRIPTION

Accurate field identification of soil types and accurate, complete descriptions of soils encountered in the field are necessary and very vital parts of any soils investigation. This information forms the basis for preliminary screening of field samples for laboratory testing, reduces the logs, and aids in the extrapolation of boring profiles to soil profiles over the investigated area. The experienced field engineer or engineering geologist can also determine a field classification of the soil material, which provides important information to the laboratory technician and design engineer. Incomplete or erroneous field identification always multiples the required amount of laboratory work and may lead to a fallacious interpretation of the soil profile at the site or incorrect assessments of design criteria.

Almost every organization engaged in site investigations has its own system of soil-type descriptions or soil nomenclature; however, all have a great deal in common, and most base the nomenclature on the same arbitrary grain-size definitions of soil components. Accurate identification of characteristic physical features, such as feel, grittiness, and consistency, are the primary components of a complete soil description. In some professions, like agronomy, the ultimate product from a field study is a complete description, while in others (engineering, for example) the description provides the basis for the field classification of the soil material. Recording a complete soil description may appear to be wasted effort if the desired result is a soil classification, particularly if the classification scheme is a simple two-letter (Unified Soil Classification) or letter-number (American Association of State Highway and Transportation Officials [AASHTO]) system. The description, however, provides far more information and better conveys the

CASE IN POINT

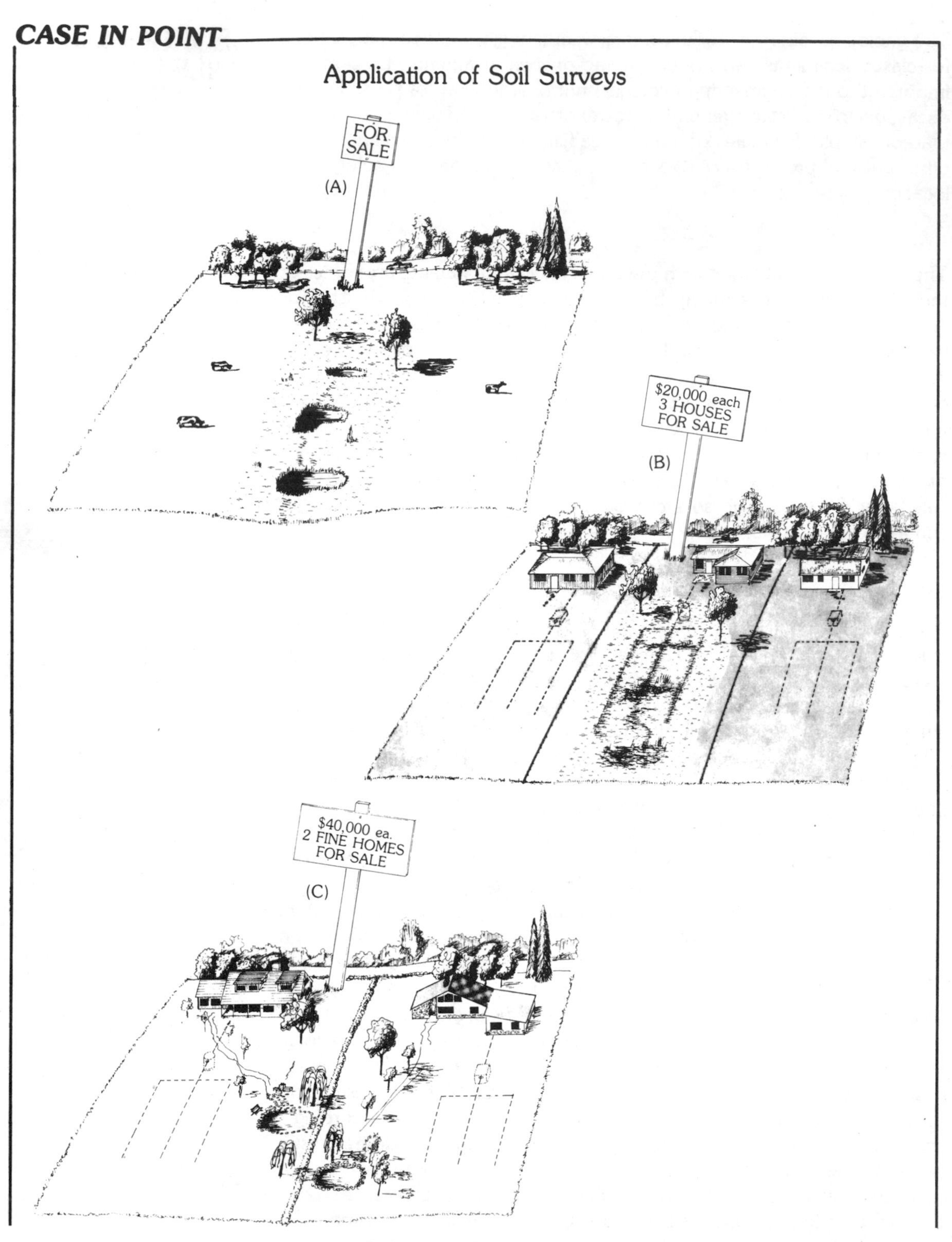

Soils surveys by the U.S. Department of Agriculture Soil Conservation Service are highly accurate and valuable sources of information to the engineering geologist for any project that will be constructed within the upper 3 m of the earth's surface. Since soils reflect the combined influence of the five soil-forming factors, the soil survey identifies soils that are wet, clayey, susceptible to floods, or affected by other factors. In the diagrams in this box there are two soils on the tract—a dry, well-drained pasture and a wet, poorly drained marshy area (A). The developer has many options available in any project. In this case the development is a rural, single-family project with either three small homes (B) or two larger homes (C). In case B, the middle homesite will have serious septic tank problems that could adversely affect the nearby homes. A design (C) that uses the wet area as an advantage to the entire project can often recover the additional cost by offering such a unique feature. The site investigation necessary to provide information to maximize the design of the site is available from the Soil Conservation Service at no cost to the developer.

interpretations of the field classification from the field to the laboratory to the design engineer.

SOIL COLOR

The simplest and least ambiguous descriptive terms are those of the primary colors and some binary ones (such as purple and green) modified by the adjectives "light" and "dark." Medium shades require no modifying adjective. Terms like "mauve," "beige," "orchid," "tan," and the like should be avoided; they are more suitable for interior decorating than for engineering geology.

Soils change color, however, with changes in moisture content. In many soils, the dry color is significantly different from the moist color. As a result, both moist and dry color should be recorded in the description whenever possible. In agricultural descriptions, charts are used to define the soil color; however, this accuracy is not necessary for most descriptions in engineering geology.

Organic soils often have dark gray to black or brownish black colors and may have a faint to strong woody or decaying odor to them. Occasionally, an organic soil shows distinct banding. Reddish colors often indicate the presence of iron, while gray colors indicate iron-poor conditions. Mottling often indicates poor or limited drainage.

SOIL TYPE

Most soil descriptions and classification systems use grain size or texture as the primary feature of soil type. A sample of a grain-size classification for soil components is as follows:

Boulders:	Larger than 15 cm
Cobbles:	5 cm to 15 cm
Gravel:	2 mm to 5 cm
Sand:	0.05 mm to 2 mm
Silt:	0.005 mm (5 microns) to 0.05 mm
Clay:	Smaller than 5 microns

In general, the use of the term "rock" should be avoided in a soil description because it is ambiguous: Does it mean particles larger than gravel size, ledge rock, or bedrock? However, the specific use of the term "rock" should be included if it is necessary. For example, "ledge rock," "boulders of granite," or "gravel-sized rock fragments" are proper uses of the term. Note also that this classification applies only to grain sizes and does not imply anything about the mineralogic composition, geological origin, or genesis of the soil. Clay-sized particles may be true clay minerals, other layer silicates (such as mica), or rock flour, with almost any mineral composition. Therefore, in a strict sense, particles in the fine-size range should always be referred to as silt-sized or clay-sized and not as silt or clay. In practice, however, soil descriptions use these terms and a grain-size classification to indicate that the nomenclature is based on grain size only.

Soil type is usually described by two words: the noun is *always* the dominant soil component, and the adjective is *always* the component that is next in quantity. Thus, there are clayey sands, sandy clays, silty sands, clayey silts, and so on.

MODIFYING TERMS

Modifying terms provide additional information that make the description unique to the particular soil being analyzed. In agricultural descriptions, soil structure, consistency, and reaction with hydrochloric acid (HCl) are the most common modifying terms. In engineering, the modifiers generally fall into two basic types: (1) clauses that follow the basic soil type, such as "silty clay *with fine gravel*," and (2) adjective modifiers that precede the color, such as "*varved* blue silty clay." The combination of the agricultural and engineering descriptions provides significant advantages for the interpretation of the soil and the soil classification.

SOIL STRUCTURE

Soil structure is a description of the shape or geometry of clusters of individual soil particles, soil aggregates, or **peds.** Soil peds are separated from one another by surfaces of weakness. The aggregate type refers to the shape and arrangement of the predominant soil peds, if any (Figure 7–4).

Soil structure is related to the physical characteristics of the soil material. Coarse-textured soils, like sands, lack cohesion and therefore have a single-grained structure. A clayey sand, however, may have sufficient

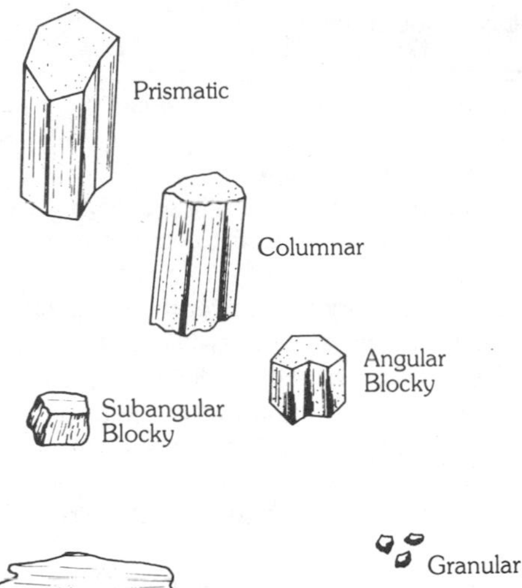

FIGURE 7–4 Common soil structures used in soil descriptions.

cohesion to be blocky or prismatic. Silty clays or clays have a blocky or subangular blocky structure, depending upon the clay mineralogy and geological history. A highly plastic, expansive clay will usually have an angular blocky structure.

SOIL CONSISTENCY

Soil consistency is the tendency of a soil aggregate to hold together when stress or pressure is applied to an undisturbed mass. Soil consistency provides descriptive data about the effect of changing moisture content on the physical characteristics of the soil. Consistency should be described at three soil moisture states: air-dry, natural field moisture content (or moist), and wet. These three moisture states approximately correspond with the shrinkage limit, plastic limit, and liquid limit (see Concept 2–3). Table 7–1 gives commonly used terms to describe soil consistency.

SOIL IDENTIFICATION TESTS

Accurate and complete soil descriptions are often impossible without the assistance of some indicator test. Coarse-grained soils are usually described correctly because the grain size of fine sand is near the limit of resolution for the average person. These soils can therefore be described by visual observation. Silt-sized and clay-sized soils, however, must rely on identification tests as part of the process of soil description.

There are numerous tests to differentiate between silt sizes and clay sizes or to determine if small quantities of fine sand-sized material are in a soil. Some of the common tests are given here, with more detail in Appendix 2. The **feel test** detects the presence of sand when you forcefully push your thumb through a soil pat in your palm. If sand particles are present, they will

TABLE 7–1 Soil consistency terms

Loose	Noncoherent even when pressed together.
Very friable	Soil material crushes under very gentle pressure but coheres when pressed together.
Friable	Soil material crushes easily under gentle to moderate pressure between thumb and forefinger and coheres when pressed together.
Firm	Soil material crushes under moderate pressure between thumb and forefinger, but resistance is distinctly noticeable.
Very firm	Soil material crushes under strong pressure; barely crushable between thumb and forefinger.
Hard	Soil material cannot be crushed.
Very stiff	Soil material deforms without crushing under strong pressure.
Stiff	Soil material deforms without crushing under moderate pressure.
Plastic	Soil material behaves like modeling clay.
Soft	Soil material is easily deformed.
Sticky	Soil material is soft and sticks to hand.
Very sticky	Soil material is very soft, gooey, and sticks well to the hand, requiring large volumes of water to wash it off.

Note: Other terms may be used when necessary, provided that they have a common meaning and are descriptive.

grind the palm. The **taste test** differentiates between silt size and clay size by its grittiness or smoothness, respectively, when the soil is ground between your front teeth.

Two tests are based on the consistency of the soil. In the **molding test,** a wet soil sample is rolled into a thread. If it breaks apart when lifted from one end, the soil is a silt, while if it has wet strength (cohesion), it is a clay-sized soil. The **crushing test** is run on an air-dried sample. Silts will crush or crumble under light to moderate pressure, while clays require moderate to high pressure and in some cases will not crush at all.

COMMON TERMINOLOGY AND ABBREVIATIONS

Since most soil descriptions are made in the field immediately after samples have been collected, the engineering geologist or engineer must develop both accurate and rapid techniques to record the data. As a guide, a list of common terms and abbreviations is presented in Table 7–2. Keep in mind, however, that different engineering departments and organizations probably use different terms and/or abbreviations; therefore, this list is only a guide.

SOIL CLASSIFICATION

After the engineering geologist has completed the soil description, usually as the sample is collected, the soil can be classified by any desired system. Engineering geologists use knowledge of the geology of the parent material,

TABLE 7–2 Terminology and abbreviations used in soil descriptions

Colors		
white (wh)	blue (bl)	red (r)
black (blk)	yellow (y)	purple (p)
gray (gr)	brown (br)	green (grn)
Adjectives: light (lt), dark (dk)		
Basic Soil Types		
bedrock (bdrk)	cobbles (cobs)	silt(y) (si)
ledgerock (ldgrk)	gravel(ly) (grav)	clay(ey) (cl)
boulder(s) (bldr)	sand(y) (sa)	
Adjectives: large (lge), medium (med), small (sm); coarse (co), medium (med), fine (fi); inorganic (inorg), organic (org)		
Modifying Terms (general)		
with (w/)	rock fragments (rk frags)	lens(es) (lns)
with some (w/so)	varved (v)	alternate (alt)
with scattered (w/sc)	partings (ptgs)	
Modifying Terms (coarse-grained soils)		
angular (ang)	rounded (rd)	poorly graded (p.g.)
subangular (sbang)	well-graded (w.g.)	uniform (uni)
subrounded (sbrd)		
Modifying Terms (fine-grained soils)		
soft (s)	loose (lo)	compressible (comp)
medium (med)	very friable (v. fri)	plastic (pl)
stiff (st)	friable (fri)	nonplastic (npl)
very stiff (v. st)	firm (fir)	sticky (stk)
hard (hd)	very firm (v. fir)	very sticky (v. stk)
Modifying Terms (soil structure)		
single-grained (s. grd)	blocky (bloky)	platy (plty)
granular (gran)	subangular blocky (sbang. bloky)	massive (mas)

the effects of the soil-forming factors, and the engineering properties of the soil in soil classification. The most important classification skill is translating from one classification system to another and communicating the information to other scientific fields.

Soil classification in agronomy originated as a system to describe soil as a medium for plant growth. That is, soils are either "good" or "poor" producers. Even today, a soil may be classified as a good corn, cotton, or beet soil. This original concept of soil was modifed after the science of geology developed, and such soil classes as limestone soil or sandy soil developed. With continued scientific research on the genesis of soil and the recognition of the soil-forming factors, the U.S. Department of Agriculture developed a genetic soil classification system. The problem of defining exactly which genetic process or soil-forming factor was responsible for the soil led to the final establishment of the **Comprehensive Soil Survey System,** which is in use today.

The Comprehensive Soil Survey System is very similar to the biological system used to classify plants and animals. A series of six categories, from general to specific, was established: **order, suborder, great group, subgroup, family,** and **series.** Order is based on soil morphology with some genetic base; suborder is based on genetic factors, such as slope or moisture conditions. Great groups are defined by diagnostic horizons, with subgroups being subdivisions of great groups. Plant growth properties establish the family category. The Soil Survey Staff (1960) defines a series as a "collection of soil individuals essentially uniform in differentiating characteristics and in arrangement of horizons."

Soil scientists in their surveys describe the soil at different depths below the surface and establish **soil horizons** (Figure 7–5). The characteristics of the soil and the horizons are the basis for soil classification; therefore the engineering geologist should have a basic understanding of the ten orders in the Comprehensive Soil Classification System (Table 7–3). Understanding the morphological characteristics (order) of the soil helps the engineering

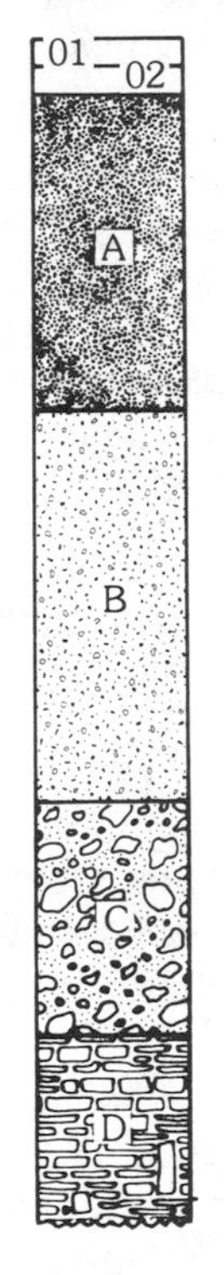

FIGURE 7–5 Example of an agricultural soil profile.

TABLE 7–3 Comprehensive Soil Classification System

Soil Order	Description
Entisol	Recent soil without genetic soil horizons
Vertisol	Soils susceptible to shrink-swell, usually having mixed horizons
Inceptisol	Young soils; horizons result from alteration of parent material without extreme weathering
Aridisol	Soils formed in desert areas; frequently contain cemented horizons and salts
Mollisol	Excellent agricultural soil with a thick, dark upper horizon
Spodosol	Forest soil, subsurface horizon rich in organic matter; frequently form on permeable parent materials and show extreme weathering
Alfisol	Excellent agricultural soil; gray to brown surface horizon and a well-developed clay subsurface horizon. Moderate weathering
Ultisol	Deeply weathered soils that form in moist, warm climatic zones. Commonly red in color
Oxisol	Severely weathered soil with deep oxidized zone
Histosol	Organic soils

geologist to use and interpret soil surveys and soil maps which map the soil by soil series.

Generally, geologists do not have the same kind of soil classification system that soil scientists do. The geology of soils, however, is important in the interpretation of a soil, because the parent material is a geologic (rock) material. Many soillike materials, particularly entisols, are actually uncemented to weakly cemented clastic sedimentary rocks. Some geologists differentiate between *modern soils,* which exist on the surface, *ancient soils* or *paleosoils,* which are buried soil horizons, and *recent sedimentary* rocks.

Engineering classifications of soil materials are based on their mechanical and material properties. The engineer is usually dealing with soil as a building material and is concerned about the performance of the soil when subjected to some stress. In many cases, the extensive amount of information contained in a soil survey report has little if any significance to an engineering project because the foundation, subgrade, or other engineering use is placed below the modern or agricultural soil profile. In other projects, where the structures are light or something is to be buried, such as utilities, the soil survey report can be invaluable. The engineer is therefore interested in both the agricultural soils and in the unconsolidated to lightly cemented sedimentary rocks.

Unified Soil Classification System						
					Typical Names	Identification Procedures
Coarse-grained Soils More than half of material is *larger* than No. 200 sieve size. The no. 200 sieve size is about the smallest particle visible to the naked eye.	Gravels More than half of the coarse fraction is larger than no. 4 sieve size	(For visual classification, the ¼ in. size may be used as equivalent to the no. 4 sieve size.)	Clean gravels (little or no fines)	GW	Well-graded gravels, gravel-sand mixtures, little or no fines	Wide range in grain size and substantial amounts of all intermediate particle sizes.
				GP	Poorly graded gravels or gravel-sand mixtures, little or no fines	Predominantly one size or a range of sizes with some intermediate sizes missing.
			Gravels with fines (appreciable amount of fines)	GM	Silty gravels, gravel-sand-silt mixtures.	Nonplastic fines or fines with low plasticity. (For identification procedures, see ML in Fig. 7–8.)
				GC	Clayey gravels, gravel-sand-clay mixtures.	Plastic fines. (For identification procedures, see CL in Fig. 7–8.)
	Sands More than half of the coarse fraction is smaller than no. 4 sieve size.		Clean sands (little or no fines)	SW	Well-graded sands, gravelly sands, little or no fines.	Wide range in grain size and substantial amounts of all intermediate particle sizes.
				SP	Poorly graded sands or gravelly sands, little or no fines.	Predominantly one size or a range of sizes with some intermediate sizes missing.
			Sands with fines (appreciable amount of fines)	SM	Silty sands, silt-sand mixtures.	Nonplastic fines or fines with low plasticity. (For identification procedures, see ML in Fig. 7–8.)
				SC	Clayey sands, sand-clay mixtures.	Plastic fines. (For identification procedures, see CL in Fig. 7–8.)

FIGURE 7–6 Unified Soil Classification System for coarse-grained soils.

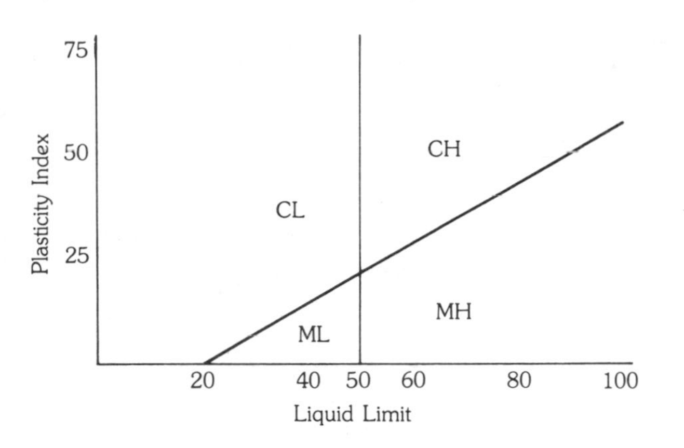

FIGURE 7–7 Relationship between plasticity index and liquid limit used to classify fine-grained soils in the Unified Soil Classification System.

There are many different engineering soil classification systems. Among the common ones are the **AASHO,** now the **AASHTO** (American Association of State Highway and Transportation Officials); **FAA** (Federal Aviation Administration); and **USCS** (Unified Soil Classification System). The AASHTO and FAA classification systems are keyed to the performance of the soil for roads or runways. Casagrande (1948) developed the Unified Soil Classification System during World War II to classify soils for airfield construction, and this system has become the most widely used. The AASHTO and FAA systems are still used to classify soils for some design procedures.

The Unified Soil Classification System is based on a logical grouping of the soil's material properties. The first division is based on the grain size of the soil sample. Since grain size influences the soil's cohesion, the grain-size division is then subdivided according to cohesion. For noncohesive soils, the modifier is based on grain-size distribution, while for cohesive soils the modifier describes the influence of changes in moisture content (soil plasticity). This system can therefore be directly related to the soil descriptions of grain size, grading or sorting, and soil consistency. Soil consistency is quantitatively represented by the **Atterberg limits** of the fine-grained or cohesive soils.

Coarse-grained soils are those in which more than 50 percent of the aggregates are retained on a 200-mesh screen. The Unified Soil Classification System subdivides coarse-grained soils into two classes: gravel (G) and sand (S), based on a grain size of ¼-in. (6.35-mm) or number 4 screen. Modifiers are related to the grading: well-graded (W) (poorly sorted in geology) or poorly graded (P) (well-sorted in geology), or the existence of fines (materials that pass a 200-mesh screen), such as silt (M) or clay (C). Figure 7–6 shows the Unified Soil Classification System for coarse-grained soils.

Fine-grained soils, those that pass a 200-mesh screen, are silt (M) or clay (C), and then modified by the Atterberg limits. If the liquid limit is less than fifty, the soil is classified as low plasticity (L). If it is greater than fifty, the soil has high plasticity (H) (Figure 7–7). Figure 7–8 shows the Unified Soil Classification System for fine-grained soils.

				Identification Procedures on fraction smaller than no. 40 sieve size		
				Dry Strength (crushing characteristics)	Dilatancy (reaction to shaking)	Toughness (molding test)
Fine-grained Soils More than half of material is smaller than no. 200 sieve size The no. 200 sieve size is about the smallest particle visible to the naked eye.	Silts and Clays Liquid limit is less than 50	ML	Inorganic silts and very fine sands, rock flour, silty or clayey fine sands or clayey silts with slight plasticity.	None to slight	Quick to medium	None
		CL	Inorganic clays of low to medium plasticity, gravelly clays, silty clays, sandy clays, lean clays.	Medium to high	None to very slow	Medium
		OL	Organic silts, and organic silty clays of low plasticity.	Slight to medium	Slow	Slight, feels weak and spongy
	Silts and Clays Liquid limit is greater than 50	MH	Inorganic silts, micaceous or diatomaceous fine sandy or silty soils, elastic silts.	Slight to medium	Slow to medium	Slight to medium
		CH	Inorganic clays of high plasticity, fat clays.	High to very high	None	High
		OH	Organic clays of medium to high plasticity, organic silt.	Medium to high	None to very slow	Slight to med. spongy
Highly Organic Soils		Pt	Peat and other highly organic soils.	Readily identified by color, odor, spongy feel, & frequently by fibrous texture.		

FIGURE 7–8 Unified Soil Classification System for fine-grained soils.

PHYSICAL PROPERTIES OF SOIL MATERIALS

The complex nature of soil materials, which is related to the five soil-forming processes and the environment of deposition of uncemented sediments, gives them highly variable physical properties. Relationships between the solid, liquid, and gaseous phases play an important role in controlling the behavior of a soil material when subjected to stress or strain. Cohesive soils undergo changes in their physical properties because their consistency changes with the moisture content.

As Figure 7–7 shows, the Atterberg limits are primary criteria in the engineering classification of soils. Since the engineer needs a means of relating the expected engineering properties of a soil to its classification, extensive research relating the Unified Soil Classification to engineering properties has been conducted. Table 7–4 gives the results of some of this research.

The physical properties of some soils can be predicted from a knowledge of the geology of the soil or the environment of deposition of the parent material. Such soils include **glacial soils, windblown soils, alluvial soils,** and **organic soils.** Glacial soils are derived from the deposition of rock materials that were eroded and transported by glaciers or glacial meltwaters during the Pleistocene. Windblown aeolian soils are found where wind processes have been active. They are fine sand and silt-sized soils that are highly permeable, occasionally weakly cemented, and highly susceptible to erosion. Alluvial soils are associated with major rivers where thick deposits of silts, sands, and gravels have accumulated. These soils are highly productive for agriculture as well as being potential sites for sand and gravel quarrying. Soils deposited in swamps and marshes are rich in organic matter (organic soils); they may in fact be composed of thick beds of it. These low-strength soils are highly compressible and usually saturated.

TABLE 7–4 Properties of soils based on the Unified Soil Classification System

Unified Soil Classification	Permeability*	Shear Strength*	Compressibility*
GW	High	Very high	Very low
GP	Very high	High	Low
GM	Low to medium	High	Low
GC	Very low to medium	Medium	Medium
SW	High	Very high	Very low
SP	High	High	Low
SM	Low to medium	Medium	Medium
SC	Very low to low	Low	High
ML	Low	Low	High
MH	Very low to low	Very low	Very high
CL	Low	Very low	Very high
CH	Very low	Very low	Very high

*Determined on compacted, saturated samples.

FIGURE 7–9 Distribution of glacial till deposits in the United States.

The study of glacial deposits is a field of specialization within geology, and is far too complex to cover in this text. The advance and retreat of the glaciers has formed unique landforms and sediments. Sediments derived from these processes are the glacial soils, or **till.** They are clastic sediments and are widely distributed throughout the northern portion of the continent (Figure 7–9). These deposits usually fall into the engineering classification of soil. Tills are deposited without further transport as the ice melts. Outwash sediments are deposited in front of the glacial ice by meltwater, and aeolian, windblown sediments are deposited far from the ice front. The mechanism of deposition of the material plays a significant role in establishing the characteristics of the till. Till derived from the melting of a stagnant glacier

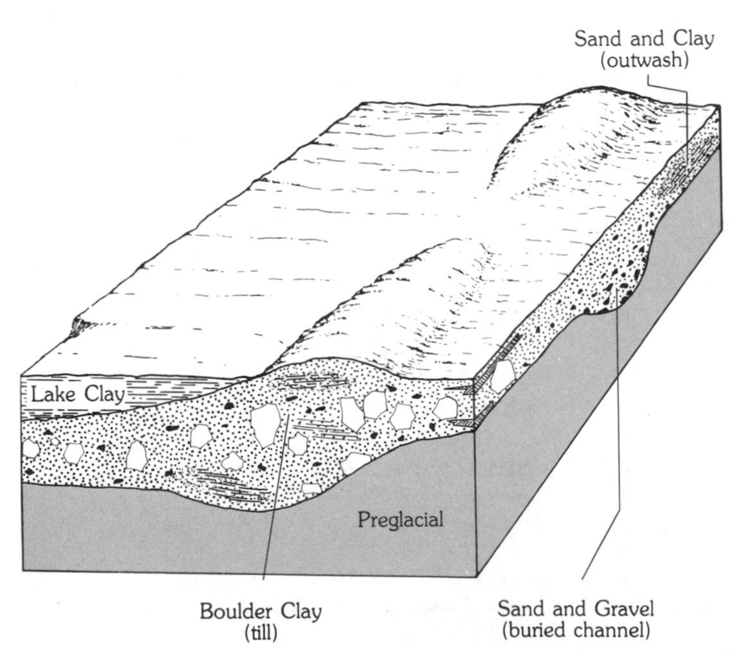

FIGURE 7–10 Block diagram of the environments of deposition of glacial sediments.

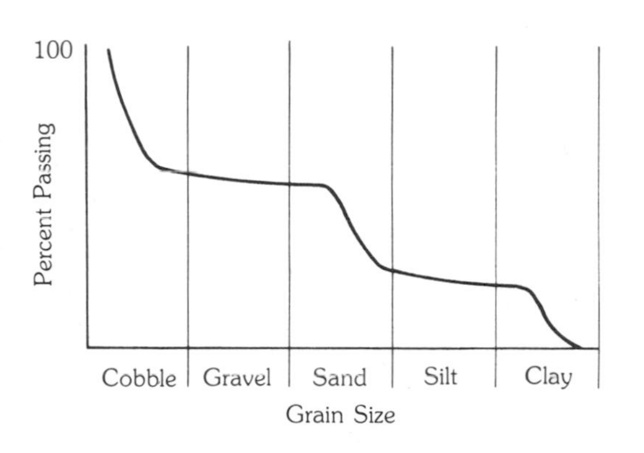

FIGURE 7–11 Step grading or multimodal size distribution common to glacial sediments.

contains a great mixture of particle sizes, ranging from very large boulders to clay-sized particles. Sediments deposited by meltwater, on the other hand, are similar to clastic sedimentary rocks, in which the size and degree of sorting reflect the energy of the meltwater streams. The deposits range from gravel to clay size (Figure 7–10). Tills are frequently mineralogically heterogeneous and have a multimodal size distribution (Figure 7–11). The characteristics of a till to a large degree reflect the lithology of the terrain that was scoured and eroded by the glacier (Table 7–5).

The most significant feature of glacial tills for the engineering geologist is the variability of the till with depth. Since the advance and retreat of a glacier is controlled by variations in the climate, the rate of melt and therefore the available sediment transport energy varies greatly. During the late summer, meltwater streams are at peak flood, while during the late winter they are at minimum flow and may even be frozen solid. As a result, the sediments at any one station may range from coarse gravel, which was deposited during each flood period, grading upward to clays, which were deposited during the low-flow periods. In addition, these meltwater deposits may overlie deposits of boulders through clay that were laid down as a boulder till when the stagnant or retreating ice melted (Figure 7–12).

Windblown or aeolian soils can be subdivided into two basic classes: **dunes,** which are predominantly sand-sized materials and are found in modern and recent coastal zones and deserts, and **loess,** which is a silt-sized

TABLE 7–5 Relationship between till texture and terrain lithology

Texture of Till	Lithology of Terrain
Granular	Gneiss, granite, quartzite (very high-strength rocks)
Well graded–poorly sorted	Limestone, slate, sandstone (high–medium-strength rocks)
Stoney clay	Shale, soft limestone, clays (low-strength rocks)

Source: After Derbyshire et al., 1976.

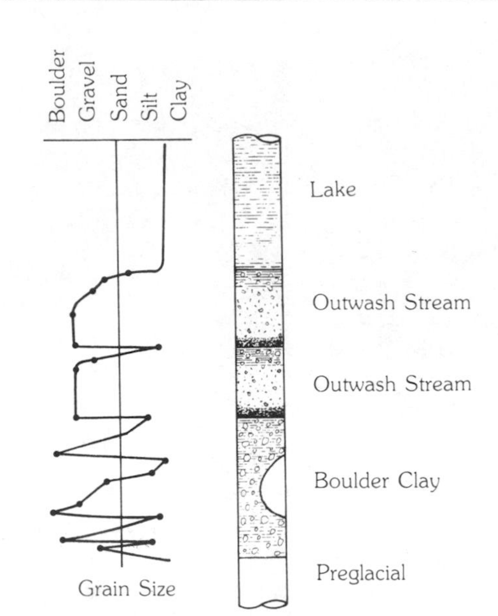

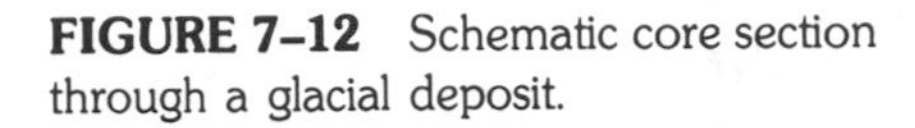
FIGURE 7–12 Schematic core section through a glacial deposit.

deposit that blankets much of the central United States (Figure 7–13). These deposits tend to be well sorted (poorly graded) and on a regional scale, to be coarser closer to their source. Aeolian soils are highly erodible by either running water or wind if they are not protected. Shallow roadcuts in loess deposits are frequently cut vertically in order to control slope erosion. In this way, rainfall soaks into the permeable soils rather than running off, thereby reducing erosion problems.

Alluvial soils are deposited by streams and are frequently shown on geologic maps by the symbol Qal, which represents Quaternary alluvium. These soils support about one-third of the world's food production (Soil Survey, 1951). Alluvial soils also form on deltas and as alluvial fans, where

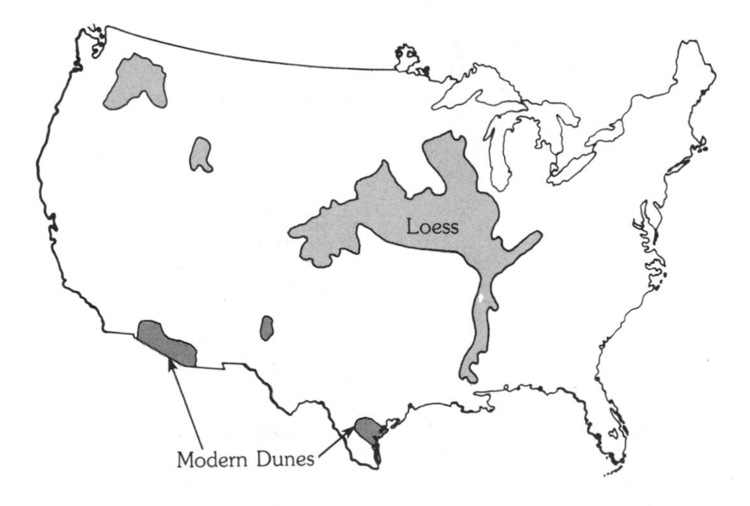

FIGURE 7–13 Distribution of aeolian (windblown) sediments in the United States.

stream energy decreases as it enters either a body of water or the open plain at the mouth of a mountain stream.

Some soils are best grouped by their characteristic behavior, rather than by geologic origin. Three such soils are significant to the engineering geologist. **Expansive soils** change their volume as the soil moisture conditions change, and **collapsing soils** decrease abruptly in volume when saturated. **Quick clays,** or **sensitive clays,** and soils susceptible to **liquefaction** appear to be stable, solid materials but suddenly liquefy and flow.

EXPANSIVE SOILS

For a soil to be labeled *expansive,* it must contain a significant amount of smectite (montmorillonitic clay minerals). Expansive soil becomes a problem when variations in the ambient environment produce changes in the soil moisture content, in turn causing a volume change.

Expansive soils are a worldwide problem; Donaldson (1969), for example, lists nineteen countries where swelling and shrinking soils cause serious engineering problems. In the United States, expansive soils are responsible for 2.3 billion dollars damage annually (Jones and Holta, 1973), with more than a billion dollars damage to highways and streets (Figure 7–14). Expansive soils are, in fact, the single most costly natural disaster; the average yearly loss from earthquakes, hurricanes, tornadoes, and floods combined amounts to only half that caused by expansive soils.

Expansiveness is directly related to the ability of a clay mineral to give up water (shrink) and take on water (swell). The greater the intra-crystalline charge imbalance, the greater is the ability of the minerals to respond to moisture changes, and therefore the greater its expansiveness. Because expansive clay minerals can absorb water onto their crystal lattices, they can undergo significant moisture content changes and therefore have a high plasticity index. These materials require large volumes of water to change the physical characteristics from a semisolid to a viscous liquid.

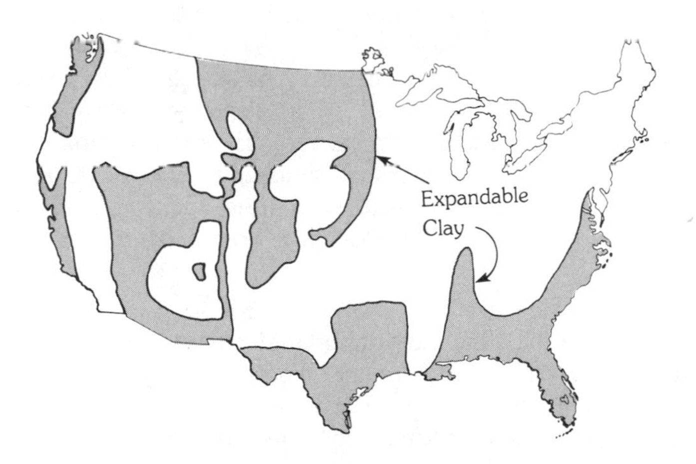

FIGURE 7–14 Distribution of rocks containing expandable clay minerals. (After Tourtelot, 1974. Reprinted by permission of the Association of Engineering Geologists.)

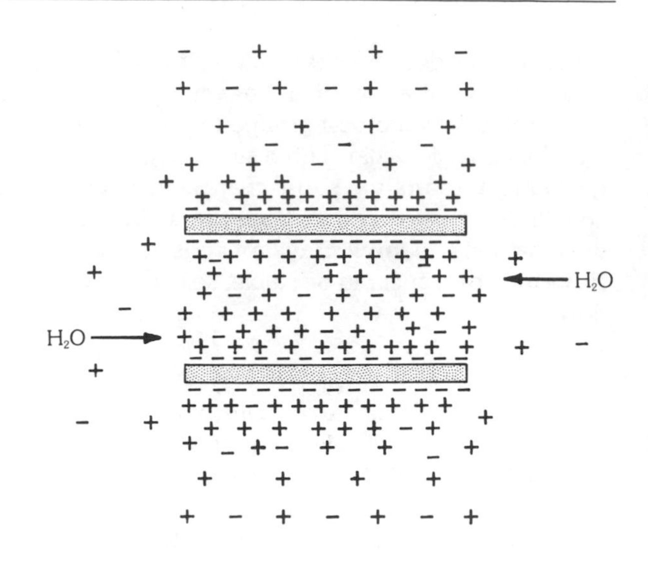

FIGURE 7–15 Movement of water into the space between expandable clay minerals to satisfy the existing electric charge field.

The adsorption of water into the interlayer involves two mechanisms (Gillott, 1968). Water, a dipolar molecule, will tend to orient itself according to the charge of the atoms or molecules that it surrounds; the result is ion hydration. On the surface of the clay platelets, the external oxygen atoms of the tetrahedra attract the positive ends of the water molecules. Hydrogen bonding maintains a very strong grip on this moisture layer, known as the absorbed or "solid water" layer. The other mechanism involved in occupation of the interlayer by water is a result of the osmotic potential between the clay interlayers and the surrounding pore fluids. Unless the pore fluid is unusually saline, there will be a concentration gradient between the water in the interlayers and the pore fluid. Since the cations are restricted from diffusing into the pore fluids, water tries to enter the clay interlayer in an attempt to equalize the concentrations. Since the interlayer bonding in smectite is very weak, the osmotic potential overcomes the bonding strength, resulting in transfer of moisture (Figure 7–15). When a smectitic soil is allowed free access to water and is unrestrained by surcharge pressure, the clays will hydrate and the soil will attain its maximum volume.

When the soil begins to dry out, capillary tension increases, and consequently so does the soil suction. If the soil dries enough, the capillary potential of its pore space exceeds the osmotic potential, and then smectite will begin to release its interlayer water and simultaneously to collapse. Thus, cycles of wetting and drying result in volume fluctuations. This fluctuation in volume is the cause of the damage from expansive soils.

OTHER FACTORS INFLUENCING SWELL POTENTIAL

The smectite content of a soil is the most important factor in determining a soil's expansive potential; however, there are others, including the texture and carbonate content.

Since sand and silt are essentially inert compared to clays, texture is an important influence on expansive potential. Clays are often colloidal in size

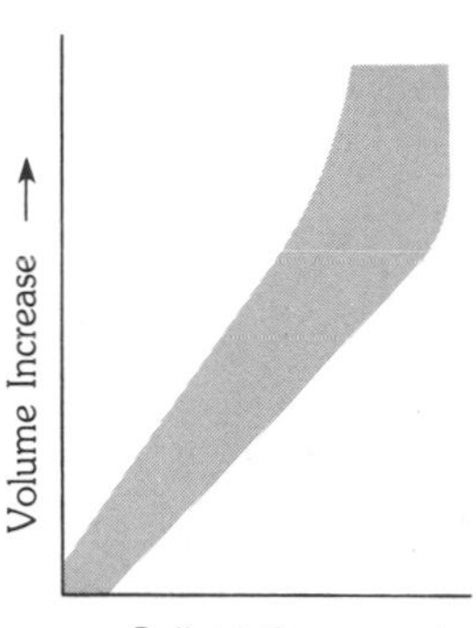

FIGURE 7–16 Generalized relationship between colloid content and soil volume increase from wetting.

and have a very high surface area per unit volume. The result is that clays have a high affinity for water just because of surface activity. Several authors have pointed out the relationships between a soil's colloid content and degree of expansion (Figure 7–16) (Holtz and Gibbs, 1956).

A soil's carbonate content affects the physical and chemical conditions of the soil environment. Carbonate in soil occurs primarily in the form of calcite, $CaCO_3$, which acts as a cementing agent, forming aggregates out of the fine clay platelets. This results not only in a decrease in the plasticity, but also in an increase in the strength of the soil. Chemically, calcareous soils tend to have excess Ca^{+2} dissolved in the soil water. Availability of Ca^{+2} has two important effects. First, the high concentration of Ca^{+2} decreases the normally high osmotic potential between the soil water and the clay interlayers. Thus, smectite's "thirst" for water is reduced. Secondly, bivalent Ca^{+2} ions displace the monovalent cations in the smectite interlayers. Thus, a highly expansive Na-smectite becomes a much less active Ca-smectite. Ca^{+2}-saturated clays also tend to flocculate, contributing to the agglomeration effect of the calcite cement.

ENVIRONMENTAL FACTORS IN EXPANSIVE SOILS

Climate is the single most important environmental factor affecting expansive soils because it largely controls the depth of the water table and the active zone of soil moisture. Problems of expansive soils may be nonexistent in humid climatic regions where the water table lies near the ground surface. On the other hand, the water table in arid regions may be at great depths and exert little influence on soil moisture conditions at the surface. The **soil moisture active zone** is that portion of the profile that experiences major variations in seasonal moisture content. The depth of this zone is important to foundation engineers because it represents a minimum depth for placement of piers and because it is the critical depth to be considered in calculating swell potential. The thickness of the active zone is usually determined through long-term measurement of heave to depths of about 5 m below the surface.

Monitoring moisture equilibrium conditions in an expansive soil profile before and after the placement of a foundation slab is of interest to the

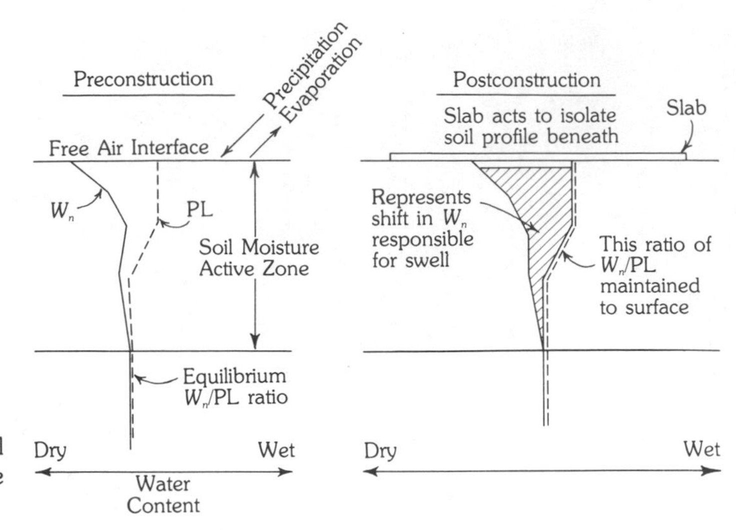

FIGURE 7–17 Alteration of the soil moisture profile after an impermeable surface is placed on the soil surface.

engineer because it indicates the magnitude of possible soil movement. In an attempt to correlate equilibrium moisture content with soil index properties, Russam (1966) suggested that after an expansive soil has been covered with an impervious layer, such as a foundation or road, the ratio of natural soil moisture content (W_n) to the plastic limit (PL) will reach a constant value (Figure 7–17).

Several climatic ratings attempt to quantify rainfall, evapotranspiration, and soil drainage characteristics (Thornthwaite, 1948: Prescott, 1949). An additional study by the Building Research Advisory Board (BRAB, 1968) produced a climatic rating system for the United States, which gives the probability that drought will not occur. These ratings, along with a Unified Soil Classification designation, constitute the Federal Housing Administration (FHA) method for selecting the proper foundation.

Vegetation, especially large trees, near a foundation may bring about large soil volume changes as it withdraws soil water (Figure 7–18). During relatively wet periods, the drawdown effect may be somewhat muted, but during drier seasons, the vegetative demand for soil water may result in pronounced, though localized, decreases in volume. Altmeyer (1956), Barber (1956), and Bozozuk and Burn (1960), among others, have studied the deleterious effects of vegetation. Felt (1953) showed that structures placed over areas with variable soil moisture content were likely to fail even after the removal of vegetation because of volume changes during the rewetting phase. Hammer and Thompson (1966) report that severe structural failures were related to planting trees during construction. This study shows that the most serious failures occurred where large elm, poplar, and willow trees were planted within 12.2 m of a structure. Trees with a 30 cm diameter or less and shallow roots produced significantly fewer failures. These authors suggested that trees should be planted no closer to the foundation than one-half their expected mature height.

Whether the problem is cyclic, as in the case of stresses imposed by vegetation, or unidirectional, as in the case of heave produced by placing a

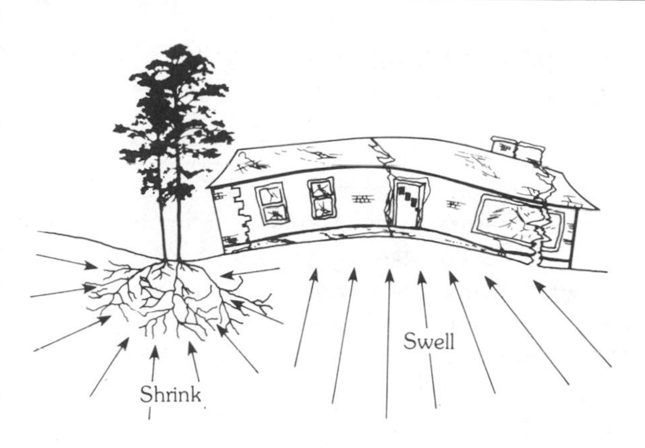

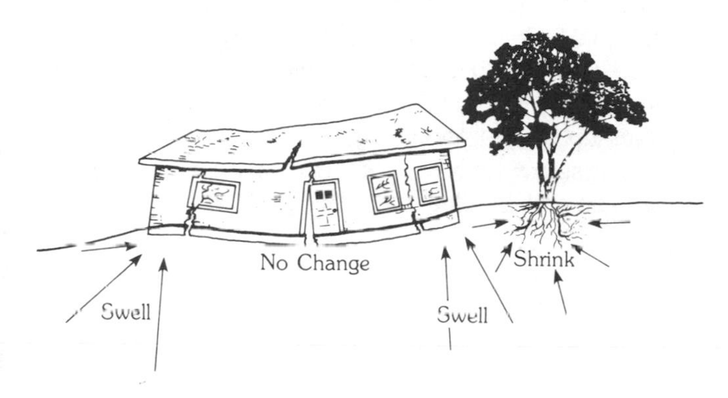

FIGURE 7–18 Influence of trees on the deformation around a foundation set on an expansive soil.

structure over a heavily desiccated clay, time is a crucial factor. Most clays have low permeabilities, which are further decreased by swelling; thus, ultimate volume changes in expansive soils and attendant deformation of the overlying structure may involve a time scale measured in years—a design consideration that the engineer cannot ignore.

Topography, especially steep slopes, is an important consideration for building on expansive soils. Expansive soil on slopes undergo a form of creep in which movements are normal to the slope during expansion and parallel to gravity during shrinkage. As a result, a lateral downslope component is generated. Poor drainage, inherited from the original topography, or poor site finishing can result in ponding of water around structures. Unless corrected, this condition may result in localized swelling and foundation movement.

Site control is important in controlling the damage that expansive soil movement can cause. Maintenance or improvement of lot drainage, the installation of gutters, landscaping, and lawn maintenance all play a complex role in causing volume changes in expansive soils. In some cases, homeowners who are aware of an expansive soil problem have watered the perimeter of the home foundation and have experienced few cracks, while others have seen little reduction in cracking. Actions by homeowners, ignorant of their impact, may result in increased damage.

COLLAPSING SOILS

Some soils abruptly decrease in volume when they are saturated. These materials usually exist in a partially saturated to dry state in nature and, like expansive soils, do not present an engineering problem until the soil moisture conditions are altered. Collapsing soils are commonly associated with fine sand-silt-clay materials deposited along the base of mountains in semiarid environments. Recent work by Schneider (1977), however, has identified this process in reclaimed surface mines in Texas.

Collapsing soils were deposited in metastable conditions that become unstable when the soil is saturated. It is believed that the soil structure in partially saturated soils is maintained by the surface tension of water held at the contacts of the grains and that saturation of the soil fills the voids and reduces the surface tension to zero. In dry soils, the bonding agent is believed to be an oriented layer of clay minerals that was drawn into the fine pores as the soil dried out and that collapse occurs as the clay layer is resuspended in the voids upon saturation (Figure 7–19).

Problems of collapsing soil can involve a wide range of areas from one portion of a foundation to an entire valley. Excessive plant irrigation or the

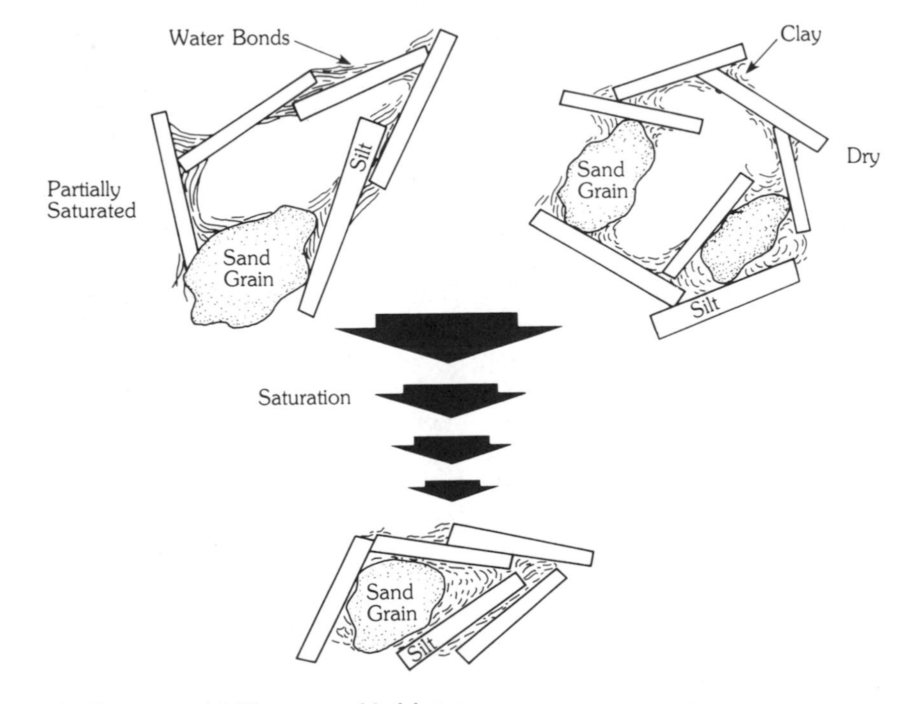

FIGURE 7–19 Mechanisms of collapse in a soil. The metastable fabric is maintained by water bonds or clay bonds that are broken when the soil is saturated.

concentration of roof drainage near a building can lead to differential collapse of the soil supporting the foundation and a corresponding foundation failure. On a larger scale, **hydrocompaction** is a problem in many irrigated valleys in the arid Southwest where continuous irrigation of agricultural lands has led to the settlement of the irrigated fields. Pipelines and roads then may fail, or the direction of land drainage may reverse.

Soils susceptible to liquefaction are low-density sands or silts that are usually saturated. When these soils are subjected to a vibrating load, be it human or tectonic activity, the soil aggregates abruptly densify. Rapid densification leads to an immediate increase in pore water pressure with a corresponding decrease in shear strength. If the conditions are right, the shear strength is decreased to near zero and the soil behaves as a liquid.

Quick clays or **sensitive clays** are clay-sized glacial deposits found in the Scandinavian countries and along the St. Lawrence River valley. The mechanism of failure for these deposits is still being debated; however, all theories depend upon some mechanism that bonds the clays together in some metastable manner. Rosenqvist (1953) proposed that the clays were originally deposited in a marine environment where ions from the seawater acted as the bonding agent. Subsequent to uplift, the bonding cations were leached out by freshwater. Pusch and Arnold (1969) regarded organic matter as the bonding agent, while Kenney et al. (1969) suggested a weak interparticle cement. Recent work by Smalley (1971) has suggested that the clay in quick clays is finely ground, clay-sized particles, or **rock flour,** bound together by a weak cement. Moon (1975) has been able to demonstrate in laboratory tests that quartz-rich rock flour can exist as either a solid or a liquid material at the same moisture content and that the property of the material is controlled by the strain rate; it is a solid at low strain rates and a liquid at rapid (shock) strain rates.

Fluid Earth Materials

eight

The most common and most significant fluid earth material that the engineering geologist must deal with is **water**. Other fluid earth materials that may have to be considered are relatively rare—such as **lava** from volcanic eruptions and a wide variety of **gases** like sulfur and dust. Because they are rarely encountered, these materials are not discussed in this text.

Water is a unique substance in both a chemical and physical sense. Chemically, water is a simple compound of oxygen and hydrogen, H_2O. The two hydrogen atoms are very small compared to the oxygen atom, and are offset at an angle of about 105°. This offset results in a dipolar molecule that can respond to any charge field (Figure 8–1). The water molecule also dissociates as:

$$H_2O \rightleftharpoons H^+ + OH^-$$

This dissociation, which produces free H^+ ions, allows water to act as the universal solvent, since the H^+ ions are available for the hydrolysis reaction.

Physically, water, unlike any other common earth material, can exist in three forms: solid (ice), liquid (water), and gas (vapor) at ambient temperatures and pressures. Moreover, water can occur in these three phases at the same temperature, 0°C at atmospheric pressure. Another unique feature of water is that it reaches its maximum density at 4°C while in the liquid phase (ice at 0°C floats in water) and expands in volume by about 10 percent as it freezes. Because water has unique physical characteristics at ambient temperatures and pressures on the earth, it is cycled through the environment. The **hydrologic cycle** is the circulation of water from the ocean (fluid), to the atmosphere (vapor), to the land (rainfall), and back to the ocean (runoff) (Figure 8–2). The hydrologic cycle can be divided into a series of processes: evaporation and transpiration, precipitation, and runoff and infiltration, which are all related through the hydrologic equation:

$$\text{Precipitation } (P) = \text{Runoff } (Q) + \text{Evaporation and Transpiration } (ET) + \text{Infiltration } (I) + \text{Storage } (S)$$

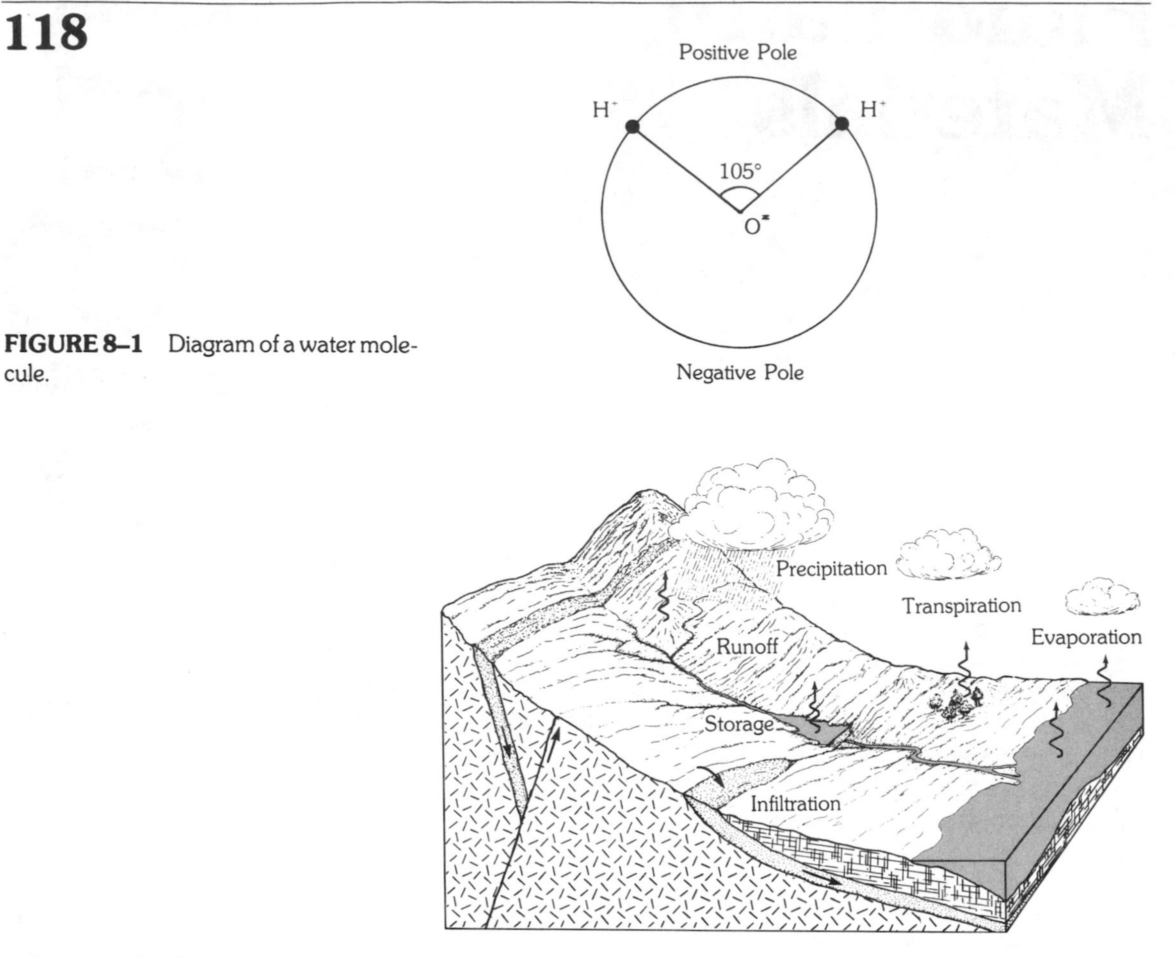

FIGURE 8–1 Diagram of a water molecule.

FIGURE 8–2 The hydrologic cycle.

This equation simply states that water cannot be created or destroyed, and therefore the hydrologic cycle must balance. Calculating this balance, however, is a complex scientific and mathematical problem that has not yet been solved.

HYDROLOGIC CYCLE PROCESSES

PRECIPITATION

The determination of total precipitation in all its forms in any area is a complex problem. Published figures and maps are based upon statistical sampling at rain-gauge stations. The validity of the figures depends on the number of stations and the characteristics of the storm. Intense thunderstorms may produce extremely complex rainfall patterns that require a very tight rain-gauge network to define them (Figure 8–3). Broad regional rainstorms associated with warm fronts, on the other hand, may be so uniform that only a few rain-gauge stations are needed to characterize the precipitation (Figure 8–4).

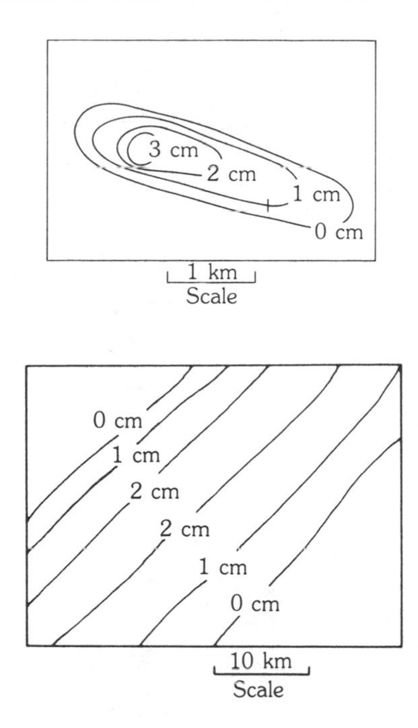

FIGURE 8–3 Map of equal rainfall (isohyetal map) of a thunderstorm.

FIGURE 8–4 Map of equal rainfall (isohyetal map) of a regional rain caused by a weather front.

RUNOFF

Measurements of runoff are simpler because runoff is simply the volume of water that leaves the drainage area in streams or rivers. A stream-gauging station measures the average velocity of flow (v) and the cross-sectional area of the stream channel (a) and determines discharge (runoff) Q as:

$$Q = av$$

Most stream-gauging stations record the water level, or **stage**, as a function of time. Since the stage is related to cross-sectional area and flow velocity is related to stage, it is possible to plot a **hydrograph** that relates discharge to time (Figure 8–5). The total discharge (runoff) from any one rainstorm is the area under the hydrograph minus the base flow.

EVAPOTRANSPIRATION

The volume of water directly returned to the atmosphere from the drainage area by evaporation (E) and transpiration (T) depends upon many climatic and environmental factors. The temperature and humidity of the air and the temperature and type of ground cover influence evaporation. This process often occurs when a summer rainstorm wets hot asphalt pavement; steam (water vapor) rises from the pavement. Transpiration is an organic process whereby growing plants release water vapor through their leaves. Transpiration depends on the growing season, type and density of the vegetative cover, temperature, humidity, and sunlight. Evapotranspiration is, therefore, another complex parameter in the hydrologic cycle.

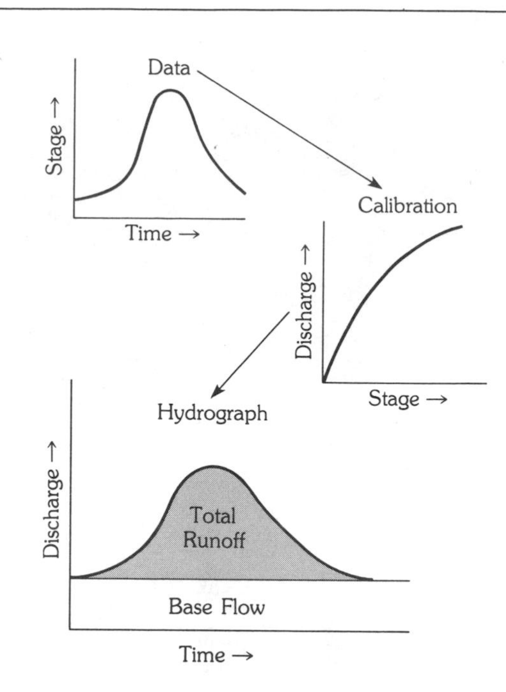

FIGURE 8–5 Determination of a flood hydrograph from a stream-gauging station.

INFILTRATION

Some of the precipitation that falls on the land surface seeps into the soil and is held in the drainage area. The amount of infiltration (seepage) depends upon the permeability of the ground surface. Coarse soils, sands and gravels, and fractured rock can accept more water than can fine-grained soils or massive rocks. In addition to the type of earth material and the precipitation intensity, previous rainfall and temperature factors influence the amount of infiltration. An intense storm will supply water at a rate that is much faster than the surface materials can accept it, and infiltration will be reduced. Water from a storm that follows earlier precipitation will be unable to infiltrate, because the available pore spaces will already be occupied by water. Rain that falls on frozen ground will also be unable to infiltrate. The slope of the land surface influences the amount of time that the water will be in contact with a particular tract of land; thus, steep slopes decrease infiltration compared to flat slopes.

STORAGE

Some of the precipitation that falls in an area is trapped in depressions and other temporary or permanent storage sites. The amount of storage depends upon the general roughness of the land surface. Stored water eventually returns to the atmosphere through evaporation or infiltrates into the ground.

THE HYDROLOGIC CYCLE

The hydrologic cycle is a complex system that has a significant impact on the engineering geologist. In order to work with the system, he or she must recognize that the components of the system vary with time and location and that any quantitative analysis is at best an educated approximation. Any analysis must deal with seasonal or annual averages and statistical predictions of extreme values.

The most important components of the hydrologic cycle to the engineering geologist are runoff (Q) *and infiltration* (I). Both components vary with time and location; however, numerous mathematical relationships have been derived to approximate the expected values.

RATIONAL EQUATION FOR RUNOFF

Runoff (Q) can be approximated by the **rational equation**:

$$Q = ciA$$

where

c = coefficient of runoff, which ranges from 0 to 1.0
i = intensity of rainfall (length/time units)
A = area of precipitation (length^2 units)

The coefficient of runoff (c) is an empirical factor that reduces the total precipitation (iA) to compensate for infiltration plus evapotranspiration plus storage (see Concept 8–1). The selection of a value for the coefficient of runoff is based on the following factors:

1. Increase c with increased slope.
2. Increase c for fine-grained or massive surface materials.
3. Increase c for periods of low evapotranspiration.
4. Increase c for storms of long duration.

Table 8–1 gives representative values for this coefficient.

TABLE 8–1 Representative values for c, coefficient of runoff

c	Characteristics of the Drainage Area
0.1–0.3	Forest
0.1–0.5	Grasslands
0.3–0.6	Row crops, plowed ground
0.5–0.8	Bare ground, smooth
0.5–0.9	Suburban lands
0.7–0.95	Urban lands
0.9–0.99	Water body, full
0.0	Water body, empty

INFILTRATION CAPACITY

The infiltration capacity (f_p) is the maximum sustained rate at which a particular soil can transmit water and is determined by soil structure, soil texture, vegetative cover, biological structures in the soil, soil moisture content, and the condition of the surface. The value of infiltration capacity varies with time during a storm. It is at a maximum when rainfall begins (f_0) and declines to a constant value (f_c) as rain packs the soil by raindrop impact,

CONCEPT 8–1

Rational Equation

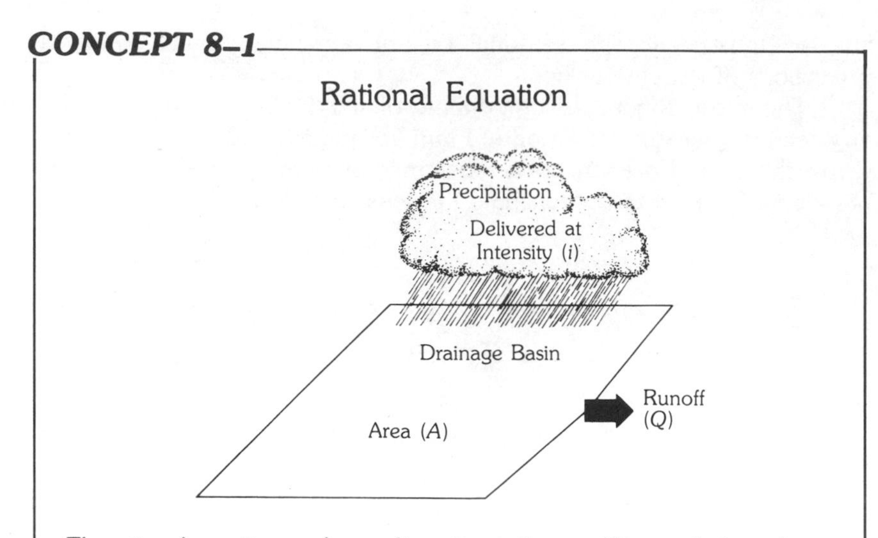

The rational equation can be used to estimate the runoff from a drainage basin and to investigate the effect of a change in land use. Given a drainage basin with an area A, which is receiving precipitation at the rate i, the runoff Q is estimated from:

$$Q = ciA$$

The coefficient of runoff, c, is a correction factor that estimates the effect of infiltration, evapotranspiration, and storage. The rational equation is a very valuable tool to estimate the impact of land use changes. For example, assume that a pasture ($c = 0.3$) is changed to a shopping center ($c = 0.8$). How much additional runoff should be expected from this new land use?

$$Q = ciA = 0.8(iA) \quad \text{shopping center}$$
$$Q = ciA = \underline{0.3(iA)} \quad \text{pasture}$$
$$\text{Increased runoff} = 0.5(iA) = 167\% \text{ more water}$$

fine soil and organic materials wash in, soil expands as the clays take on water, and the structure of the soil surface breaks down. In some expansive soil areas, desiccation (drying) cracks may be more than 5 cm wide and as much as 4 m deep after the summer dry season, and the soil will have a very large infiltration capacity. By mid-fall the soil moisture content has been recharged, closing the cracks and significantly reducing the infiltration capacity. Infiltration capacity (f_p) at any time (t) after a rainstorm begins can be expressed by:

$$f_p = f_c + (f_0 - f_c)e^{-kt}$$

where k = a constant for each soil type and vegetative cover.

At some time (t_1) after a storm begins, the e^{-kt} function will approach 1.0, and the infiltration capacity will be equal to a constant value (see Concept 8–2). Infiltration capacity must be measured in the field by using a

CONCEPT 8–2

Infiltration Capacity

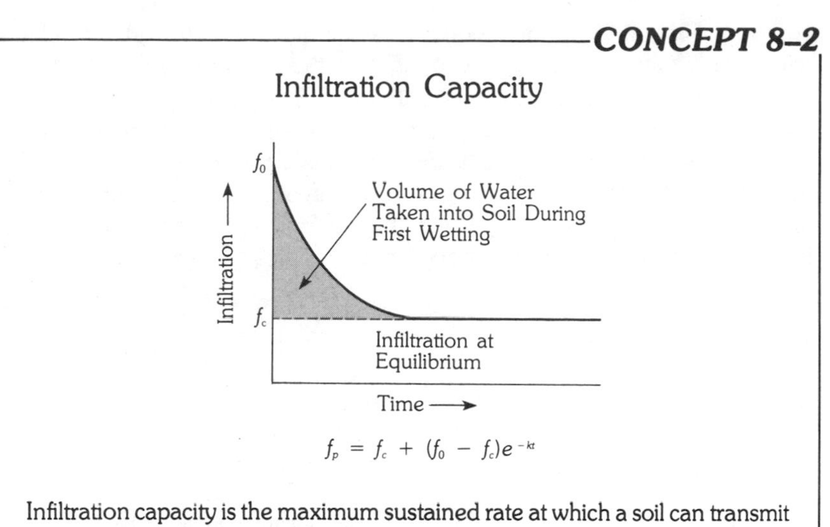

$$f_p = f_c + (f_0 - f_c)e^{-kt}$$

Infiltration capacity is the maximum sustained rate at which a soil can transmit water under the influence of gravity. A soil that is initially dry will be able to take more water than the same soil after it is saturated. The relationship between infiltration and time since rainfall started takes the form of a log decay curve that must be measured for each soil type and soil condition.

rainfall simulator producing a known intensity of rainfall over a known area. The amount of surface runoff is measured during the test, and the infiltration capacity is calculated from:

$$\text{Infiltration capacity} = \text{Precipitation} - \text{Runoff}$$

Once the infiltration versus time data are plotted, f_0, f_c, and k values for the site can be calculated. These data are then used to draw maps of equal infiltration capacity, called **isopotal maps** (Figure 8–5).

The influence of soil type on infiltration capacity is shown in Figure 8–6, where land use (row crop) and soil moisture content (wet soil) are

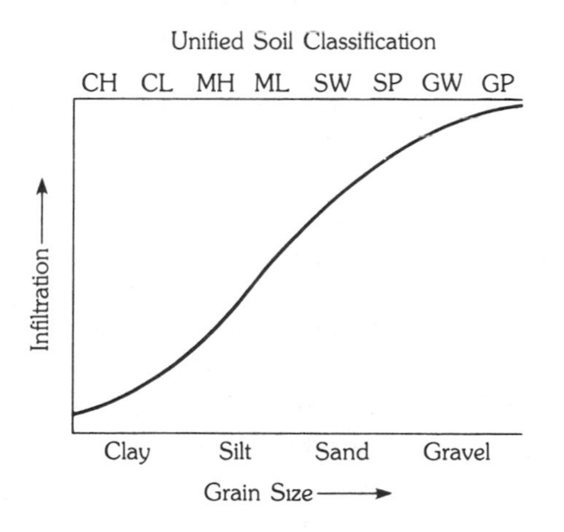

FIGURE 8–6 Generalized relationship between infiltration and the grain size and sorting of soils.

constant. Table 8–2 shows the relative effect of changes in agricultural land use on infiltration capacity, assuming that the soil type is constant.

Estimates of runoff (Q) and infiltration (f_p) values for a drainage area are based upon a knowledge of the precipitation history, obtained from the National Weather Service; the topography; vegetative cover; seasonal factors; and the soil and rock types in the area. Runoff estimates play an important role in the analysis of the surface drainage system, while infiltration estimates are important for the analysis of groundwater.

TABLE 8–2 Effects of land use on infiltration

Infiltration	Land Use
Low	Fallow, bare ground
↓	Row crops
	Poor quality pasture
	Small grain crops
	Fair quality pasture
	Poor quality woodlands
	Good quality pasture
	Fair quality woodlands
	Meadows, not grazed
High	Good quality woodlands, undisturbed

GROUNDWATER

Since ancient times, water has been known to exist underground and has been obtained from wells. From observations of **springs,** where water issues forth from the earth from no apparent source, came the belief that water flowed via tunnel-like passages. It was assumed that such water could not be derived from rainfall because rainfall was believed to be inadequate and the earth too impervious. The concept of underground rivers was perpetuated by people who claimed the ability, by methods veiled in mystery, to find water: well-witchers, dowsers, wiggle-stickers, and doodle-buggers.

A clear understanding of the behavior of groundwater in the hydrologic cycle was reached during the seventeenth century in France. However, it was restricted to the rather exclusive and small scientific community of the day. For the first time, theories were based on scientific observations and quantitative data. Measurements on the River Seine and its drainage basin showed that the rainfall in the basin was about six times more than the river discharge. With the realization that more rain fell than flowed in the river, there was strong evidence to support the theory that infiltration of rainwater was the source of groundwater.

Well-witchers notwithstanding, the basis for understanding the occurrence of groundwater grew with the establishment of the fundamentals of geology. The occurrence of groundwater could be predicted scientifically, based on knowledge of geologic materials and their continuity throughout a region. In 1905, N. H. Darton of the U.S. Geological Survey used his regional structural and stratigraphic studies of the Great Plains to predict the depth to an aquifer on the flanks of the Black Hills. After almost three years of drilling, the Burlington Railway completed a flowing artesian well yielding 400,000 gallons (1500 m^3) per day from within 30 feet (9 m) of the predicted depth.

Today, the problem of groundwater retrieval lies not in finding water, but in determining how much water is available. Since most aquifers cover a large area, they must be regarded as large, natural underground reservoirs, because when one owner uses the resource, the supply of all the others is affected. In order to maintain the resource indefinitely, there must be a hydrologic balance between all waters entering and leaving the aquifer, although it need not be continuous. Based on an understanding of groundwater flow, predictions can be made that allow an excess of water to be withdrawn during periods of need, to be balanced by periods of diminished withdrawal, allowing recovery of the supply. Just as in the case of surface water flow, the flow of water through porous and permeable rock and soil can be predicted. It is necessary to know how withdrawal will affect the overall storage as well as the state of the water table locally.

The modern concept of groundwater flow is based on the findings of Henry Darcy, a French hydraulic engineer who worked on a project to improve the water supply of Dijon, France. After investigating the flow of water through sand filters, in 1856 he reported that his experiments positively demonstrated that the volume of water which passes through a bed of sand of a given nature is proportional to the pressures and inversely proportional to the thickness of the bed traversed. This statement, that the flow rate through a porous media is proportional to the head loss and inversely proportional to the length of the flow path, is known as **Darcy's law.** It, more than any other contribution, serves as the basis for present-day knowledge of groundwater flow.

Darcy's law may be expressed as a formula by introducing a constant of proportionality to the properties he measured. Discharge per unit area (Q) is equal to the constant (K) times the head loss (h) divided by the length of flow path (L):

$$Q = K\left(\frac{h}{L}\right), \text{ or } K\left(\frac{dh}{dL}\right)$$

where K is defined as the hydraulic conductivity.

It is not always correct to think that groundwater flows downhill because it does so only when the flow is horizontal. With inclined flow, water may actually flow uphill. A simple experiment, in principle the same as those Darcy performed, will illustrate this concept (Concept 8–3).

Groundwater is contained in the interstitial pore spaces in the ground and is present virtually everywhere, including the pore spaces in sandstone

CONCEPT 8–3

Darcy's Law

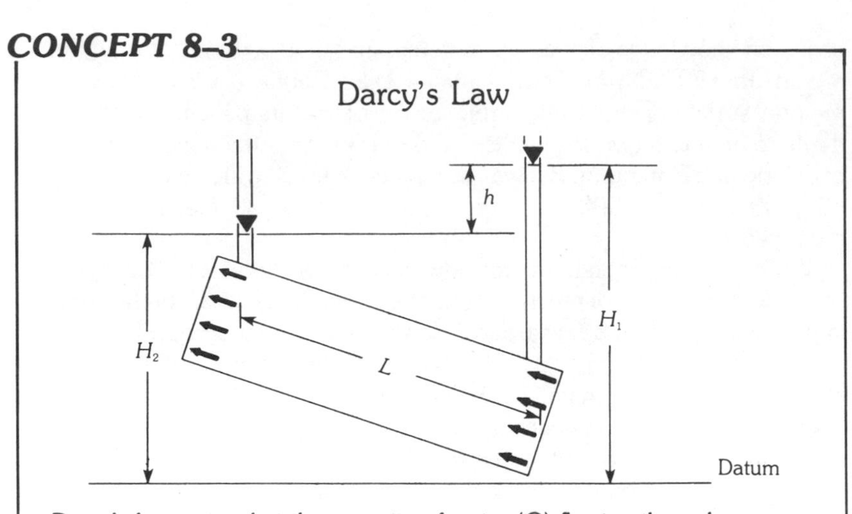

Darcy's law states that the quantity of water (Q) flowing through a porous medium is proportional to loss in head (h) and inversely proportional to the length of the flow path (L). In the diagram above, a fluid is flowing from right to left through a porous medium that is L units long and has a cross-sectional area A units in size. As the fluid moves from right to left, the pressure head H_1 is reduced to H_2; thus, the fluid is flowing down the hydraulic gradient (h/L) but up the pipe.

Darcy's law states that:

$$Q = K\left(\frac{h}{L}\right)A$$

where

Q = flow
K = hydraulic conductivity
$\frac{h}{L}$ = hydraulic gradient = i
A = cross-sectional area

and shales, solution cavities in limestones, and fissures and fractures in igneous and metamorphic rocks. The groundwater conditions at any site are controlled by the characteristics of the rocks at the site. The amount of pore space, or **porosity**, determines the maximum volume of water that can be stored in the rock. The **specific yield** is less than the porosity because some water will be retained in a rock as *hygroscopic* water, which clings to the surface of the soil or rock aggregates (Figure 8–7). Specific yield is calculated from the ratio of the volume of water drained due to gravity to total volume of rock.

The rate at which water can flow through a rock is the **hydraulic conductivity**, which is controlled by grain size and shape, fracture width, continuity, and density (Table 8–3). It is just as possible to have a rock or soil

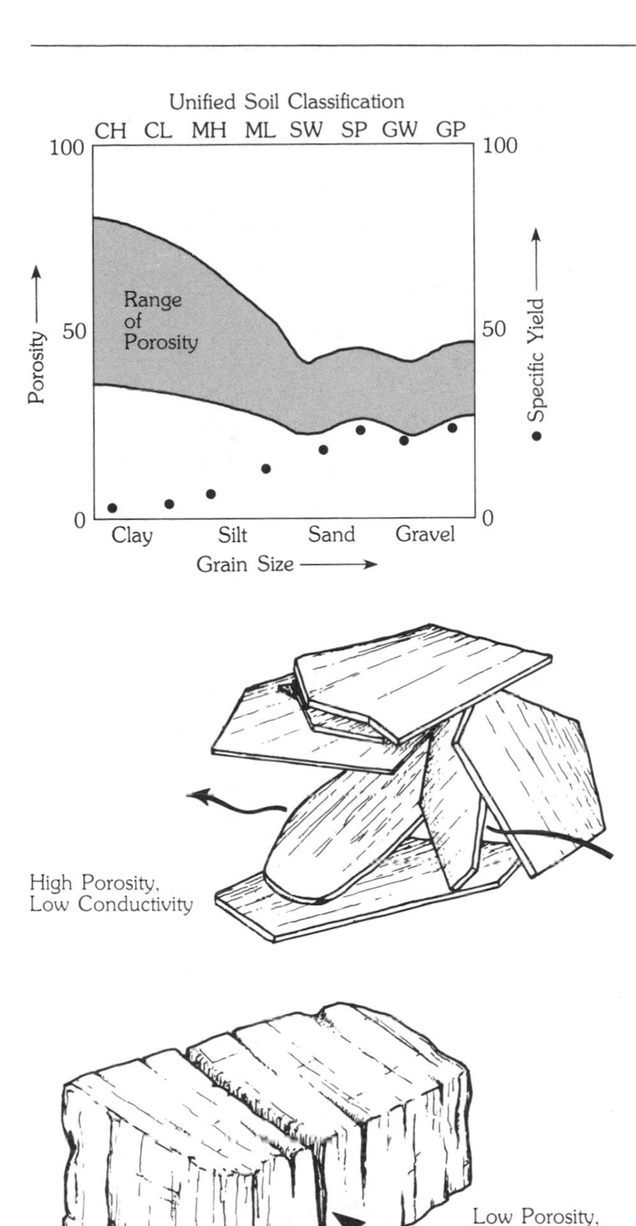

FIGURE 8–7 Generalized relationship between porosity and specific yield of groundwater in various soil types.

FIGURE 8–8 High porosity in a fine-grained material has a low hydraulic conductivity because the connections between the pore spaces are very small, while a massive rock with low porosity may have a very high hydraulic conductivity through fractures.

with 50 percent porosity and a very low hydraulic conductivity as it is to have a high hydraulic conductivity in a dense rock (Figure 8–8).

Groundwater geology, or **hydrogeology**, is the study of the hydraulic properties and three-dimensional characteristics of the associated earth materials. Groundwater flows through permeable units, or **aquifers**, is restricted in units with low permeability, or **aquitards**; it is prohibited in impermeable units, or **aquicludes**. The arrangement of the rock units, their

TABLE 8–3 Hydraulic conductivity

Material	Hydraulic Conductivity	
	cm/sec	ft/day
Clay		
Unweathered	10^{-8}–10^{-9}	10^{-3}–10^{-4}
High plastic (CH)	10^{-7}–10^{-9}	10^{-2}–10^{-4}
Low plastic (CL)	10^{-6}–10^{-8}	10^{-1}–10^{-3}
Silt		
High plastic (MH)	10^{-5}–10^{-7}	1 –10^{-2}
Low plastic (ML)	10^{-4}–10^{-7}	10 –10^{-2}
Sand		
(SP)	10^{-1}–10^{-5}	10^{-4}– 1
Well sorted, fine	10^{-3}–10^{-5}	10^{-2}– 1
Well sorted, medium	10^{-2}–10^{-4}	10^{-3}–10^{-1}
Well sorted, coarse	10^{-1}–10^{-3}	10^{-4}–10^{-2}
(SW)	10^{-1}–10^{-4}	10^{-4}–10^{-1}
Poorly sorted, fine	10^{-2}–10^{-4}	10^{-3}–10^{-1}
Poorly sorted, medium	10^{-1}–10^{-3}	10^{-4}–10^{-2}
Poorly sorted, coarse	10^{-1}–10^{-3}	10^{-4}–10^{-2}
Silty sand (SM)	10^{-4}–10^{-6}	10 –10^{-1}
Clayey sand (SC)	10^{-5}–10^{-7}	1 –10^{-2}
Gravel		
(GP)	10^{-3}–1	10^{-8}–10^{-5}
Well sorted	10^{-3}–1	10^{-8}–10^{-5}
(GW)	10^{-2}–1	10^{-7}–10^{-5}
Poorly sorted	10^{-2}–1	10^{-7}–10^{-5}
Silty gravel (GM)	10^{-3}–10^{-6}	10^{-2}–10^{-1}
Clayey gravel (GC)	10^{-4}–10^{-7}	10 –10^{-2}

hydraulic properties, and the groundwater levels establish the hydrogeology of any region. There are two possible basic hydrogeologic conditions: The aquifer has direct access to surface recharge and no excess fluid pressures; it is then **unconfined** (Figure 8–9), or the aquifer is confined at the top with excess fluid pressures and is termed **confined**, or **artesian** (Figure 8–10). The relationship between the base of the aquifer and the regional groundwater system determines whether or not an unconfined aquifer is perched (Figure 8–11).

Groundwater is recharged into the hydrogeologic system where permeable units, or aquifers, crop out at the surface (Figure 8–12). These sites usually have a higher infiltration capacity. In some cases, the system is completely filled, making additional recharge impossible, and runoff is increased by **rejected recharge**. Groundwater discharge is the natural removal of groundwater through seepage into lakes or streams, flow from springs, and evapotranspiration through plants. Artificial discharge through wells or drainage works is groundwater **withdrawal** (Figure 8–13).

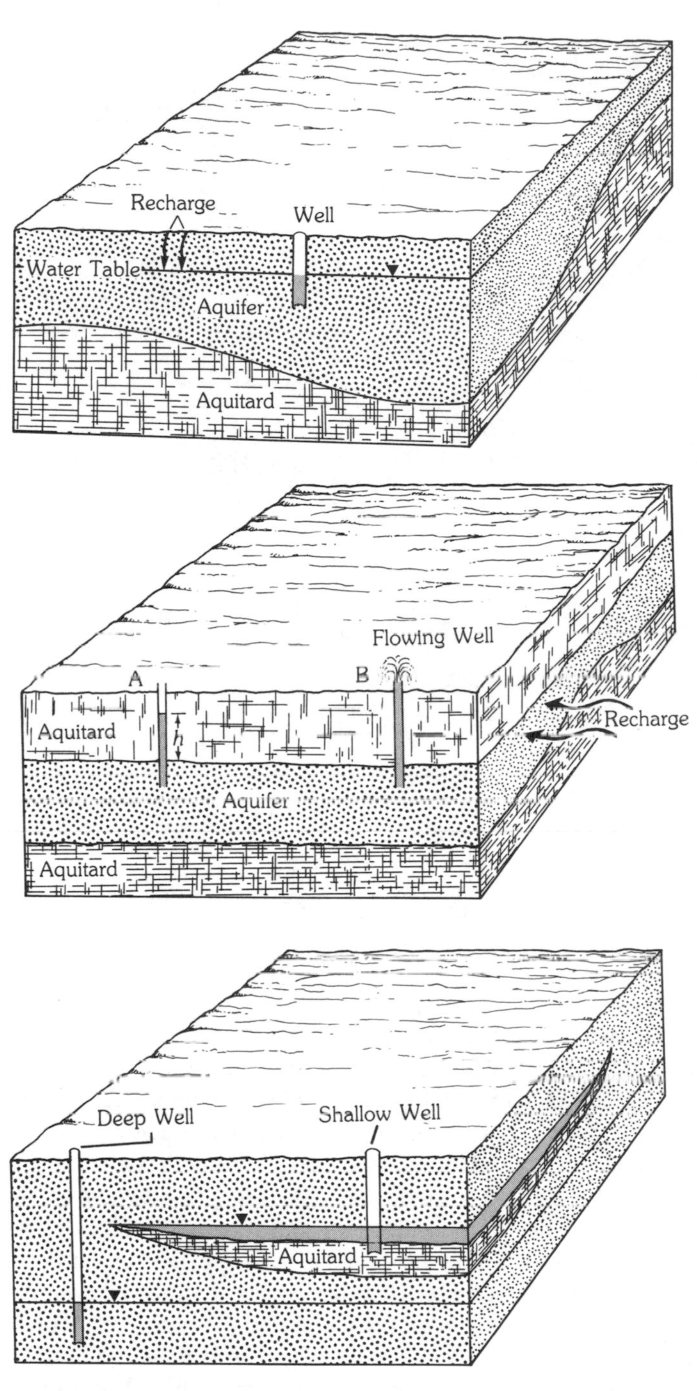

FIGURE 8–9 Unconfined aquifer. Notice that the water level in the well is within the aquifer and that recharge occurs on the site.

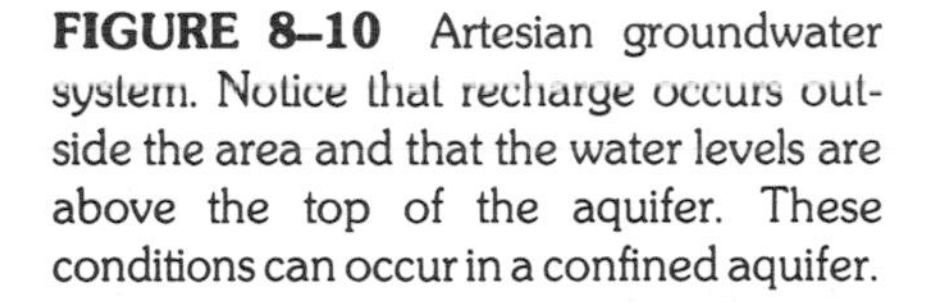

FIGURE 8–10 Artesian groundwater system. Notice that recharge occurs outside the area and that the water levels are above the top of the aquifer. These conditions can occur in a confined aquifer.

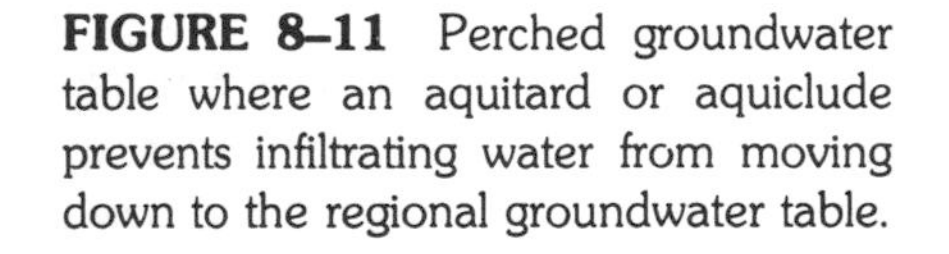

FIGURE 8–11 Perched groundwater table where an aquitard or aquiclude prevents infiltrating water from moving down to the regional groundwater table.

Numerous factors control the chemical quality of groundwater. Rainwater is naturally slightly acidic due to the formation of carbonic acid:

$$H_2O + CO_2 \rightleftharpoons H_2CO_3 \rightleftharpoons H^+ + HCO_3^-$$

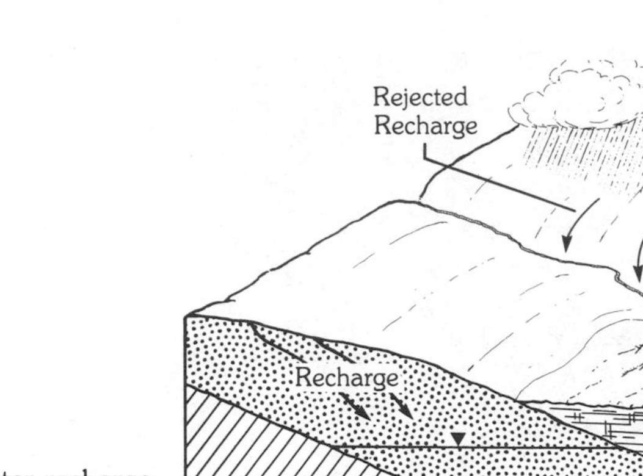

FIGURE 8–12 Groundwater recharge and rejected recharge. Recharge is rejected when the aquifer cannot accept additional water, and the water flows off as runoff.

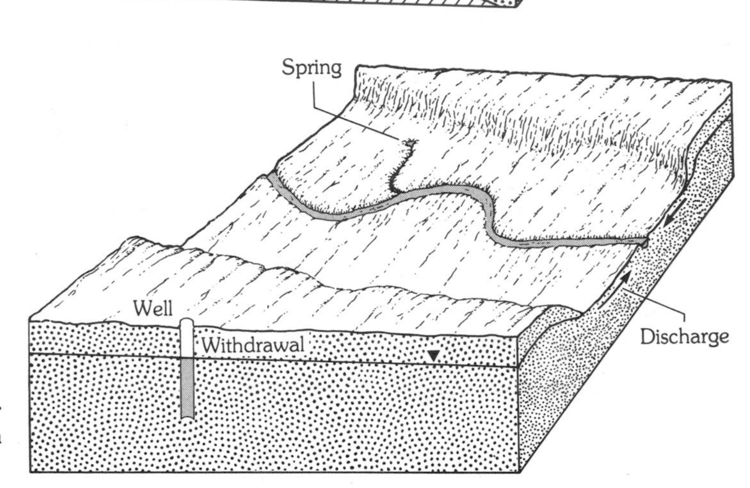

FIGURE 8–13 Groundwater **discharge** from a spring or bank flow into a river and **withdrawal** from a well.

In urban areas the atmospheric concentration of CO_2 is increased by combustion of organic-based fuels, such as wood, coal, oil, or gas. This higher CO_2 concentration can lead to **acid rainfall**. Other gases, like SO_2, also contribute to problems of acid rainfall. When the acidic rainwater reaches the ground, the free H^+ ions can react in the hydrolysis reaction, leaching ions from the soil and carrying them into the groundwater system. In anaerobic environments of deposition, sulfur is retained as H_2S or other forms, which, along with many other ions, can be leached into the groundwater. Groundwater **contamination** is the natural alteration in water quality by the reaction between the water and the rocks with which it comes in contact. In other cases, contaminated groundwater may be trapped marine water. Groundwater **pollution** is the artificial contamination of the system through the introduction of organic or inorganic materials.

GROUNDWATER HYDROMECHANICS

Groundwater may be an important natural resource or a serious problem for an engineering project. In both cases, the quantitative analysis of the groundwater conditions, or groundwater hydromechanics, requires a thorough knowledge of the hydrogeology of the site. The hydrogeology establishes the boundary conditions for the analysis of groundwater flow problems. Once the boundary conditions are defined and the necessary simplifying assumptions made, the problem becomes solvable. The important boundary conditions are that:

1. Water level or pressure head is defined.
2. Boundaries to the aquifer are known.
3. Hydraulic properties have been defined.

and the important simplifying assumptions are that:

1. Aquitards and aquicludes are impermeable.
2. Darcy's law is valid.
3. Flow through the aquifer is laminar flow.
4. Aquifers are infinite and flow is steady (uniform).

From a practical standpoint, being able to predict the yield from a well is most desirable. Darcy's law is directly applicable, with certain modifications, according to the type of aquifer.

Confined aquifers offer the simplest case because the aquifer is enclosed by impermeable material and is completely saturated. The water level in a penetrating well will be above the top of the aquifer because the water is under pressure. Water levels in several wells can be used to describe a **piezometric surface**, or potential water table. Darcy's law can now be used to determine the hydraulic conductivity (K) by measuring the discharge from a well having a known area through which water enters the well and by knowing the difference in elevation between the water in the well and the piezometric surface. As the well is pumped, the piezometric surface will be drawn down as water near the well is removed. In three dimensions the shape of this drawdown describes a surface known as the **cone of pressure relief**. The outer limit of the cone defines the area of influence of the well. Darcy's law may be rewritten so that it describes the shape of this cone in a uniform aquifer that has a horizontal piezometric surface and impermeable boundaries (Concept 8–4).

The case of an unconfined aquifer is harder to solve because the piezometric surface is within the aquifer. If the upper boundary is the ground surface, the piezometric surface is the groundwater table. Difficulties arise in the determination of the cone of depression around a pumping well and the radius (r_0) to the outer limit of the cone. Since Darcy's law requires that the cross-sectional area of the aquifer be known, and since the cross-sectional area changes as an unconfined aquifer is drawn down, the problem is more complex. In most sand aquifers, the effect of large changes in r_0 (150 to 300 m) has only a slight impact on the value of Q, and a large r_0 value is usually selected in practical applications (Concept 8–5).

Flow conditions near a well may be turbulent with very high values for the hydraulic gradient, thereby violating the initial assumptions. The two numerical solutions given in Concepts 8–4 and 8–5 can be used either to

CONCEPT 8–4

Pumping a Confined Aquifer

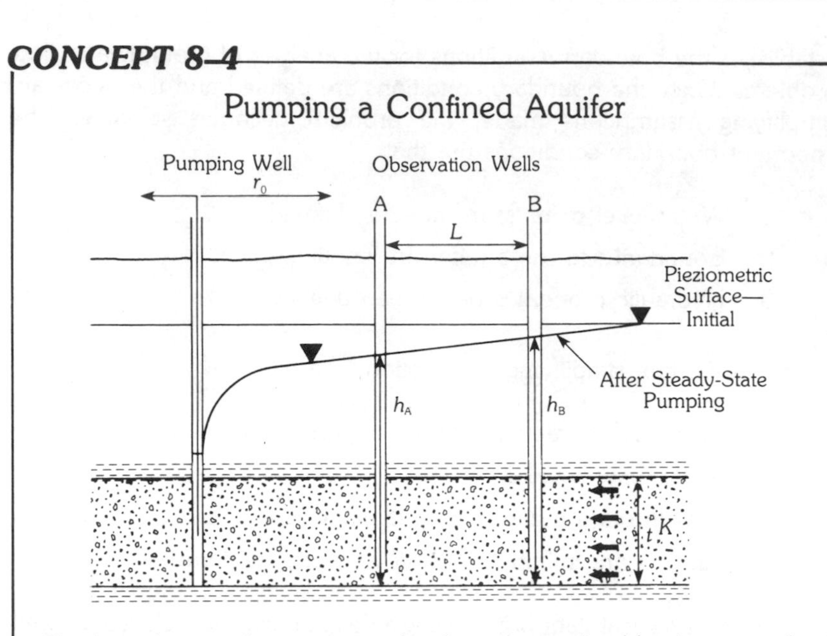

The boundary conditions of a confined aquifer are fixed by the aquitards; thus, if steady-state flow is established in the aquifer, the hydraulic gradient is the slope of the piezometric surface. Since the geometry of the aquifer is fixed, the cross-sectional area is constant, and groundwater flow can be calculated from Darcy's law:

$$Q = KiA = K\left(\frac{h_B - h_A}{L}\right)t$$

where

Q = flow per unit width of the aquifer
K = hydraulic conductivity
h_B, h_A = head at observation wells
L = distance between wells
t = thickness of the aquifer

Frequently hydraulic conductivity (K) × thickness (t) is used as the aquifer transmissivity (T).

If an aquifer test, or pump test, is performed to determine the hydraulic conductivity of the aquifer materials, and steady-state flow conditions toward only one pumping well are established, the hydraulic conductivity equals:

$$K = \frac{Q}{2\pi(h_B - h_A)t} \ln\left(\frac{r_B}{r_A}\right)$$

where

Q = steady-state pumping rate
h_B, h_A = piezometric surface
r_B, r_A = distance from pumping well to observation wells

CONCEPT 8–5

Pumping in an Unconfined Aquifer

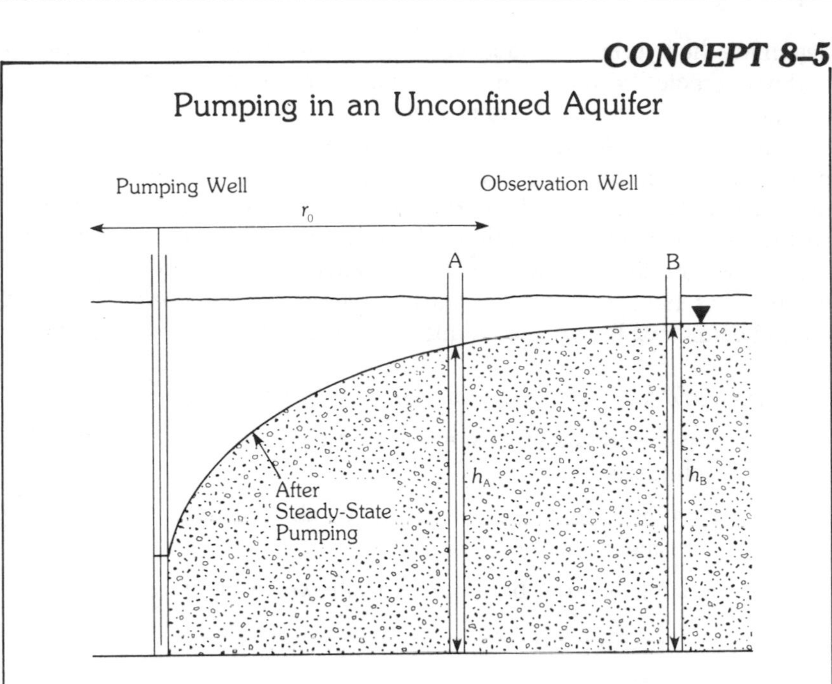

In an unconfined aquifer, the water table forms one boundary of the groundwater flow system; thus, the cross-sectional area (a) changes with both location from the well and time. This variability in the area, plus possible losses from (by evapotranspiration) or gains to (by infiltration) the system, makes the mathematical analysis of unconfined aquifers a complex problem. If steady-state flow conditions are established, changes in area with time become zero; however, the groundwater surface still forms a curved boundary, and the area changes with distance. An approximation of the hydromechanics of the unconfined aquifer system can be made if the following simplifying assumptions are made:

1. The groundwater table represents the hydraulic gradient.
2. The groundwater system is a steady-state system, and the quantity pumped is constant.
3. Flow is in only one direction, toward the well, and no other pumping wells influence the established cone of depression.
4. There are no evapotranspiration losses or infiltration gains.

If these assumptions are reasonable, the hydraulic conductivity of the aquifer based on a pump test can be calculated from:

$$K = \frac{Q}{\pi(h_B^2 - h_A^2)} \ln\left(\frac{r_B}{r_A}\right)$$

where

Q = steady-state pumping rate (withdrawal)
h_B, h_A = head measured at two observation wells
r_B, r_A = distance from pumping well to observation wells

determine the hydraulic conductivity (K) of an aquifer by running a **pump test** or to predict the impact on the groundwater resource of a pumping well producing a known groundwater yield.

In some complex hydrogeologic cases, such as in seepage, drainage, or well selection problems where the quantity of water withdrawn is unknown, the groundwater flow conditions can be estimated with the use of a **flow net**. Flow nets can be drawn in plan (map) view or in section view, depending on the problem to be solved (Figure 8–14). The basic assumptions in a flow net solution are the same as those used to calculate groundwater flow, but additional rules of flow net construction require that:

1. Flow lines are perpendicular to equipotential lines.
2. Equipotential surfaces (lines) are perpendicular to impermeable boundaries.
3. Flow lines and equipotential lines always define a network of curvilinear squares.
4. Flow lines cross boundaries as shown in Figure 8–15.

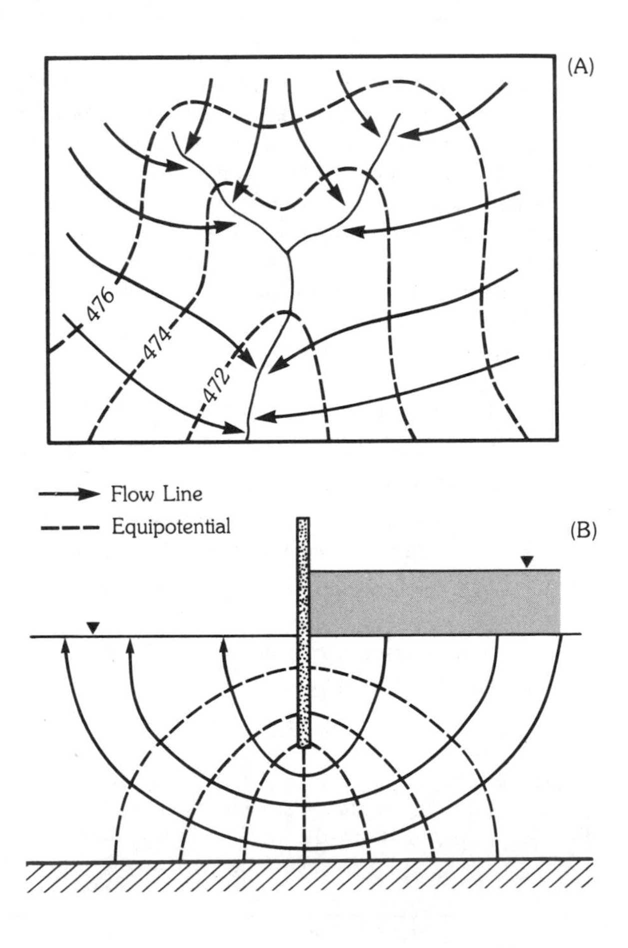

FIGURE 8–14 Flow nets. Part (A) is a map and (B) is a cross section.

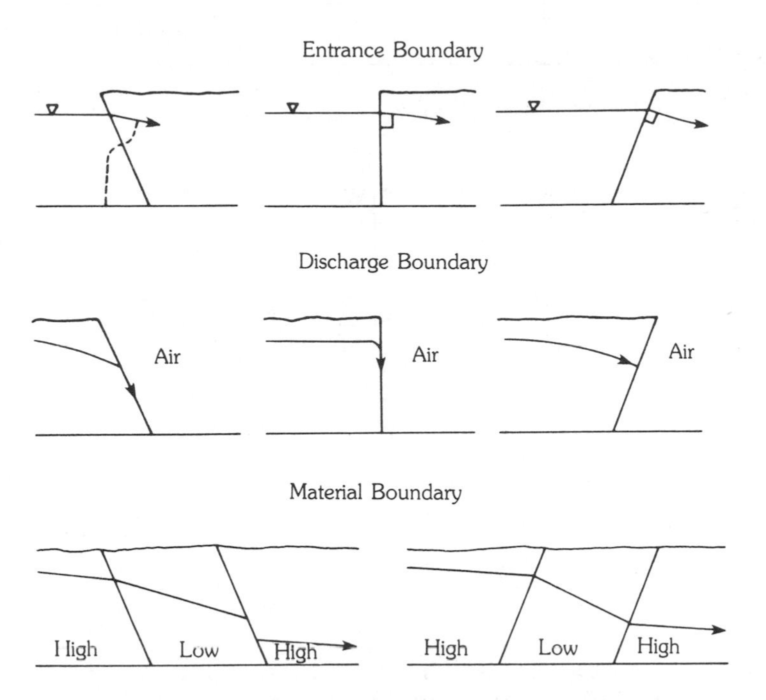

FIGURE 8–15 Construction rules for flow nets as they cross boundaries.

Flow nets can be constructed to show the hydrogeology of an area in map view in order to assist in well selection and the identification of areas of high or low hydraulic conductivity. In this case, the elevation of the piezometric surface defines the existing equipotential lines, and flow lines are drawn in compliance with the rules. Since flow nets are similar to wave ray diagrams in that they both show the characteristics of energy flow, *no flow crosses a flow line,* and the quantity of water flowing between any two flow lines is always equal and constant. Thus, flow lines are impermeable, frictionless boundaries, and flow conditions along any flow path can be analyzed independently of all others. In areas where the flow lines are nearly parallel, the ratio of the hydraulic conductivity (K) at any two points is equal to the ratio of the hydraulic gradients at the same two points (see Concept 8–6). In areas of a uniform regional groundwater system, sites with high hydraulic conductivities will have a wide contour spacing on a piezometric surface elevation map. In areas with low hydraulic conductivities the contour spacing will be close.

Flow net analyses are valuable for predicting the amount of seepage into an excavation or through and under dams and levees. In the analysis of seepage into an excavation, the hydrogeology of the site establishes the boundary conditions and preconstruction flow net. In dam and levee analyses, the engineering project changes the existing hydrogeologic conditions, and the new boundary conditions must be supplied. Anytime flow is increased, the risk of **seepage erosion** or **quick conditions** is also increased; therefore, a flow net analysis is important to project safety. In these types of analyses, the orientation of the flow net changes from a map

CONCEPT 8–6

Hydraulic Properties and Flow Nets

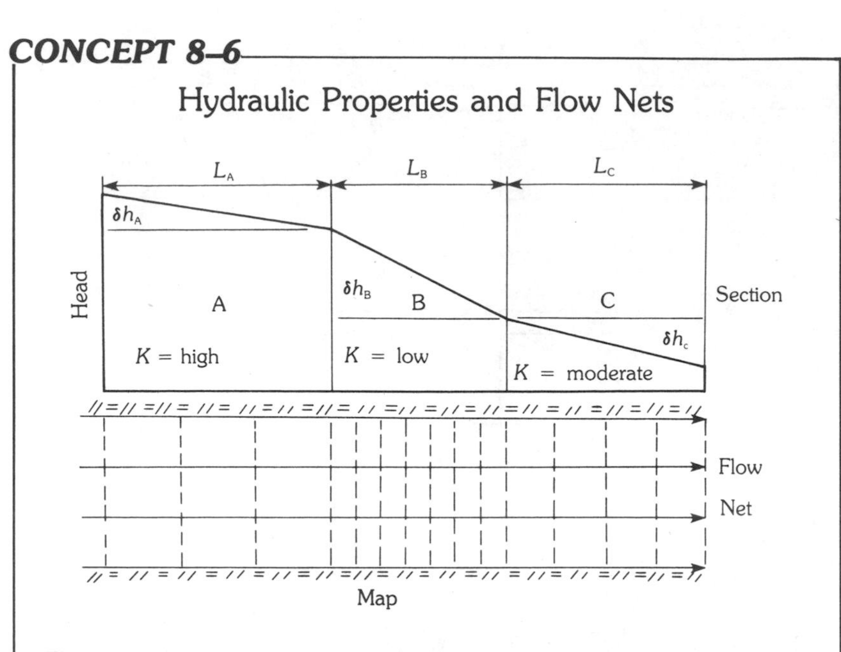

Flow nets are a graphical representation of the hydraulic properties of a groundwater system. In the groundwater system shown above, three different units (A, B, C) have three different hydraulic conductivities. Since a flow line is a flow boundary, the quantity of water flowing between any two flow lines must be constant. Using Darcy's law, $Q = KiA$, the relationship between K and i must be constant:

$$Ki = \frac{Q}{A}$$

If hydraulic conductivity increases, hydraulic gradient must decrease. This relationship makes it possible to estimate the hydraulic conductivity of geologic units once a flow net or equipotential map is drawn. In the map above, the close spacing of the equipotential lines in unit B demonstrates that unit B has a low hydraulic conductivity.

view to a cross-sectional view, and some of the boundary crossings are different because there is an air-soil and/or water-soil boundary (Concept 8–7).

Flow-net construction procedures also change because the equipotential lines are not provided as they were for map studies. A hydrogeologic cross section that defines the boundaries, piezometric surface, aquifer unit, and aquitards is first constructed. Flow lines are then constructed, followed by the determination of the equipotential lines as guided by the rules. Since the flow lines control the net and therefore affect the final calculation of seepage, the lines should be drawn with great care, meeting the requirement that the flow net is a network of squares when drawing the equipotential

CONCEPT 8–7

Flow Nets

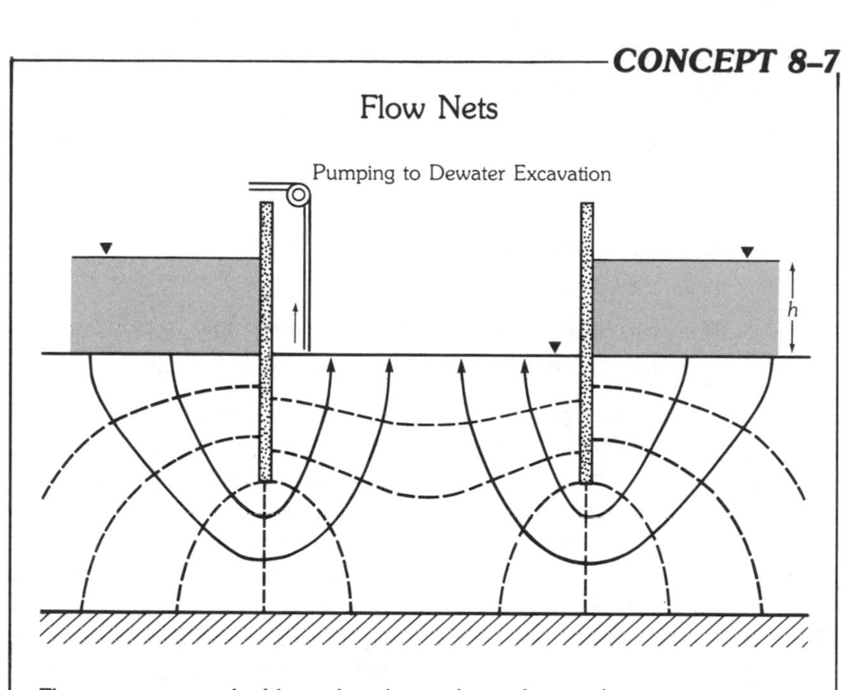

Flow nets are a valuable tool in the analysis of groundwater impacts on a construction project. For example, the sheet pile bulkheads in this diagram are proposed to keep water out of an excavation to be cut into the bottom of a lake. The flow net can be used to estimate the amount of water that will seep into the excavation from below and will need to be removed by pumps. Flow volumes are determined from the flow net, using Darcy's law and the geometry of the net. Darcy's law states:

$$Q = K\left(\frac{h}{L}\right)A$$

where

h = head at both ends of the net
L = length of the flow path
K = hydraulic conductivity

Since the flow net is a planer net, $A = 1$.

Flow through a flow net is calculated from

$$Q = Kh\left(\frac{n}{m}\right)$$

where

n = number of flow paths in the net
m = number of equipotential lines

In this example, $n = 6$ and $m = 8$.

We can then determine seepage per unit length:

$$Q = Kh\left(\frac{6}{8}\right)$$

lines. Total seepage through the net is calculated from the geometry of the net and the following equation:

$$Q = Kh\left(\frac{m}{n}\right)$$

where

K = hydraulic conductivity
h = total head loss
m = the number of channels formed by flow lines
n = the number of squares between two flow lines

Flow nets are graphic models that represent the groundwater flow conditions. In complex hydrogeologic cases or for three-dimensional problems, other flow models have been developed. **Physical scale models** of the conditions can be used for both groundwater and surface water problems. **Electric analog models,** using either a network of resistors and capacitors or electric conducting sheets, can be used because the groundwater flow follows the same physical laws as the flow of current. **Analytical** and **numerical models** are computer solutions to groundwater flow problems. In these models, the boundary conditions, properties of the aquifers and aquitards, and hydraulic gradients and pressures are placed on a network of finite cells. The computer then calculates the equipotential at each location within the network, using the basic equations of fluid mechanics (Figure 8–16).

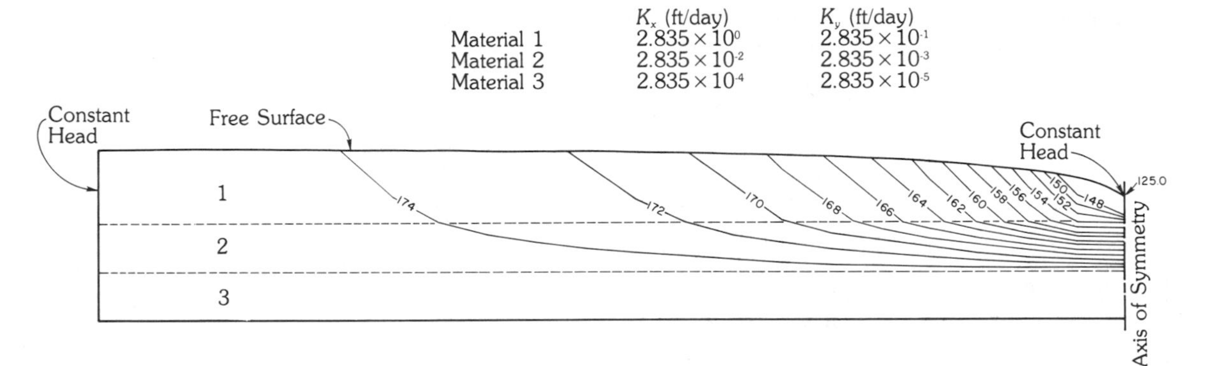

Flow Quantities

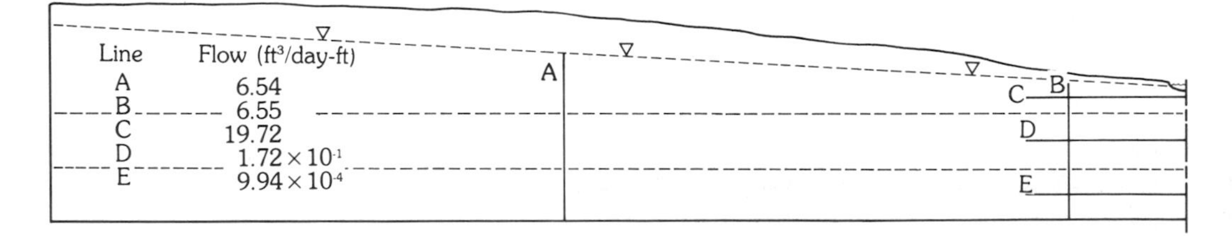

FIGURE 8–16 Computer generated flow net and flow quantity calculations. (From J. L. Kennedy, III, 1980.)

GROUNDWATER INTERACTIONS

Since groundwater is directly associated with other water bodies in the hydrologic cycle and with the solid earth materials that make up the aquifers, aquitards, and aquicludes, groundwater interactions are important to the engineering geologist. Three such interactions can be significant: **quick conditions, saltwater intrusion,** and **land subsidence.** Quick conditions have the same impact as quick clays or liquefaction but are caused by seepage forces that flowing groundwater can produce (Concept 8–8). In many coastal areas the shallow fresh groundwater resource is being threatened by saltwater intrusion, the encroachment of saline marine waters (Concept 8–9). Groundwater withdrawal from artesian aquifers reduces the pore water pressure, which can lead to subsidence of the land surface (Concept 8–10).

CONCEPT 8–8

Quick Conditions

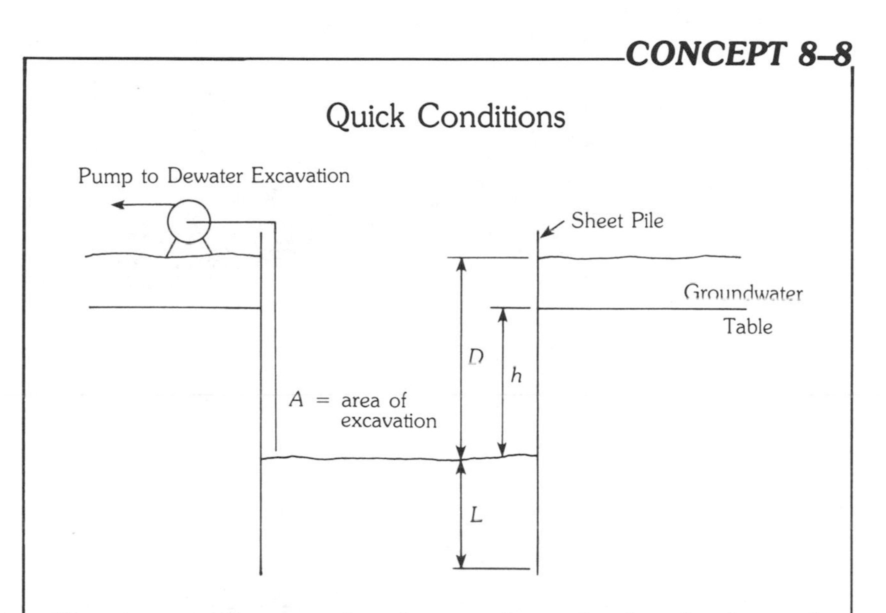

There is an equilibrium condition between the weight of a soil and upward directed seepage pressures, so that when the two are equal, the soil will behave like a fluid. This situation creates *quick conditions.* The submerged unit weight (γ_b) of the soil material above the sheet pile in the bottom of the excavation equals the pressure gradient of the water flowing into the excavation:

$$\gamma_b = \frac{G - 1}{1 + e}\gamma_w A$$

$$i = \frac{h}{(h + 2L)}\gamma_w A$$

For quick conditions $\gamma_b = i$:

$$\frac{G - 1}{1 + e} = \frac{h}{h + 2L}$$

The minimum depth that the sheet pile must extend below the excavation is:

$$L = \frac{1}{2}\left[\frac{h(1 + e)}{G - 1} - h\right]$$

where

G = specific gravity of the soil
e = void ratio of the soil
γ_w = unit weight of water

CONCEPT 8–9

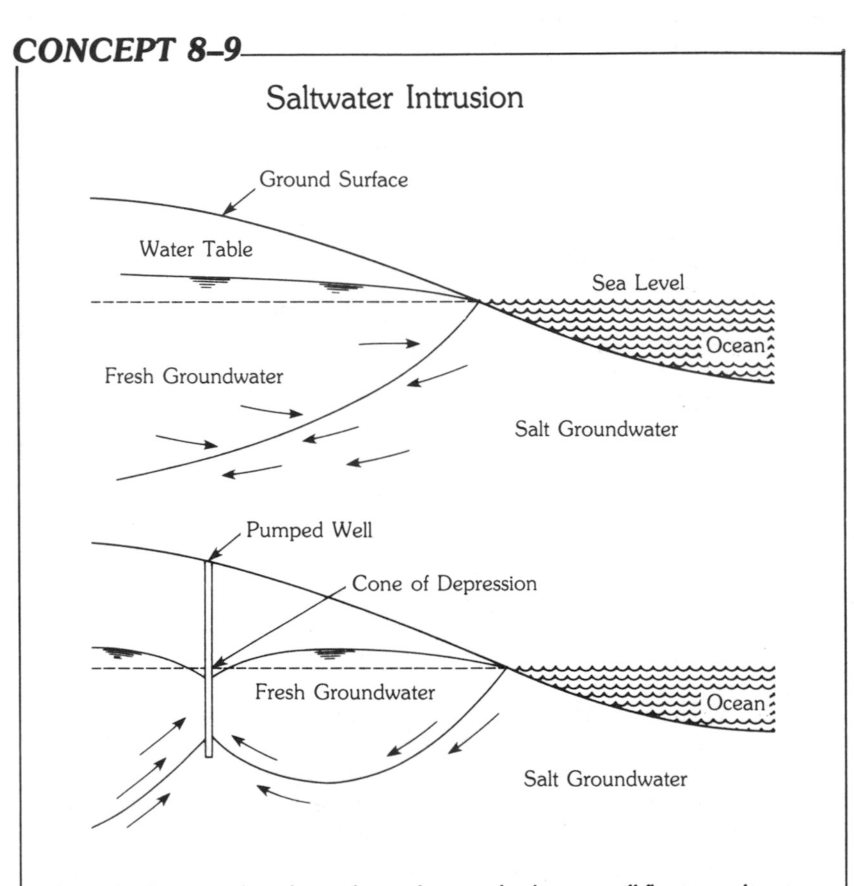

Since freshwater is less dense than saltwater, freshwater will float on saltwater. The density relationships between them establishes the elevation of the freshwater surface above sea level. Lowering the freshwater level through groundwater withdrawal causes a corresponding rise in the saltwater level. If the saltwater elevation rises to the well, groundwater quality deteriorates, and the water supply is lost.

CASE IN POINT

Barriers to Seawater Intrusion, Los Angeles

The Los Angeles area has experienced an extremely rapid growth rate that has caused excessive withdrawals of fresh groundwater from coastal aquifers. Many of these aquifers are hydraulically connected to the Pacific Ocean, and a decline in the piezometric surface of the freshwater has allowed intrusion of seawater, thereby degrading water quality. The Los Angeles County Flood Control District has instituted a series of seawater barrier projects to protect the freshwater supplies in the coastal aquifers.

Each saltwater barrier project is designed specifically for the geology and hydrogeology of the project site. The geometry of the aquitards and aquifers and the hydrologic properties of the units is determined. These data form the basis for a network of pumping wells that intercept the intruding saltwater and freshwater recharge wells that build a freshwater mound. The combined effect of reducing the saltwater head and increasing the head of the freshwater system produces a hydraulic barrier to saltwater intrusion.

CONCEPT 8–10

Subsidence Due to Groundwater Withdrawal

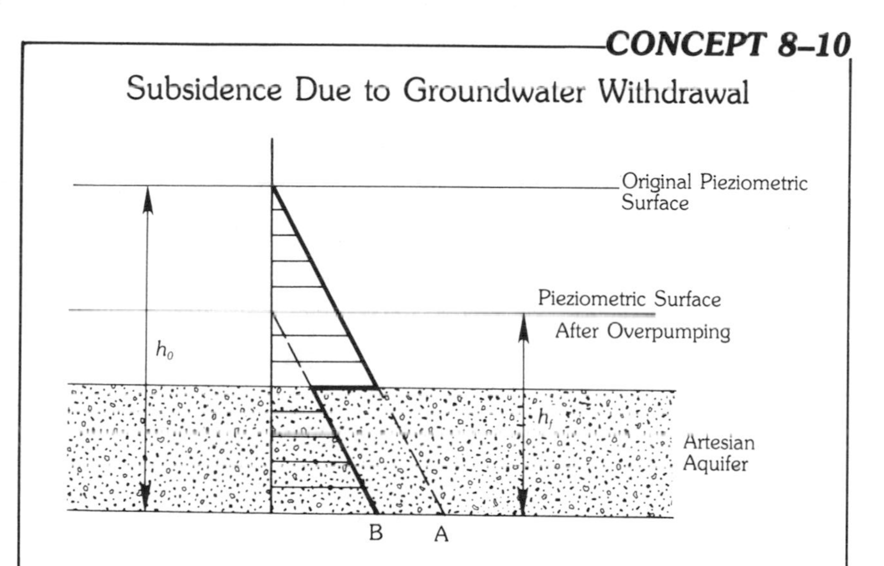

A reduction in the elevation of the piezometric surface in an artesian aquifer causes clays in the overlying and underlying aquitards to give up pore water to the aquifer because a hydraulic gradient develops between the clays and the aquifer. In A, the pore water pressure in the clay and the aquifer is hydrostatic, and there are no excess pore pressures. After pumping (B) the pressure head in the aquifer is reduced, but the low hydraulic conductivity of the aquitard prohibits the free drainage of water into the aquifer, and excess pore pressures result. As these pressures are released, the aquitard consolidates and the surface subsides.

part II

Geologic Processes and Engineering Geology

Geologic Processes and Time

nine

Geologic processes, particularly the active ones, are the most important geologic aspects that the engineering geologist has to evaluate. If the engineering geologist is to translate geologic data into engineering information, the professional must be able to predict the *environmental impacts,* the influence of the proposed engineering activity on the active geologic processes, and the *engineering impacts,* the influence of the active geologic processes on the construction performance and maintenance of the proposed engineering activity.

One of the most difficult differences to grasp between the sciences of geology and engineering is that of time. To the geologist, time is expressed in earth-time, and processes of great magnitude, like the erosion of a mountain, can be understood because the process of erosion has been operating for hundreds of millions of years. In contrast, engineering works must deal with processes and interactions in human time. The engineer does not often design for time periods very much longer than 100 years.

This difference in philosophy of time-dependent phenomena often leads to difficult communication problems between the two professions. In the geologist's more than 4 billion years of earth-time, almost anything can change. This long time period does create some problems for the geologist, in that the record of events that occurred in the past is often incomplete and assumptions are necessary. The basic assumption of geology, that of **uniformitarianism,** is that all past changes in the physical characteristics of the earth's crust are the result of processes controlled by the same physical laws in effect today. For example, we assume that water runs downhill now and in the past. Or, as we frequently say, "The present is the key to the past."

Geological history was originally divided by fossil evidence, which established relative time and a close approximation of time to and beyond the Paleozoic era. Geological time in years was added after the development of radioactive dating techniques. Thus there are four basic history blocks called eras: Precambrian, Paleozoic, Mesozoic, and Cenozoic. The Precambrian was originally described as lacking hard-shell life forms and fossil remains. The Paleozoic era is dominated by marine life forms; the Mesozoic by reptile forms; and the Cenozoic by mammals. Table 9–1 is an abbreviated form of the geological timetable.

Nongeologists often have difficulty translating geological time (earth-time) into a meaningful time scale. We can see and understand human time—the span of human recorded history or the birth and death of a tree, for example; but 100 million years—how long is that? If we select a total age of the earth of 3.65 billion years (actually, data now give a total age of more than 4 billion years) and define the birth of the earth at midnight, December 31, then earth-time is compressed into one year of 365 days. Using these numbers, one day equals 10 million years. Table 9–2 shows the geologic eras in terms of one year.

Since one day is 10 million years of earth-time, then one hour equals 833,334 years, one minute equals 13,889 years, and one second equals 232 years. Subdividing the Cenozoic era, the last seven days of December, into recent historical events yields Table 9–3, in which all the events occurred on December 31. This time scale should help you understand geological time and its relationship to human history.

The engineering geologist must relate to engineering problems that in earth-time may, as designed, have useful lives of less than one-half second. Geologic processes such as earthquakes, floods, erosion, or sediment deposition are "rare" occurrences to people but are frequent in earth-time. For example, a 100-year flood would occur every 0.43 second during our earth-time year, or about twice as often as the normal person's heartbeat.

TABLE 9–1 Geologic time scale

Geologic Eras	Million Years (before present)
I. Cenozoic era	
A. Quaternary period	
1. Holocene epoch	0.01
2. Pleistocene epoch	2.5
B. Tertiary period	
1. Pliocene epoch	7
2. Miocene epoch	26
3. Oligocene epoch	38
4. Eocene epoch	54
5. Paleocene epoch	65
II. Mesozoic era	
A. Cretaceous period	136
B. Jurassic period	190
C. Triassic period	225
III. Paleozoic era	
A. Permian period	280
B. Pennsylvanian period	325
C. Mississippian period	345
D. Devonian period	395
E. Silurian period	430
F. Ordovician period	500
G. Cambrian period	570
IV. Precambrian era	4000(?)

TABLE 9–2 Earth-time expressed in one year

Era	Geological Time (years before present)	Human Time
Cenozoic	65 million	December 25
Mesozoic	225 million	December 8
Paleozoic	570 million	November 1
Precambrian	3.65 billion	January 1

TABLE 9–3 Recent human time on the earth-time clock

Event	Human Time (before present)	Earth-time
Modern technology	75 years	11:59.59.68 P.M.
American Revolution	200 years	11:59.59.14 P.M.
Roman Empire	2,500 years	11:59.49.22 P.M.
Bronze Age	5,500 years	11.59.60 P.M.
Stone Age	10,000 years	11:59.28 P.M.
Ice Age	1 million years	10:48 P.M.

To most people, such events are irregular in intensity and intermittent in frequency and are usually the basis for federal or state regulations. They are recognized in engineering design considerations as a risk of loss. The engineering geologist must be able to relate these processes or their geologic evidence to the economic and social aspects of any engineering design and in so doing will face statements like "I have never seen a landslide on that mountain!" or "I've lived here thirty years and that creek has never flooded!" or have to *prove* that a fault is, in fact, inactive and has not moved during the past 1 million years.

The engineering geologist also deals with geologic processes that in human time are nearly uniform and regular in character, such as daily winds, climatic zones, and seasonal rains, but continually change in earth-time. These processes are often forgotten because they are so familiar. But, for example, along parts of the Texas coast uniform, regular daily winds have been shown to move massive quantities of sand in dunes. These dunes relentlessly advance over roads, forcing their abandonment and relocation—a costly engineering failure.

The concepts of earth-time and human time and their interaction with the geologic characteristics of a site to produce adverse or favorable conditions for the safe and economical completion of an engineering project play a significant role in the professional life of an engineering geologist. Much of your professional effort will be spent evaluating, describing, and defining past geologic processes and events in an attempt to predict future events or impacts. It is just as important to predict the impact of human activity on the physical environment as it is to predict the impact of the physical environment on people.

Erosion Processes

ten

Erosion is a geologic process of mixed emotions. Soil erosion removes topsoil, leading to books like *The Rape of the Earth* (Jacks and Whyte, 1939), but it is also a primary component of the rock cycle as the source of all clastic sediments. The engineering geologist must understand the processes of erosion in order to assure a continued supply of stream sediment without losing too much soil or overloading stream systems.

Erosion agents include moving water and wind and, on a more local scale, ice and gravity, which drives all the other agents as well. Erosion is therefore the transformation of potential energy toward a lower energy state. The most powerful and significant erosion agent is flowing water.

EROSION BY WATER

Anytime water flows over the land surface, soil or rock particles are subjected to a boundary shear between the particle and the flowing water. Soil erosion involves two sequential events. First a soil or rock particle must be detached and then it must be transported. Detachment is frequently accomplished through the dissipation of raindrop energy—**raindrop erosion**—or to a lesser degree through the dissipation of the kinetic energy of turbulent flow. Sediment transport is accomplished by the kinetic energy of the surface flowing water.

Early workers noticed that there was a direct relationship between raindrop energy and the amount of soil or rock particles carried by the runoff. This work has led to the realization that the primary factor in soil erosion is raindrop energy. Erosion studies must recognize the characteristics of the precipitation in the area of concern. The important characteristics are drop mass, size, size distribution, direction of impact, rainfall intensity, and the terminal velocity of the drops. The kinetic energy ($\frac{1}{2}mV^2$) and momentum (mV) of each drop can be calculated and statistically summed to obtain the total raindrop energy for a particular storm. Raindrop or splash erosion of the soil is related to the kinetic energy of the raindrops:

$$Q_R = K(E)^a$$

where

Q_R = raindrop erosion
K = coefficient of proportionality
E = kinetic energy of raindrops
a = constant for the type of soil

Using this work, Smith and Wischmeirv (1962) developed a **rainfall-erosion index,** or EI_{30} index, which is the product of raindrop energy (E) and the 30-minute intensity of rainfall (I_{30}).

The **universal soil loss equation** defines the most important factors that control surface erosion:

$$A = RK(LS)CP$$

where

A = average soil loss (tons/acre)
R = rainfall factor ($EI_{30}/100$)
K = soil-erodibility factor (f[soil type])
LS = slope length–steepness factor (f[topography])
C = cropping factor (f[land use])
P = conservation practice

In this equation, R is the eroding mechanism—rainfall energy and intensity. All other factors are related either to natural characteristics of the soil (K) or

TABLE 10–1 Influence of variables in the universal soil loss equation

Variable	Influence
Soil Characteristics (K)	
Clay soils	Low K value
Silt and fine sand	High K value
Sand	Low K value
Gravel	Low K value
Slope Length–Steepness (LS)	
Increase length of slope	Increase erosion
Increase steepness	Increase erosion
Cropping Factor (C)	
Barren ground, fallow	1.0
Row crops	0.5–0.9
Coarse grasses	0.1–0.6
Fine grasses and mature woodlands	0.01–0.1
Conservation Factor (P)	
No erosion control	0.9–1.0
Contour plowing	0.5–0.9
Terracing	0.2–0.4

site (LS) or to human actions, such as land use (C) or soil erosion control (conservation) (P). The value for K is the amount of soil loss (tons/acre) at the site under a rainfall factor (R) on a 9 percent slope that is 22.13 m long, having bare ground (fallow) exposed. The slope length–steepness factor (LS) is the ratio between the actual slope length and the standard 22.13 m slope length. The cropping factor (C) is the ratio of soil loss from the actual vegetative cover to soil loss from a slope under standard conditions and ranges from 0 to 1. The conservation practice (P) ranges from 0 to 1; total soil loss lessens as soil conservation practices are improved. P is assigned the value of 1 when no contour practices are used and cultivation is straight up and down the slope (Table 10–1).

The universal soil loss equation can be used to estimate the **wash load,** or amount or increase in soil erosion from a site (see Concept 10–1). However, the determination of the various factors used in the equation is a complex task. The Soil Conservation Service of the U.S. Department of Agriculture can assist in this task.

CONCEPT 10–1

Application of Soil Loss Equation

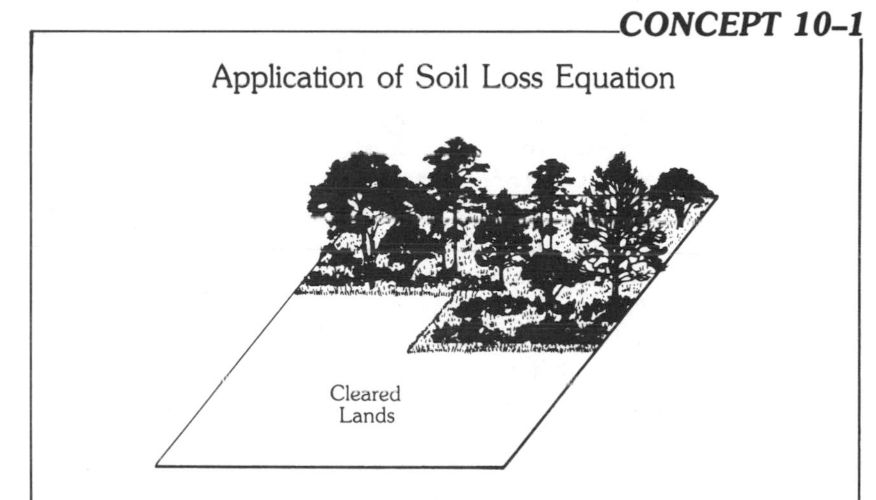

Given a tract of land that will be cleared for a construction project, what soil erosion problems can be expected? The soil loss equation states:

$$\text{Erosion} = RK(LS)CP$$

Since the construction project will not change the rainfall factor (R) or the soil-erodibility factor (K), the equation simplifies to:

$$\text{Erosion} = (LS)CP$$

The slope length-steepness (LS), the cropping (C), and the conservation (P) factors may change. If a cut is to be made on the tract, the slope length–steepness factor will increase by the ratio of cut slope angle to the precut slope angle and the ratio of slope length. Changes in vegetative cover are also evaluated.

EXAMPLE

Condition	Ratio to standard
(LS) Precut slope = 5°	5°/9° = 0.56
(LS) Cut slope = 12°	12°/9° = 1.34
(C) Precut vegetation	C = 0.2
(C) Cut slope bare ground	C = 1.0

Increase in erosion:

Erosion before construction = $RK(.56)(.2) = 0.113RK$

Erosion after construction = $RK(1.34)(1.0) = 1.34RK$

The results of this example suggest that erosion from the site will increase from 0.113 to 1.34 or by about 10 times more material. Erosion control measures (soil conservation factor, P) should be instituted to reduce this effect. If mulching and string nets are used on the slope until a grass cover is established, the estimated erosion is reduced ($P = 0.3$) to 0.40, or erosion is increased only by about 4 times.

EROSION BY WIND

Wind erosion of soil and rock particles is a more complex problem than erosion by water because the energy needed to transport particles is derived from the turbulent energy and boundary shear of the moving air mass. The variables in the wind erosion system include surface wind and climatic characteristics plus the surface material and its characteristics. Surface winds that move sediment dissipate energy as they transport materials; as a result, surface wind velocities decrease as the quantity of sediment in transport increases. The determination of the amount of energy driving the system therefore is a complex problem. The critical property of the wind is the velocity, since velocity and kinetic energy are related.

Climatic characteristics, such as wind turbulence, direction, frequency and duration, humidity, and evaporation potential of the air, all interact to influence the amount of sediment in transport. As the air becomes drier, the potential for wind erosion increases. The size and density of the surface materials play a significant role in controlling erodibility. The most erodible particle sizes range from 0.1 to 0.15 mm for a density of 2.65 (quartz). Surface characteristics that influence the amount of wind erosion include particle bonding agents, soil moisture content, surface roughness, vegetative cover, and topography.

The complex mathematical and statistical solutions necessary to define and calculate wind erosion have not as yet been found. A wind erosion equation has been proposed, and the equation provides a list of important physical factors but not a quantitative solution to the problem:

$$E = f(I, K, C, L, V)$$

where

E = annual wind erosion (tons/acre)

I = soil and slope erodibility

K = soil ridge roughness factor
C = local wind erosion climatic factor
V = equivalent quantity of vegetative cover

All factors used in the wind erosion equation are complex in themselves; however, their general influence on wind erosion can be qualitatively assessed. As Table 10–2 shows, an increase in wind velocity will cause a corresponding increase in wind erosion, while an increase in vegetative cover decreases wind erosion.

TABLE 10–2 Influence of variables in wind erosion system

Variable (increasing)	Influence on Erosion
Wind	
Velocity	Increase
Frequency	Increase
Duration	Increase
Intensity	Increase
Particle size above 0.15 mm	Decrease
Soil cementation	Decrease
Organic matter in the soil	Decrease
Vegetation	Decrease
Soil moisture content	Decrease
Surface roughness	Decrease
Surface length (fetch)	Increase

EROSION BY ICE AND GRAVITY

Present-day active erosion by moving ice is limited to sites where active glaciers exist; however, landforms produced by glaciers in the geologic past are visible throughout the Northern Hemisphere. Erosion by continental glaciation, which advanced into the central plains of the United States, has, for the most part, been buried by glacial sediments (Figure 10–1). In the Arctic terrain of Canada, much of the landscape shows the effects of continental glaciation (Figure 10–2). Alpine glaciation has left erosional evidence in the form of oversteepened valley slopes and other landforms (Figure 10–3).

Direct gravitational erosion occurs in the form of downslope movement of solid earth materials. This erosional process will be discussed in Chapter 13.

EROSION CONTROL

Erosion usually becomes a problem when people alter the land surface in some manner for a new land use (Figure 10–4). The best known erosion problems have come with the rapid development of agriculture after the invention of mechanized equipment. Large tracts of land could be cleared and planted, leading to changes in the vegetative cover—woodlands and prairies were planted in row crops. Serious erosion problems developed in the Western mountains when the forest industry introduced clear-cutting. Erosion problems triggered by forest or grass fires are somewhat less in degree. Land clearance for construction projects is probably the single most

FIGURE 10–1 Till plains of the central United States.

FIGURE 10–2 Subdued topography caused by continental glaciation, Bathurst Island, Northwest Territories, Canada.

FIGURE 10–3 Topographic expression of alpine glaciation.

significant cause of erosion as vegetation and vegetative debris are cleared away from a building site for fire prevention (Figure 10–5).

The engineering geologist plays a significant role in erosion prediction and control. The first step is to evaluate the soil loss equation (water or wind) and identify the factors that will be altered by the engineering project. Design changes and erosion control methods that will decrease soil losses are then suggested. Many of the solutions to erosion control problems are reached by empirical methods; a treatment is applied, the results are monitored, and the treatment is modified. This process is continued until the desired results are obtained.

FIGURE 10–4 Severe rill erosion on a highway cut slope.

FIGURE 10–5 Clearing vegetation from a construction site increases erosion from the site.

River Processes

eleven

GEOMORPHOLOGY

A river is a system that constantly interacts with a much broader system including the lithosphere, hydrosphere, atmosphere, and biosphere. The water carried by the channel is part of the hydrologic cycle. Channel configurations and the different landforms that the river manifests are the net result of the system attempting to reach a state of equilibrium with all the different parameters of its environment. As a river makes its way to its eventual destination, any change in the environment that affects the river or any portion of its drainage basin will necessitate a readjustment, varying in magnitude, of the shape and pattern of the river. Leopold and Wolman (1957a) cite the following variables that enter into the consideration of a stream channel: discharge, sediment load, size of sediment, channel width, channel depth, flow velocity, stream bed roughness, and slope or gradient. Obviously such things as climate, bedrock type and structure, vegetation, and human projects such as dams or changes in land use patterns can have an impact on one or more of these parameters.

An idealized path of a river starts with headwaters in the mountains and discharge into the sea (Figure 11–1). The most obvious factor is a constantly decreasing gradient toward the ocean (Figure 11–2). As streams flow from mountainous areas, where gradients can be extremely high, large amounts of clastic debris can be carried and deposited at the foot of the mountain range as the stream energy decreases. This deposition forms large, roughly conical-shaped masses known as **alluvial fans** (Figure 11–3). Alluvial fans are best developed in arid areas where rainfall is infrequent but intense. A slower, relatively continuous stream flow, characteristic of more humid areas, tends to inhibit the growth of alluvial fans because of constant erosion. Farther away from the mountain range, gradients become lower and the river develops a braided pattern (Figure 11–4). Meandering river patterns typically develop closer to the coast, where gradients become even lower (Figure 11–5).

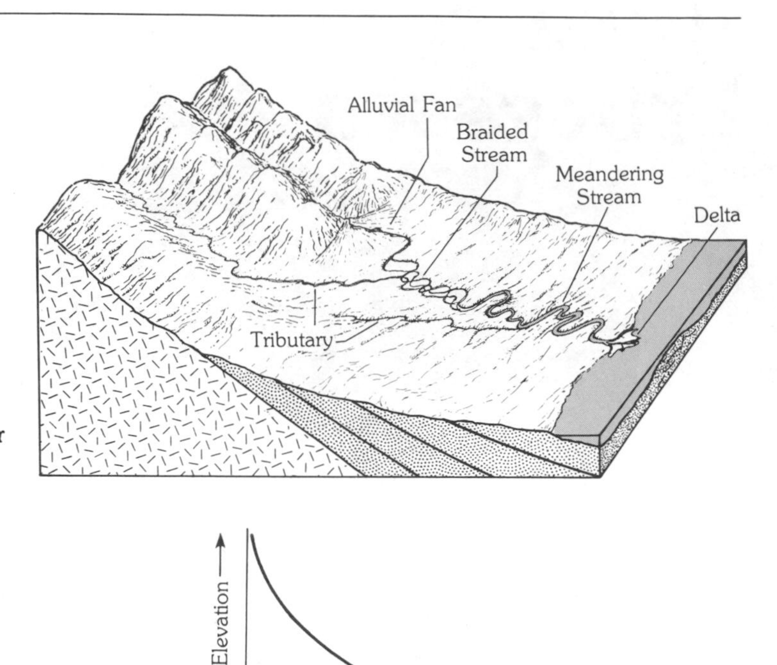

FIGURE 11–1 Block diagram of a river system.

Elevation →
Sea Level
← Distance from Coast

FIGURE 11–2 Long profile of a river. The long profile is the profile of the bed of the stream **(thalweg)** from the head of the river to the coast.

FIGURE 11–3 Alluvial fan.

As the river reaches the ocean, the gradient decreases to near zero, and the flow spreads out. There is an immediate decrease in velocity, resulting in the deposition of much of the sediment load. This mass of

FIGURE 11–4 Braided river.

FIGURE 11–5 Meandering river.

deposited material forms a delta (Figure 11–6). The type and extent of delta development depends on (1) the amount of sediment and water discharged by the river, and (2) the effects of marine processes such as waves, tides, and longshore currents. River processes tend to build the delta outward, whereas marine processes work as an opposing force to inhibit their growth. In the case of the Mississippi River delta, the river predominates over marine processes.

Most river valleys have **floodplains, natural levees,** and **terraces** (Figure 11–7). As a river flows through its valley, it creates and then flows in a low, flat area within the valley that is susceptible to flooding. The floodplain is composed mainly of sediments deposited within the stream as it migrates

FIGURE 11–6 Delta.

laterally within the valley. A smaller contribution comes from the deposition of very fine-grained sediment when floodwater spills over stream banks. Leopold and Wolman (1957b) estimate the contribution of such overbank deposition to be only 10 to 20 percent of the total sediment in the floodplain.

Natural levees are broad, low ridges paralleling the stream channel. They are formed during floods when the stream overflows its banks. As the floodwater escapes the confines of the channel banks, there is an immediate decrease in velocity resulting in the deposition of some of the coarser parts of the sediment load. In some cases, these levees can be constructed to such heights that floodwater can no longer completely overtop them. The river may then exploit and breach low areas in the levee, creating fan-shaped deposits of sediment on the floodplain known as **crevasse splays**. In some cases, these breaches can lead to the sudden or gradual abandonment of the stream channel downstream from the break.

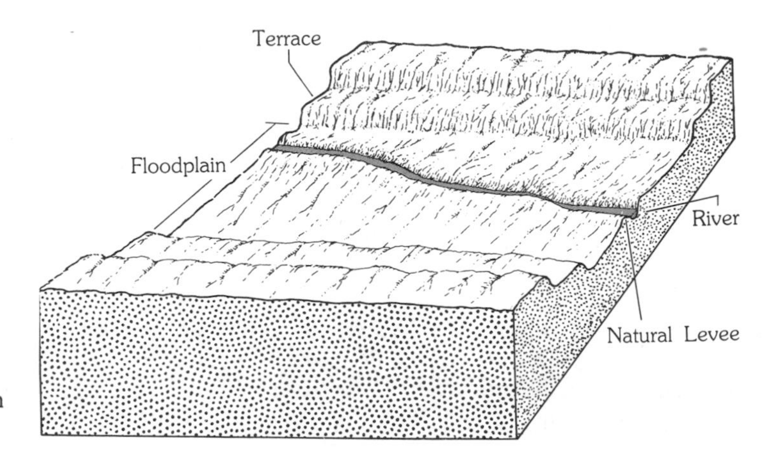

FIGURE 11–7 Geomorphology of an alluvial river valley.

FIGURE 11–8 River terrace (designated by arrow).

River terraces are topographic features within the stream valley that mark the position of former floodplains or valley floors. They are generally created by tectonic or climatic changes that cause the river to entrench deeper into the established floodplain (Figure 11–8).

RIVER MORPHOLOGY

Rivers fall into three basic classifications: *braided, meandering,* and *straight.* The natural occurrence of rivers that are straight for any significant distance has been determined to be so exceedingly rare that they will not be discussed here.

A **braided river** is composed of a series of multiple, relatively shallow, intertwining channels separated by sediment bars (Figure 11–9). Morphologically, these sediment bars can be classified into two types: **longitudinal bars**, whose long axis is parallel to the stream, and **transverse bars**, which are at an angle to the stream. Overall, the channels, which migrate laterally very rapidly during flood stage, have a high width-to-depth ratio.

Braided rivers often exhibit extremely variable rates of sedimentation, high and variable gradients, and large fluctuations in discharge within short time spans (LeBlanc, 1972). Flash-flood discharges, which can affect both slope and sedimentation rate, occur in arid areas where rainfall is infrequent but intense. Seasonal melting of snow and glaciers in mountainous regions can also produce the same flash-flow effect. The net result is an environment of rapidly changing discharge conditions to which the river system must constantly adjust. A highly fluctuating discharge is apparently very instrumental in maintaining the braided pattern.

FIGURE 11–9 Braided stream channel. Notice that the stream can follow numerous channels. (Photo by M. P. Wilson.)

Formation of bars in a river channel plays an important role in the initiation of a braided stream. Bar formation, according to Leopold and Wolman (1957a), takes place by deposition of some of the coarser part of the sedimentary load as a result of some local hydrologic condition during

FIGURE 11–10 Stream channel stabilized by extensive vegetation cover.

high stream flow. This initial deposition traps some of the finer material and functions as a nucleus for further deposition and construction of the bar in a downstream direction. The bar then diverts stream flow laterally, in turn allowing the stream to widen its overall channel configuration by eroding its confining banks.

The ease with which banks are eroded affects the rate of lateral migration of the channel. Banks composed of unconsolidated sediments and/or sparse vegetation facilitate the rapid lateral migration rates characteristic of braided streams. McGowen and Garner (1970) point out that the Amite River of Louisiana and the Colorado River of Texas, both prone to flash-flood conditions, would probably be braided were it not for the dense vegetation that serves to stabilize their banks (Figure 11–10).

Meandering rivers, unlike braided streams, flow through a single channel that migrates within its floodplain in a winding, sinuous path (Figure 11–11). They also differ in that the width-to-depth ratio of the channel is lower than that of braided streams. Factors that aid in the maintenance of the meandering pattern are a more continuous and uniform year-round discharge and abundant vegetation, which serves to stabilize the banks.

In plan view, the most obvious features are the **point bar** and the **concave bank**. The **inflection point**, or crossover, is the shallowest portion of the channel whereas the area within the **meander bend** is the deepest. Leopold and Wolman (1960) note that during high flow, the bend areas are scoured, and the inflection points fill with sediment. Conversely, during low-flow periods, the inflection points erode and the bend areas tend to fill.

A meandering stream migrates across its floodplain as material is eroded from the concave bank and deposited on the point bar. Friedkin

FIGURE 11–11 Channel of a meandering stream. Notice that the stream follows a single channel.

FIGURE 11–12 Recent neck cutoff of a meander. Sediment will be deposited at the abandoned channel ends, forming an oxbow lake.

(1945) has shown by model experiments that material eroded from the concave bank is deposited on the point bar that is immediately downstream on the same side. He observed very little material crossing to the opposite side of the channel.

Individual meander loops can be cut off from the main stream channel to form **oxbow lakes** (Figure 11–12). These oxbow lakes will eventually fill with clay or silty sand. The significance of these "clay plugs" within the floodplain shows up at a later time when the river shifts back over a former oxbow lake. Since the clay is much more cohesive than the surrounding coarser-grained deposits, they are more difficult to erode and will thus have an impact on the stream pattern. The nature of the deposits that fill these cutoffs depends upon their orientation to the main stream channel (LeBlanc, 1972). The downstream end of a meander cutoff will fill predominantly with clays, whereas the upstream end will receive an occasional influx of sediment from the main channel, resulting in coarser sediments overall.

Meander cutoffs are of two types: **neck cutoffs** and **chute cutoffs** (Figure 11–13). Neck cutoffs are created by bank erosion of opposing meander loops. Chutes are created in extreme floods when the thread of maximum stream velocity tends to straighten out and shifts from the

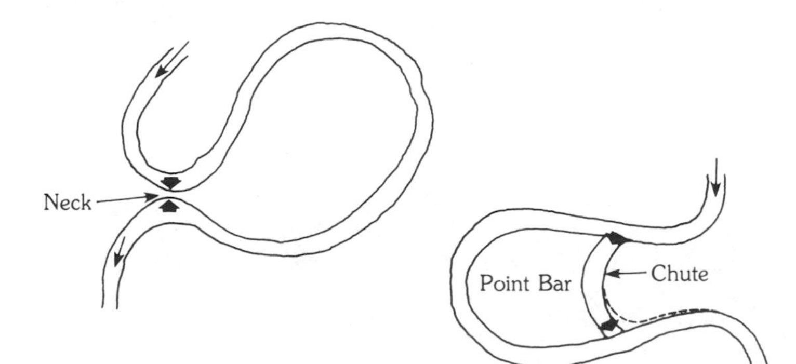

FIGURE 11–13 A neck and a chute in a meandering channel.

CASE IN POINT

Cut Off at the Neck

Meandering rivers in alluvial valleys frequently change their channels through the processes of bank erosion. A large meander loop on the Brazos River in Texas was cut off at the neck during high spring flows. These natural processes can cause significant land ownership problems whenever property lines are defined by a river. When searching past deeds and property surveys in alluvial valleys, it is important to remember that the boundary line—the river—may have moved since the last survey. These same processes can have a significant impact on the riparian rights of riverfront property, particularly if the river abandons the channel that fronts on the property.

concave bank toward and over the point bar (McGowen and Garner, 1970). Whether or not the chute results in a cutoff depends on the depth and extent to which it erodes into the point bar.

Why a river meanders rather than taking a straight path is still poorly understood. Several lines of evidence suggest that it is a phenomenon of fluid dynamics rather than simply being a result of the river winding its way around resistant material in its path. According to Leopold and Wolman (1960), most river meanders have a ratio of *radius of curvature* to *channel width* that ranges from 2 to 3. This is not what one would expect if the ratio were the result of random occurrence of materials of differing resistance within the valley. Furthermore, Leopold and Wolman have observed that meltwater on glacier surfaces and the main current of the Gulf Stream in the North Atlantic, both free of sediment, flow in a meandering path similar to that of rivers.

Not all rivers flow in alluvial valleys where erosion and deposition of sediment control channel shape. Rivers flowing in crystalline terrain (igneous and metamorphic rock) or in indurated sedimentary rock terrain are frequently controlled by terrain features. The stream will first erode along zones of weakness or easily weathered materials. As a result, it is trapped by these terrain features. In crystalline rocks, the control is established by fractures or faults in the rock. In folded rocks the relative erodibility of the rocks tends to control the stream; it cuts into the more erodible rock. In areas that have experienced recent uplift the stream may be entrenched, as in the Colorado River in the Grand Canyon.

RIVER MECHANICS

Every river or stream system may be considered to be a distinct entity unlike any other. No two streams possess the same physical configuration throughout their entire length. However, the morphology of all rivers is similar enough so that each river, or portions of each, may be classified according to Leopold and Wolman's (1975) system as straight, braided, or meandering. In order to understand why any river possesses a certain physical configuration, the river must be investigated as a complex system involving the interaction of numerous physical processes: varying channel dimensions, changes in flow regimes, and the size and quantity of material transported within the channel. The study of the interaction of these parameters and their relationships to the characteristics of a river is **river mechanics.**

In attempting to make a physical analysis of a river system, it becomes immediately evident that many of the parameters can only be approximated in part because the methods of making measurements are somewhat inadequate and because knowledge of the science is far from complete. Fundamental relationships that appear to be applicable to one river system often do not apply to others. For this reason, most of the derived quantitative relationships are only generalizations and should be regarded as such.

Every river has a unique set of **hydraulic parameters** that define the physical characteristics of the channel and the water flowing in it (Figure

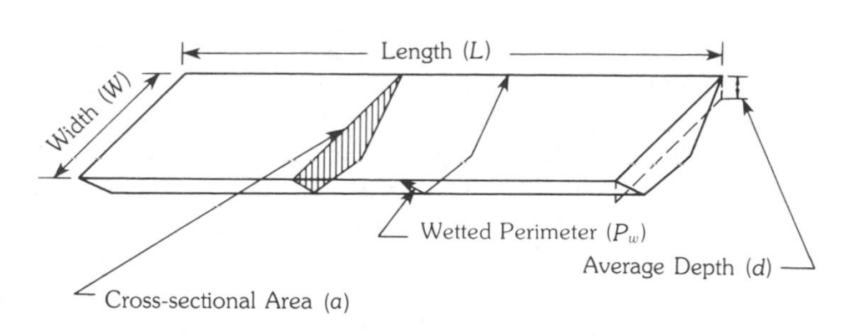

FIGURE 11–14 Geometry of a channel.

11–14). The **length** (l) of any portion of the channel is an arbitrarily chosen value. The **width** (w) of a channel at any time is the distance between the points of intersection of the water surface and the channel sides on either side of the river. The **depth** (d) is usually taken as the distance between the water surface and the normal bottom of the channel at its greatest value. The average depth across the channel is generally much less than the greatest depth in meandering rivers; however, in very shallow rivers and artificial channels, the two values closely approximate one another. The **cross-sectional area** (A) of channel is the width (w) of the channel multiplied by the depth (d). The **wetted perimeter** (P_w) is the measured distance of contact between the water and the channel cross section, measured perpendicular to the direction of streamflow. The **hydraulic radius** (R) is the ratio of the cross-sectional area to the wetted perimeter:

$$R = \frac{A}{P_w}$$

Remember that the values of these dimensions vary from point to point along the course of a river. Hence establishing proper values to use in calculations for any segment of the river is very difficult. To add chaos to confusion, the values of these parameters change daily at any given location because the amount of water in the channel varies, and erosion or deposition of material within the channel brings about changes.

Figure 11–15 diagrammatically illustrates **hydraulic nomenclature** for certain segments of a river. The **meander loop, meander width,** or **amplitude** (M_w) is the areal distance between the outer edges of two successive meander loops. **Radius of curvature** (R_c) is the length of the radius of an arc that most closely approximates the centerline of the channel. The **bend** of a channel is that portion within the looping meander. A

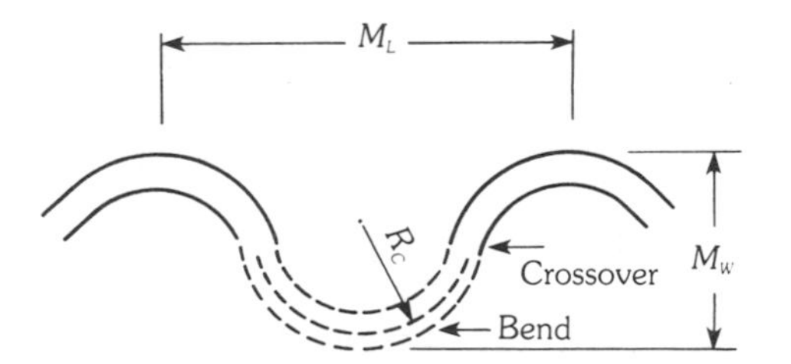

FIGURE 11–15 Geometry of a meandering channel. M_L is meander length, R_C is radius of curvature, and M_W is meander width.

crossover is the straighter portion of the channel between meander loops. The cross-sectional profile of these two regions is distinctly different and becomes of prime importance in analyzing water flow and distribution of sediment within the channel. **Slope** or **gradient** is the vertical drop in elevation over some horizontal distance, usually expressed in meters per kilometer. Slope is the tangent of the angle that the water surface makes with the horizontal.

The **velocity** (v) of a stream is simply the measure of the speed with which the water flows through the channel. A typical profile of a stream shows that the water velocity varies with depth (Figure 2–12). Flow velocity is at a maximum at the surface and perhaps extending down a short distance into the water. With increasing depth, velocity diminishes approximately as a parabolic curve. At the water-channel boundary, velocity is zero.

The volume of water that passes through any given cross-sectional area of a stream in some unit of time is the **discharge** (Q), usually recorded in cubic feet or cubic meters per second. Discharge is calculated by multiplying the cross-sectional area (A) by the average, or mean, stream velocity (v):

$$Q = Av$$

The discharge of a stream may vary widely from day to day at a given locality depending on the amount of precipitation that falls within the drainage basin. Throughout the length of the river, the discharge also varies from location to location at any given time, but in almost all cases discharge will increase at points progressively downstream, because water from tributaries increases the volume of water in the stream.

HYDRAULIC RELATIONSHIPS

River width, depth, and velocity all increase with discharge at any particular station (Figure 11–16). Farther downstream the relationships between width, depth, and velocity with **mean annual discharge** also change. Mean annual discharge is the average of all mean daily discharges for a year. Since mean annual discharge in most rivers increases downstream, the

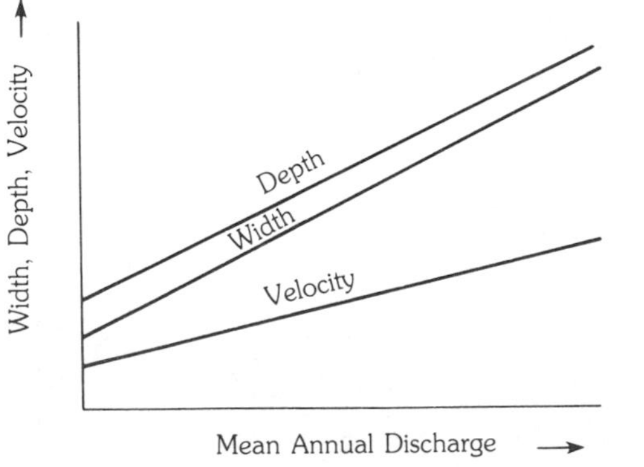

FIGURE 11–16 General relationship between width, depth, and velocity of a river in an alluvial valley to the mean annual discharge.

width, depth, and velocity of a river also increase at stations progressively farther downstream (Langburn, 1962). Of particular interest is the velocity relationship, since most people are under the impression that stream velocity is greater in the upstream area where the slope is greatest, because shallow mountain streams tumbling over boulders and other channel obstacles give an impression of high velocity.

FLOW CONDITIONS

Laminar and **turbulent flow** represent the two conditions of fluid movement. Laminar flow is a smooth condition in which water layers flow parallel to one another with no exchange of water particles from one layer to another. In turbulent flow water particles are exchanged across flow layers in a random and confused pattern (Figure 11–17). Turbulent flow conditions probably exist in all natural streams; however, it is believed that a thin layer of laminar flow, the **laminar boundary layer,** exists for some finite distance above the river bed.

Turbulent flow in a river can easily be detected by casual observation as the swirling and bubbling motion of the water. Because of this motion, the water particles travel much farther than the actual distance of the channel length. Turbulence is generated as a result of the interaction of two materials of different viscosity near the stream bottom. Random turbulent motion of individual water particles is superimposed upon a high-order flow pattern within the stream channel. The high-order flow pattern can be visualized as a corkscrew motion, called **helical flow,** and is most pronounced in the bends of a river where the water impinges against the concave bank (Friedkin, 1945). In the bends, the water on the surface flows toward the concave bank, is forced down at the sides, and turns back toward the channel center along the bottom. Because the force of the water is greatest at the surface, the concave bank is eroded while deposition occurs on the convex bank. In a straight or crossover, the single, assymetrical helical flow pattern gives way to a dual symmetrical pattern. An upwelling occurs near the center of the channel, and flow is outward in both directions from the channel center. Again, the surface flow is directed downward at the channel sides, and water returns to the center of the channel along the bottom.

Hydraulic engineers and geologists have used numerical relationships between the hydraulic parameters to predict river conditions. The **mean velocity** (v) is a significant parameter that affects discharge and the amount of energy in the river system. It includes increasing velocity, increasing

FIGURE 11–17 Laminar flow versus turbulent flow.

discharge, and energy. The **Manning equation** is an empirical equation that relates hydraulic radius (R) and slope (S) to the mean velocity:

$$V = \frac{1.49R^{2/3}S^{1/2}}{n}$$

where n = Manning's number.

Manning's number is a factor that relates all other river characteristics, including channel roughness, sediment in transport, turbulent losses, water-sediment density, and channel shape, among others. The Manning equation was originally developed for the design of artificial channels, and n was only a roughness factor. As Table 11–1 shows, the variability of Manning's number increases as the stream channel characteristics become more natural.

Sediment erosion is another significant parameter in a river system that affects channel migration, erosion and siltation, sediment infilling in reservoirs, coastal processes, and the entrainment of sediment in pumps and drainage projects. Hjulstrom (1939) related velocity to the erosion,

TABLE 11–1 Representative values for Manning's number

n	Channel Characteristics
0.010	Very smooth—glass, plastic
0.011	Very smooth concrete or wood
0.012	Smooth concrete
0.013	Normal concrete
0.014	Wood construction
0.015	Vitrified clay
0.017–0.020	Shot concrete, excellent earth
0.020–0.025	Smooth bare earth in good condition
0.025–0.035	Earth channels with some vegetation
0.035–0.040	Natural streams
0.040–0.050	Rough mountain channels or rivers
0.011–0.035	Alluvial channels, transporting sand
?	Streams with rough beds and vegetation

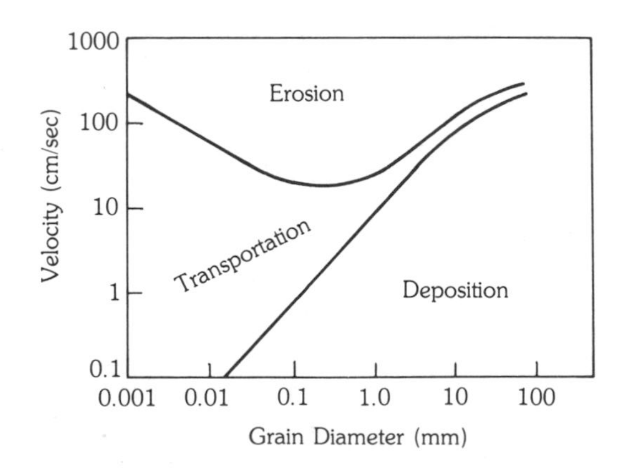

FIGURE 11–18 Hjulstrom's diagram.

transport, or deposition of loose sediments through laboratory experiments (Figure 11–18).

For grain sizes greater than 0.5 mm, a continual increase in the velocity is required to move larger and larger grains. For grain sizes less than 0.5 mm, the stream velocity required to move the grains becomes greater because cohesion of the finer-sized material must be overcome. Once grain motion is initiated, less energy is required to keep the grain in motion. Coarse-grained sediments require almost as much energy (velocity) to keep them in transport as was necessary to erode the particles, whereas finer-grained material requires a significantly lower velocity for transport than for erosion. Grains moving as bed load or suspended load will be deposited when the velocity decreases below the critical value for that particular grain size. The approximate value of this critical velocity for any grain size can be found from the Hjulstrom diagram. In the diagram, notice that grains smaller than 0.01 mm can stay in transport at very low velocities, because any upward turbulence in the stream is great enough to overcome the gravitational forces attempting to bring the grain back to the channel bottom.

Another approach to the problem of sediment transport has used the concept of **boundary shear** (τ) between a moving fluid body and the bottom of the stream channel (see Concept 2–8), in which:

$$\tau = \gamma d S$$

where

γ = density of the fluid
d = depth of water one unit wide
S = slope of the channel as a ratio

The boundary shear developed by the moving fluid is then compared to the **critical tractive force** (τ_c):

$$\tau_c = \frac{\pi}{6} c D (\rho_s - \rho) g \tan \alpha$$

where

c = packing coefficient determined by the arrangement, number, and size of the grains in a unit area
D = mean grain diameter
ρ_s = density of the grains
ρ = density of the fluid
α = angle defined by the packing coefficient which represents the angle of repose of the grains

The critical tractive force is the boundary shear that causes just the most exposed particle to move and is the initiation of sediment transport. For

CASE IN POINT

Scour at the Bridge

During February and March 1980 three major storm cells moving inland from the Pacific Ocean dumped heavy rains on the Salt River near Phoenix, Arizona. The Salt River Project has a series of water supply reservoirs in the Superstition Mountains. Heavy runoff from the storm and a lack of flood storage in the reservoirs led to serious flooding along the Salt River in Phoenix. Low water crossings and bridges were damaged by scour. Scour holes on the upstream side of the bridge piers, reaching 2 m or more in depth, undercut bridge support. The high boundary shear conditions during the flood can be seen by the scour holes and the size of the sediment transported during the flood. The cobbles shown in the photograph are 10 to 20 cm in diameter. Also notice the debris that was trapped by the bridge pier. (For scale, the man is 1.8 m tall.)

sand and fine gravel forming a smooth bed, the critical tractive force is approximated by:

$$\tau_c = 4D$$

where D = median grain size in feet.

These two approaches to the problem only define the initiation of sediment transport and indicate that erosion, transport, or depostion will occur. In the design of irrigation or drainage ditches where clear water flow, or no erosion, is desired, these approaches will work. However, in most river projects, the system is already carrying sediment, and it is necessary to predict the amount of sediment transport (Q_s).

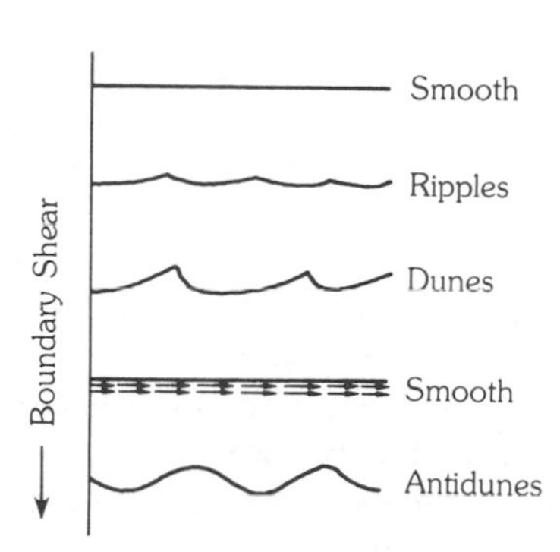

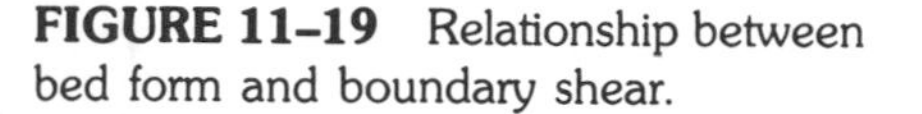
FIGURE 11–19 Relationship between bed form and boundary shear.

Laboratory and field studies of **bed form** of sand beds have determined that the bed changes form as the velocity or boundary shear increases (Figure 11–19). Starting with a smooth sand bed, no sediment transport occurs until the boundary shear equals the critical tractive force. As the boundary shear increases, the sediments move along the channel bed as **bed load** until dunes begin forming. At this point, particles begin to become suspended in the fluid and move as **suspended load.** Since most rivers have sufficient boundary shear to have dunes, the sediment transport problem is one of separating bed load from the suspended load.

In considering all the processes acting in a channel, you might intuitively feel that the process of picking up grains from the bottom of the channel and entraining them in the fluid flow as suspended load is rather simple. However, it is not. Several theories concerning the entrainment process have been proposed. All the conditions that have been noted in experiments must be considered in trying to explain how this phenomenon occurs. Foremost among these conditions is the existence of the laminar boundary layer and the velocity distribution near the channel bottom. According to current theory in fluid mechanics (for example, Swancon, 1970), the velocity of a fluid flow over a stationary flat plate must be zero at the boundary and increase in a direction perpendicular to the plate.

If the velocity is zero at the channel surface, how do the grains ever get started moving? Consider Figure 11–20, a hypothetical axial profile along a stream channel showing random particle distribution and the laminar

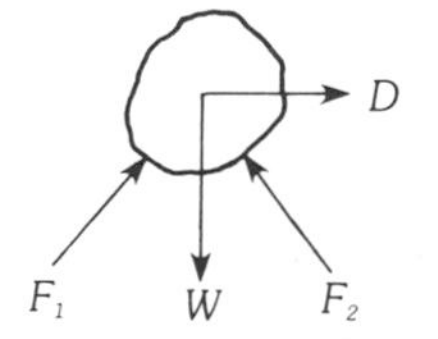

FIGURE 11–20 Flow conditions along the bed of a stream. Some of the larger particles protrude into the turbulent flow zone and are subjected to the loading shown. As the boundary shear increases, the drag (*D*) on the particle increases and the particle eventually begins to roll.

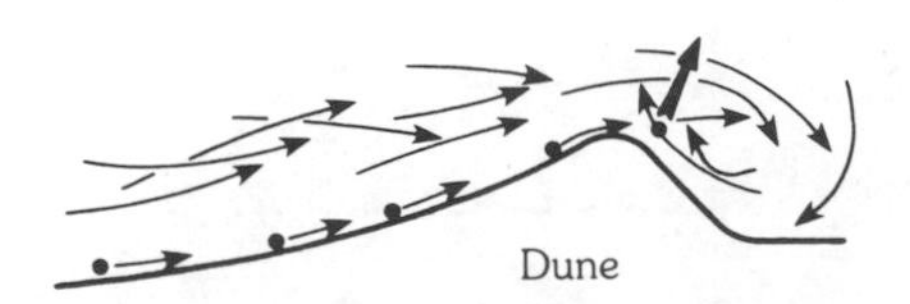

FIGURE 11–21 Launching of bed load into a stream from the crest of a dune.

boundary layer. The larger particles project above the laminar boundary layer, and therefore experience turbulent flow conditions. These grains may roll along the bottom as bed load within the laminar flow layer.

When larger ripples and dunes are formed, they modify the shape of the boundary layer, and some particles may be swept off the crest and out of the boundary layer by turbulent action. More important is that grains are launched into the fluid flow as they pass over the crest toward the trough. Water flow over the dune crest and into the trough produces an eddy, which entrains the grain and carries it into suspension (Figure 11–21).

Sutherland (1967) has proposed a method for grain entrainment which allows grains to be put into suspension from a flat bed. He observed that grain motion on a bed occurs in random bursts. These bursts are of short duration, but with increasing flow velocity the motion of the grains becomes more violent, and eventually the grains are projected into the fluid flow as suspended load. He believes that the cause of the bursts is individual eddies that collide with the bed surface, penetrating and disrupting the boundary layer. As the eddy approaches the bed surface, the grain begins rolling along the surface because the eddy induces a velocity increase. Once the velocity vectors in the eddy become inclined to the horizontal with respect to the position of the grain, the grain is propelled from the bed and into suspension. The same process is also dominant in the stagnation region of ripples and dunes.

The movement and launching of particles from the bed into suspension is a difficult problem to analyze quantitatively. Methods to calculate bed load sediment transport have existed for many years; however, due to the complexity of the system, the sediment transport equations are empirical rather than theoretical.

The **DuBoys equation** is an empirical equation developed by an English hydraulic engineer for irrigation canals in India; it relates sediment transport (Q_s) in lbs/sec/ft channel width as:

$$Q_s = K_\tau(\tau_0 - \tau_c)$$

where K_τ = the transport coefficient, which is related to mean sediment size.

Calculation of suspended load, Q_s, is a complex empirical solution similar to **Laursen's equation**:

$$\bar{c} = \Sigma p\left(\frac{d_m}{y}\right)^{7/6}\left(\frac{\tau'_0}{\tau_c} - 1\right) f\left(\frac{\sqrt{\tau_0/\rho}}{\omega}\right)$$

where

$\bar{c}$ = concentration in percent weight = Q_s/Q
τ'_0 = particle shear = $(v^2 d_m^{1/3})/(30y^{1/3})$
v = flow velocity

d_m = mean grain size of each size fraction
y = depth of water in ft
$\sqrt{\tau_0/\rho}$ = shear velocity
$\tau_0 = \gamma y s$
ρ = density of fluid
ω = fall velocity of the particle in each size fraction

Laursen's equation is solved by determining the grain size distribution for the sediment, dividing it into units of similar sizes by percent (p = unit value of each fraction), solving for $\bar{c}$ in each size fraction, and summing. Numerous graphical and computer techniques have been developed to calculate the total sediment load; however, their discussion is beyond the scope of this text.

STREAM SYSTEM PATTERNS

The complex characteristics of flow patterns in a river channel lead to the question of how these patterns affect the configuration of the stream system as a whole. If we consider that the flow of water in all meandering streams follows the helical model, then some other factors must be present to exert control over the configuration of the stream system. From among all the variables that go into producing the shape of any particular stream, there emerge five that may be considered independent: **geology, topography, river load, bank material,** and **discharge.** All other factors are dependent upon these five for their effect upon stream configuration.

Geology is considered an independent variable because the stream pattern does not control the geology. In fact, the geology definitely controls the morphology of the stream. Figure 11–22 shows numerous stream patterns and the related geologic environments where these patterns may be expected. Control of the stream pattern by the structural elements of the area is often dramatic.

Topography is also independent because the stream is forced to flow through the existing topography. Topography controls the slope of the stream, thus influencing the hydraulic parameters of boundary shear and velocity. The size and quantity of material that the river carries have a great influence on the shape of the stream (Leopold and Maddock, 1953). Many experiments with model streams under laboratory conditions have shown how stream geometry may change with a change in the river load. The material that comprises the banks of the stream is extremely important in determining channel dimensions and the meander width of the stream (Lane, 1937; Friedkin, 1945; Schumm, 1963). As the grain size of the bank material becomes finer, the banks become more difficult to erode, as shown in the Hjulstrom diagram (Figure 11–18).

The discharge is very important in stream configuration. Friedkin's (1945) studies have shown that each discharge, within certain limits, has a characteristic meander pattern that the stream channel attempts to follow. The stream channel also attempts to reach equilibrium with a particular discharge. Since stream discharge continually varies, equilibrium is never realized and the channel continues to shift and migrate across the valley

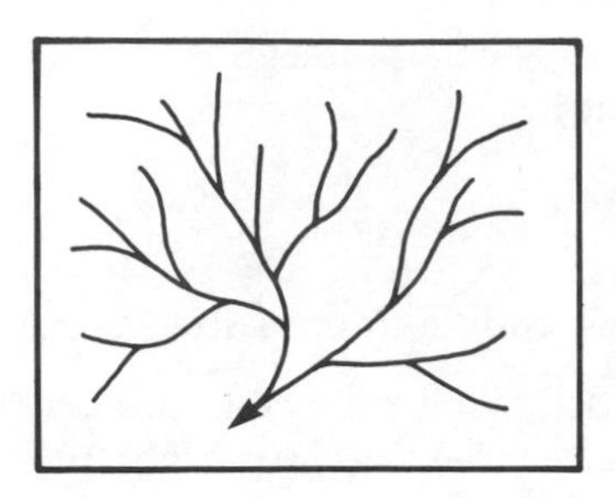

Dendritic
(no control)

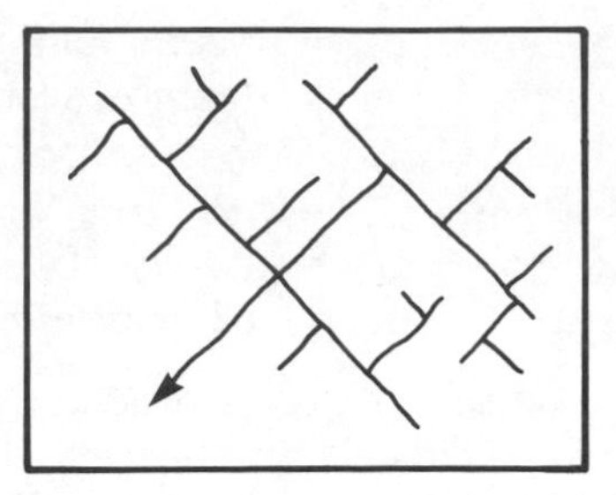

Angular
(fault/fracture control)

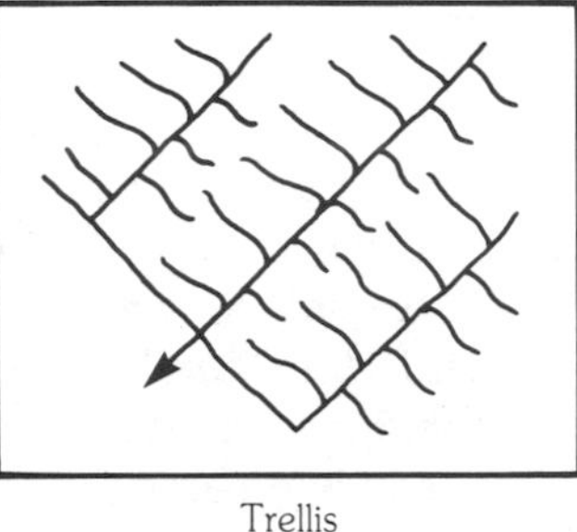

Trellis
(dipping sedimentary control)

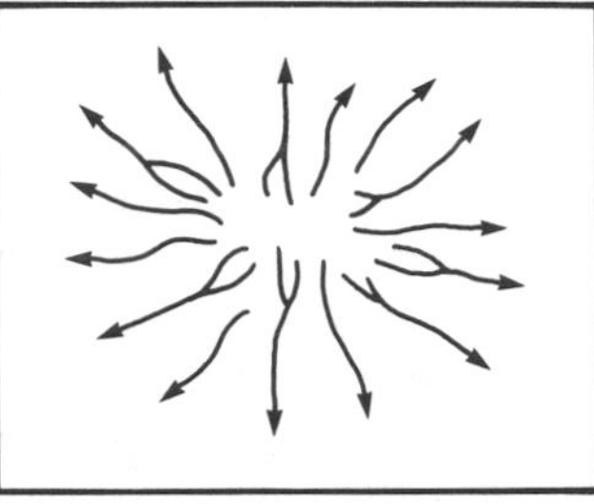

Radial
(dome/cone control)

FIGURE 11–22 Relationship between common stream patterns and the geology of the region.

floor. The discharge condition that the channel tries to approximate is usually that of mean annual discharge. Higher and lower discharge conditions create channel characteristics that are manifested upon the overall stream configuration.

HYDROLOGY AND FLOODING

Throughout this chapter, reference has been made to the fact that the discharge (Q) in a stream varies with both time and location. The amount of water that a stream receives is the amount of runoff delivered from its drainage basin. Since runoff is a complex problem, as was discussed in Chapter 8, stream discharge is also complex and is based on a statistical analysis of the rainfall-runoff characteristics of the drainage basin.

Statistical analyses of discharge are related to the **recurrence interval** to produce a **flood frequency curve** (Figure 11–23). The recurrence interval is defined as the period of time during which a given discharge will statistically be equaled or exceeded once. This factor must be based upon actual field measurements at a stream-gauging station and then applied to that location only. The flood frequency curve can now be used to determine the **statistical probability** of the amount of discharge (flood) that could be expected during any given time interval. The recurrence interval is actually a definition of the intensity of a flood event (100-year flood) and *not* the exact number of years between floods. For example, a flood having a 10-year recurrence interval defines channel flow conditions of a discharge of 500,000 cu ft/sec that have a statistical probability of

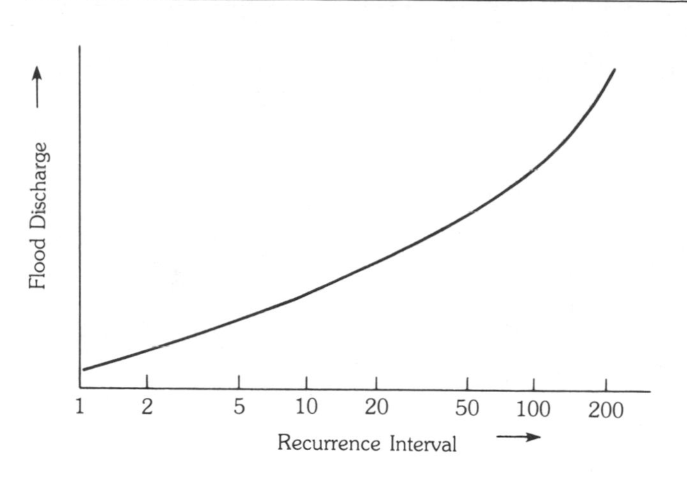

FIGURE 11–23 Statistical determination of flood discharge as a function of the recurrence interval. This relationship is based on measurements of discharge at a stream-gauging station and then used to predict the magnitude of rare events.

occurring or being exceeded during *any* 10-year period or approximately a 10 percent chance of occurring in any one year.

The time relationship between a rainstorm and the passage of the floodwater at a gauging station is shown in a **flood hydrograph** (Figure 8–5). For all drainage conditions, the water level **(stage)** in the stream does not begin to rise at the onset of the rainfall because water must travel from the place where it landed to the stream. This time difference is known as the **lag time.** Lag time is an important parameter in any **flood routing** or **flood forecasting** problem. However, the physical characteristics of the basin, which control runoff, and the stream channel, which affects the rate of the flood's downstream movement, make calculating lag time a complex problem that must be based on a thorough knowledge of the basin and the stream. Once sufficient field data on rainfall and runoff versus time have been collected and statistically analyzed, graphs and charts can be prepared to relate discharge (flooding) to time and location along the river.

A major problem with the prediction of discharge, time, and location relationships is that the characteristics of many drainage basins are drastically changed by human activities. Engineering projects can affect the drainage system in two basic ways: altering runoff and altering lag time. As discussed in Chapter 8, runoff is related to precipitation by the equation

$$P = Q + ET + I + S$$

where

P = total precipitation
Q = runoff
ET = evaporation and transpiration
I = infiltration
S = storage

and through the rational equation

$$Q = ciA$$

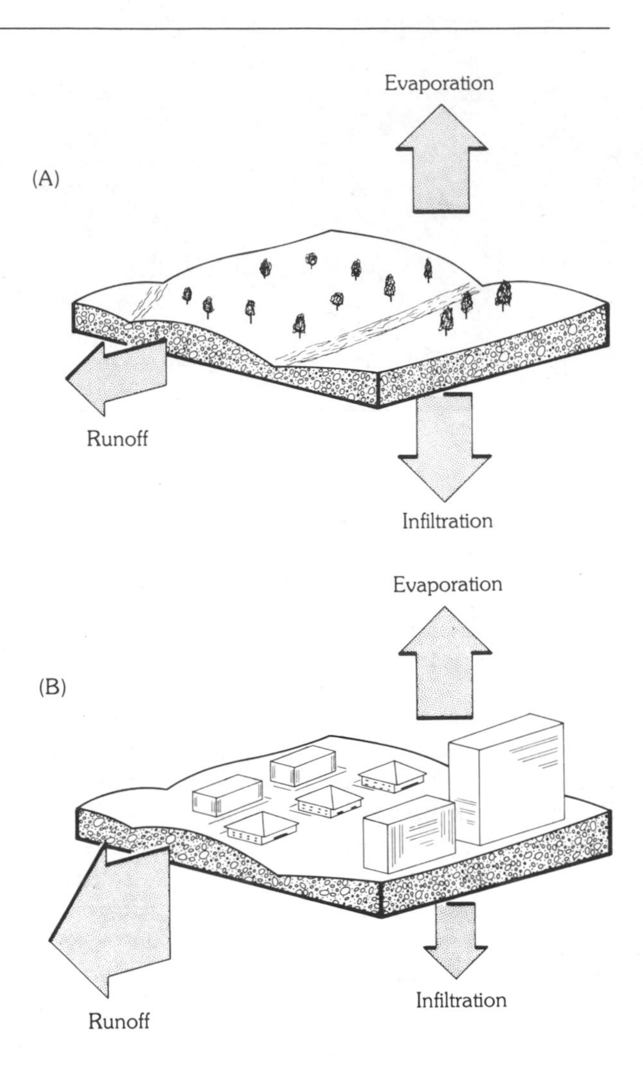

FIGURE 11–24 Schematic diagram of the influence of urbanization on the hydrologic cycle. Notice that runoff sharply increases and that infiltration decreases.

where

Q = runoff time
c = coefficient of runoff = $f(I + E + S)$
i = intensity of rainfall/time
A = area of drainage basin

Changes in I, E, or S, or in c all influence the amount of runoff (Figure 11–24).

The influence of human activities on the lag time is equally or in some cases more significant than the influence on total runoff (Figure 11–25). Drainage improvement projects, gutters, storm sewers, and channelization are all intended to reduce the lag time by allowing more runoff to leave a particular site than was possible before. In contrast, flood storage or retention projects are designed to retard runoff and increase lag time.

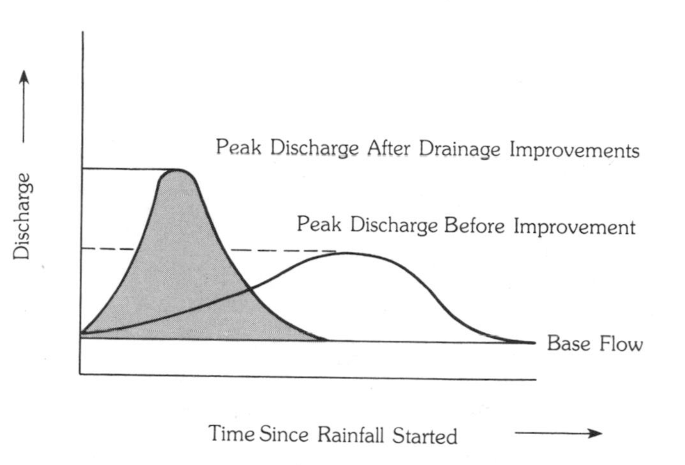

FIGURE 11–25 Effect of drainage improvements on the flood hydrograph.

RIVER ENGINEERING GEOLOGY

The engineering geologist and hydraulics engineer face a complex, constantly changing natural system whenever a river-related engineering project is undertaken. The system can be subdivided into a series of interrelated components that are themselves complex, or into a series of engineering projects and their related river processes (Table 11–2).

Channel improvements involve the alteration of the geometry of the stream channel for some engineering project, such as harbor facilities, a bridge crossing, bank erosion control, or navigation. Since the improvement does not affect the amount of water delivered to the stream system, the analysis is based on the discharge-time relationships and the sediment transport, erosion, or deposition effects of the project. The impact of the project on the river system and, more importantly from an engineering point of view, the impact of the channel alterations on the engineering project must be considered. All too frequently, sediment transport is not fully evaluated in river improvement projects. For example, navigation improvements along the Mississippi River have changed the river geometry so that some portions of the system are now hydrodynamically more stable as a braided stream than as a meandering stream. The result has been increased channel maintenance in those areas.

TABLE 11–2 River system components

COMPONENT	SIGNIFICANT FACTORS
Runoff	Meterology, climate, drainage basin geomorphology, land uses
Discharge	Stream geometry, time, runoff
Erosion/sedimentation	Stream geology, discharge
PROJECT	**RIVER PROCESSES**
Channel improvement	Discharge, erosion/sedimentation
Drainage/channelization	Runoff, discharge, erosion/sedimentation
Reservoirs	Discharge, erosion/sedimentation

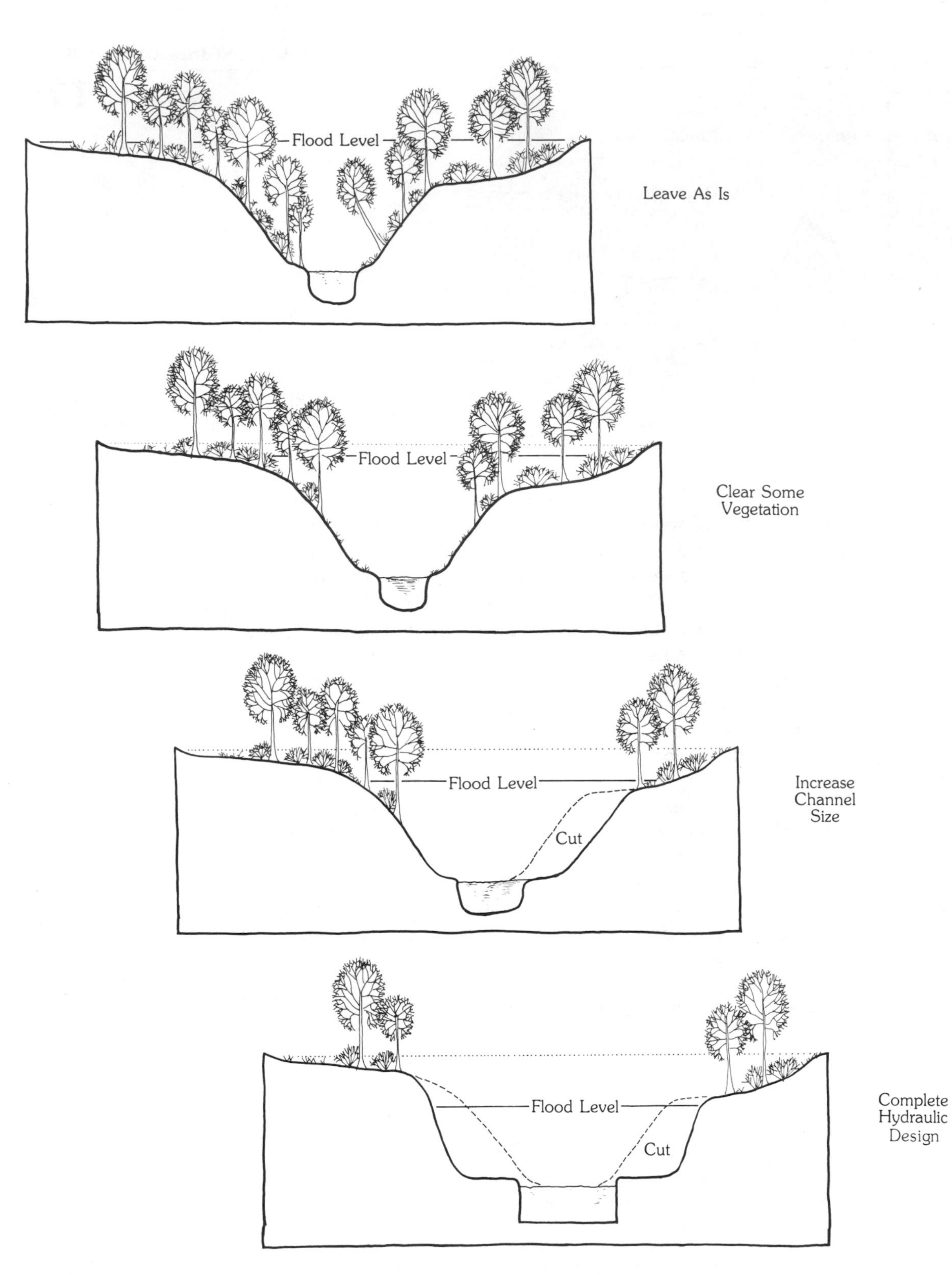

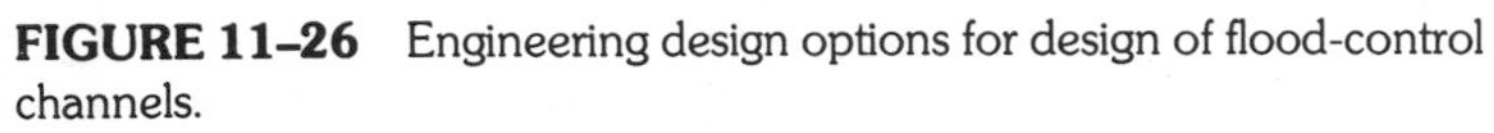
FIGURE 11–26 Engineering design options for design of flood-control channels.

Drainage and **channelization** projects are intended to improve the drainage efficiency by reducing lag time. Storm water is collected and rapidly removed from an area in order to remove excess runoff, usually related to a change in land use. Drainage projects must consider the expected amount of runoff and select a design intensity (design recurrence interval) in order to determine the size of the project. Channelization, the process of modifying a stream to improve drainage efficiency, can modify flow velocity or cross-sectional area in order to increase the stream carrying capacity, or discharge. Any change in discharge will have a corresponding effect on channel erosion and sedimentation, or sediment transport. The drainage engineer has a spectrum of design options available in any project (Figure 11–26). In urban areas where the land values are high, such projects can economically reduce the size of the flood hazard area to provide additional safe building space.

Engineering geologic considerations for channelization projects include the analysis of the effect of increased discharge and velocity, which increases stream energy and results in erosion/sedimentation problems. The higher-energy, channelized portion becomes an eroding area that brings an additional sediment load into the lower-energy, unchannelized portion. The corresponding decrease in stream energy leads to siltation in the unchannelized portion. In addition to increased sedimentation, there is the problem of increased flood levels, since more water has also been delivered to the unchannelized portion. Proper channelization project design depends upon successfully balancing the need for upstream drainage while minimizing downstream flooding and sedimentation problems.

Reservoir and **flood mitigation** projects on a river are designed for water storage, either for use during dry periods or for restraining excess floodwaters for later release. These projects can have a profound impact on the hydrodynamics of the stream or river system. On a local project scale, the hydrodynamics of the stream system can also have a significant impact on the project.

Filling rates for reservoirs are controlled by the discharge history of the river at the location of the reservoir. In most cases, the operator must allow for the continued supply of water downstream, and therefore, water available for storage is limited to excess flow. At the other extreme, once the reservoir is filled, sufficient capacity must be maintained to store the water delivered by the **design flood.** In addition to flood storage, the reservoir must be able to release safely any water that is delivered in excess of the design flood. Multiple-use reservoirs have three general levels of water storage: **conservation pool, storage level,** and **flood storage** (Figure 11–27). The conservation pool is the minimum water level necessary to maintain the lake. Storage volume varies throughout the season and must be managed so that the volume of water removed from storage **(draft)** does not deplete the supply before it is recharged during wet periods. Flood storage is the volume of a flood-control reservoir that is available to store floodwater.

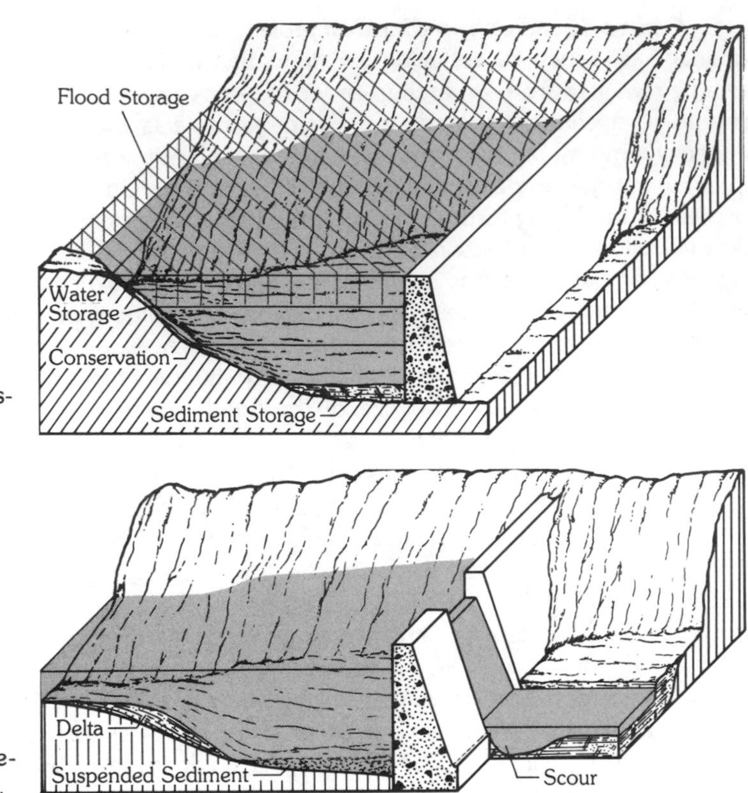

FIGURE 11–27 Design pools in a reservoir.

FIGURE 11–28 Sedimentation behind a dam and scour below the dam.

Lakes and reservoirs always have sedimentation problems, simply because the water flowing into the reservoir is carrying sediment. At the head of the lake, where the stream enters, the flow velocity decreases, coarse-grained bed load and suspended sediment are deposited, and a delta will form. Fine-grained suspended sediment will be carried into the reservoir and in many cases will be transported to the base of the dam in a **turbidity current**. Below the dam, clear water that is capable of carrying sediment is discharged. As a result, **scour** will occur in the downstream channel (Figure 11–28).

FLOODPLAIN MANAGEMENT

The U.S. Army Corps of Engineers has estimated that property loss due to flooding is over 900 million dollars a year and loss of lives more than eighty-three a year (Burton and Kates, 1964)—and these figures continue to increase all the time. Throughout human history people have settled by waterways because they provided a source of food, economical transportation for people and their goods, and a communication link to the outside world. Ancient peoples understood the risk of the places where they lived, and after being flooded, their houses were easily rebuilt with little cost or trouble. Today these needs are still important; but we are encroaching more and more upon the floodplains in direct competition with their natural uses. No one can deny that the aesthetic value of a wooded lot by a gently meandering stream makes the area a prime site for single-family dwellings and that such homes in turn encourage commercial development and multifamily housing. Even after a severe flood, people choose to settle in these areas again, believing that another flood will not happen, or certainly not to them. Franklin S. Adams (1973) of Pennsylvania State University comments:

> The feverish activity associated with rebuilding all that has been lost supports this contention. In the very best pioneering spirit, flood victims everywhere appear determined to restore every lost bridge, every damaged house and trailer, every water-logged business, to its "rightful" place on the floodplain. Their faith in themselves and nature appears limitless.

In nearly all cases, the flooding is seen as an "act of God" and not as the irregular, intermittent event that it is. Heavy rains are perceived to be the only cause of the high waters, and such climatic conditions are considered unpredictable and uncontrollable. However, floods are really caused by two things: *rain quantity* and *land use*. Floods that inundate rural areas are just an inconvenience (Figure 11–29). However, the amount and cost of damage rise considerably with increasing urbanization (Figure 11–30). Neither complete control of rivers nor complete abandonment of the floodplains is economical or practical. Floodplains are a valuable resource and will continue to be occupied. The degree of occupancy and the amount of flood protection that can be practically provided will determine the amount of risk involved.

FIGURE 11–29 Rural flood, representing an inconvenience to rural transportation and some crop damage.

FIGURE 11–30 Urban flood damage, representing loss of property and lives.

There are two major ways to prevent or reduce flood damage: (1) management of the river, and (2) management of the land, called **structural** and **nonstructural controls,** respectively. Structural controls (river management) include flood storage, drainage improvement, watershed treatment, and flood-retaining structures. Flood storage structures, usually dams across a river channel, are designed to keep incoming floods upstream from the area to be protected. Drainage improvement projects are designed to reduce the lag time and increase the drainage capacity. Channelization and storm sewers are the most common. Watershed treatment is a relatively new technique that involves structural modification of the watershed in order to increase infiltration or lag time. It utilizes such projects as small dams and lakes, permeable surfaces on parking lots, roof storage on structures, or storage areas in parking lots, parks, or other open spaces. Flood-retaining projects are designed to keep a flooding river within its banks or floodway and usually involve construction of levees or dikes. It is important that the engineering geologist realize that any structural control is designed for a particular risk (recurrence interval) based on a cost/benefit analysis and that there always exists a risk that the structure will be overtopped!

Nonstructural controls (land management) are designed to minimize the losses that would occur when the area is flooded and depend upon land use management or building regulation. **Floodplain zoning** is the legal means of protecting life and property as well as is economically possible while making the most beneficial use of the land. After an extensive study of the river system the area is divided into flood zones based on flood stages and risk. These zones are designated as specific land use areas (Figure 11–31). In the channel, no obstruction to the free flow of water is allowed: bridges must be designed to provide for the free passage of debris or ice, and no piers, fill, or jetties, which deflect currents against an unprotected bank, are allowed. Bridges must also leave a sufficient waterway underneath to prevent overtopping or be designed for overtopping with minimum

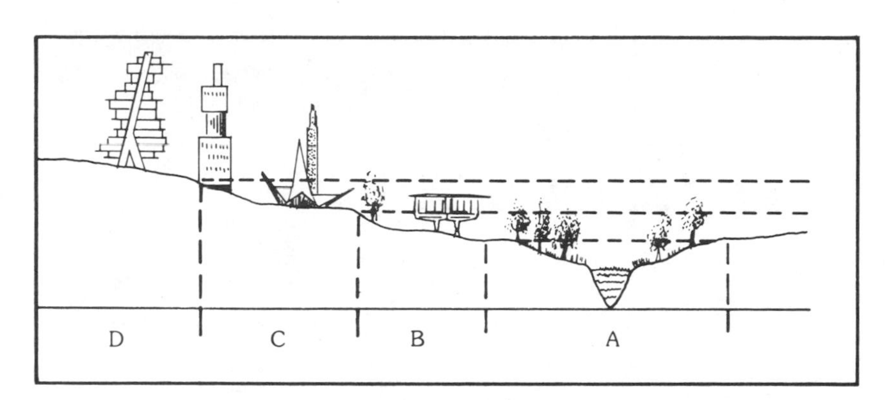

FIGURE 11–31 Flood hazard zones: (A) active floodway, which should be kept clear to allow free passage of frequent floods; (B) floodplain, where structures should be designed to be occasionally flooded; (C) low-risk floodplain, where structures need not be designed for flooding but critical land uses should be avoided; and (D) above any reasonable flood level, with no restrictions on land use or structural design needed.

backwater flooding; they must also be constructed to minimize chances of failure due to scouring around the piers.

The flood zone comprising recurrence intervals of 1 to 20 years should be left free of obstructions that would inhibit the flow of water. Permissible land uses in this area include agriculture, parks and open spaces, and roads. In the 20- to 100-year interval, buildings are permitted if the first floor is raised to above the 100-year recurrence interval flood or if the building is flood-proofed. This area can also be used for temporary and portable amusement businesses, such as circuses, rides, and shows; power and telephone lines; sewer and water pipes; and sand and gravel pits, provided that stockpiles of material are not placed where they would impede the free flow of floodwater. Prohibited in this area, besides structures not meeting flood-design requirements, is the storage of unsanitary or dangerous substances.

The federal government has rather strongly encouraged floodplain management programs. The Federal Flood Insurance Act, passed in 1969, provides for a flood insurance subsidy to individuals in communities that agree to adopt floodplain regulation guidelines. However, the Flood Disaster Protection Act, passed in 1973, states that the Department of Housing and Urban Development (HUD) must notify all communities in the nation if they have one or more special flood-hazard areas. Once notified, a community faces stiff sanctions if it does not participate in a program of federally subsidized flood insurance within one year of notification. To qualify for participation, a community must adopt and enforce "adequate land use and control measures" to reduce the likelihood and severity of flood damage.

The "adequate land use and control measures" that communities must adopt are floodplain management regulations that meet minimum safety criteria for protection against a 100-year flood. These measures include

No Flood Proofing

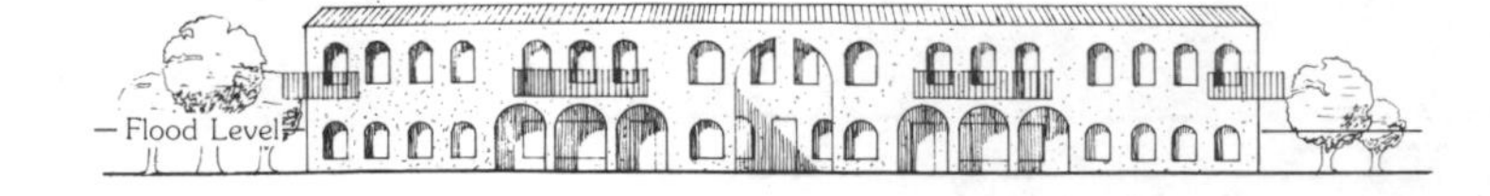

Average Annual Cost/ft²	
Construction	$1.34
Insurance	0.92
TOTAL	$2.26

Raised to +1 ft

Average Annual Cost/ft²	
Construction	$1.56
Insurance	0.02½
TOTAL	$1.58½

Raised with Watertight Closures

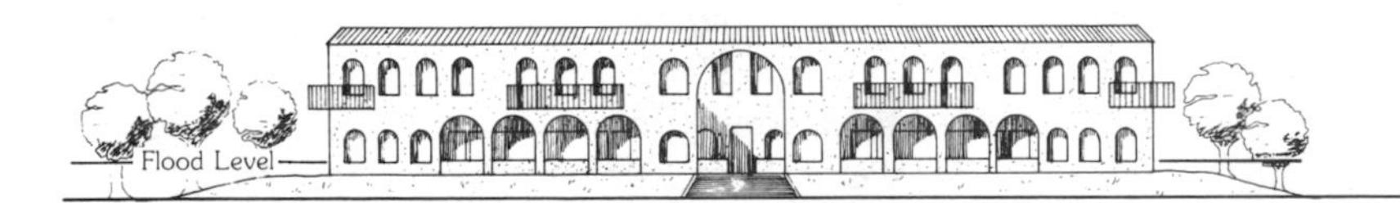

Average Annual Cost/ft²	
Construction	$1.55
Insurance	0.02½
TOTAL	$1.57½

Elevated to +6 ft

Average Annual Cost/ft²	
Construction	$1.55
Insurance	0.01
TOTAL	$1.56

FIGURE 11–32 Influence of federal flood insurance rates on the annual cost of a conceptual 22,500-square-foot project. Notice that as flood-compatible designs are incorporated, the annual insurance cost decreases.

elevating homes to a safe level, adequately flood-proofing commercial buildings, anchoring buildings, and providing sufficient drainage. The insurance provisions of this program are designed to encourage, through economic incentives, proper flood-resistant construction (Figure 11–32). Cities can also discourage urban development in the flood-hazard areas by refusing to extend utility and other public services. An official floodplain map, providing flood-hazard information, should be available to businesses and lending institutions. The erection of flood warning signs in the floodplain area or the prominent posting of previous high-water levels can be used to advise the consumer, but such signs carry no enforcement.

There are many useful guides available to a community planning to implement a floodplain management program. Each case is different because both the river system and socioeconomic system are unique to the community. The general steps to establishment of a flood management plan are as follows. Determine floodplain zones first, based on historical evidence of flooding, including relative magnitude of discharge and elevation reached; computed recurrence interval; engineering study of flood potential; effect of urbanization in the area and upstream; and anticipated growth in the area. Prepare an official floodplain map and distribute it to businesses, insurance and financial institutions, service companies, real estate brokers, and land developers. A master plan is then drawn, showing zones of most frequent flooding and the elevation of the floodwaters. Encroachment lines or zones should be established to prohibit the obstruction of flood flows. The master plan is the legal basis for the establishment of official flood zones, which have been found to be within the legal police power of a city.

Coastal Processes

twelve

The coastal zone is an extension of the geology of the land into the marine environment, where the coast is modified by erosion, transport, and deposition of the geologic materials by wave and wind processes. The geomorphology of the coast is, therefore, a modification of the geomorphology of the land. Coasts can be classified by their geology—hard rock or soft rock (Figure 12–1)—or by the geomorphic processes that formed the landmass—glacial, depositional, tectonic, and carbonate (Figure 12–2). In the first system of classification, the geomorphology of the coast is controlled by the erosional resistance of the rock materials, while in the second, it is attributed to the processes that formed the landmass.

For purposes of coastal engineering geology, classification by the characteristics of the coastal rocks is better than by the geomorphology, because the engineering geologist is concerned with predicting coastal processes. In soft-rock coastal areas the sediments are easily eroded, and the coastal form changes rapidly in the human time span. In contrast, hard-rock coastal areas undergo very slow to imperceptible changes in form during human time. Just as coastal areas can be classified by the rate of change, active coastal processes can be subdivided into those that cause gradual changes and those that cause rapid changes. The active processes that cause gradual change are regular and nearly continuous (erosion, deposition) while those that cause rapid changes are irregular and intermittent events (storms, severe erosion).

WAVES, TIDES, AND WINDS

Three primary mechanisms operate along most coasts and drive the regular, nearly continuous processes. They are **waves, tides,** and **wind.** Of these, waves are the most important along most coasts. Waves, however, are actually the result of the transfer of wind energy to the water surface through the boundary shear along the interface. Wave generation is related to the

FIGURE 12–1 Soft-rock (left) and hard-rock (right) coasts.

direction, duration, and intensity of the wind and the length of the water body across which the wind is blowing (fetch). The direction of the wind controls the direction of wave movement. Wind duration and intensity and fetch length control the height of the waves; an increase in any of these factors will cause a corresponding increase in wave height. As the wave height increases, the period (number of seconds between wave crests) also increases.

Waves are classified by period, as shown in Table 12–1. Capillary through swell waves are wind waves, the result of wind stresses on the water surface. **Surf beat** is a wave train of smaller waves that tend to form groups of higher and lower waves, and is frequently identified by surfers in their saying, "Every seventh wave is the largest" (Figure 12–3). **Tsunami** waves are generated by submarine or nearshore fault movements due to tectonic activity. **Tides** are gravitational waves generated by a sun-moon-earth interaction. Extensive research in the field of physical oceanography has led to the development of numerous charts for predicting wave heights. Charles L. Bretschneider of the University of Hawaii developed a set of deep-water

TABLE 12–1 Classification of waves

Period (seconds)	Waves
Less than 0.1	Capillary
0.1–1.0	Ripple
1–5	Chop
5–11	Sea
9–25	Swell
50–100	Surf beat
800–3000	Tsunami
8000–9000	Tide

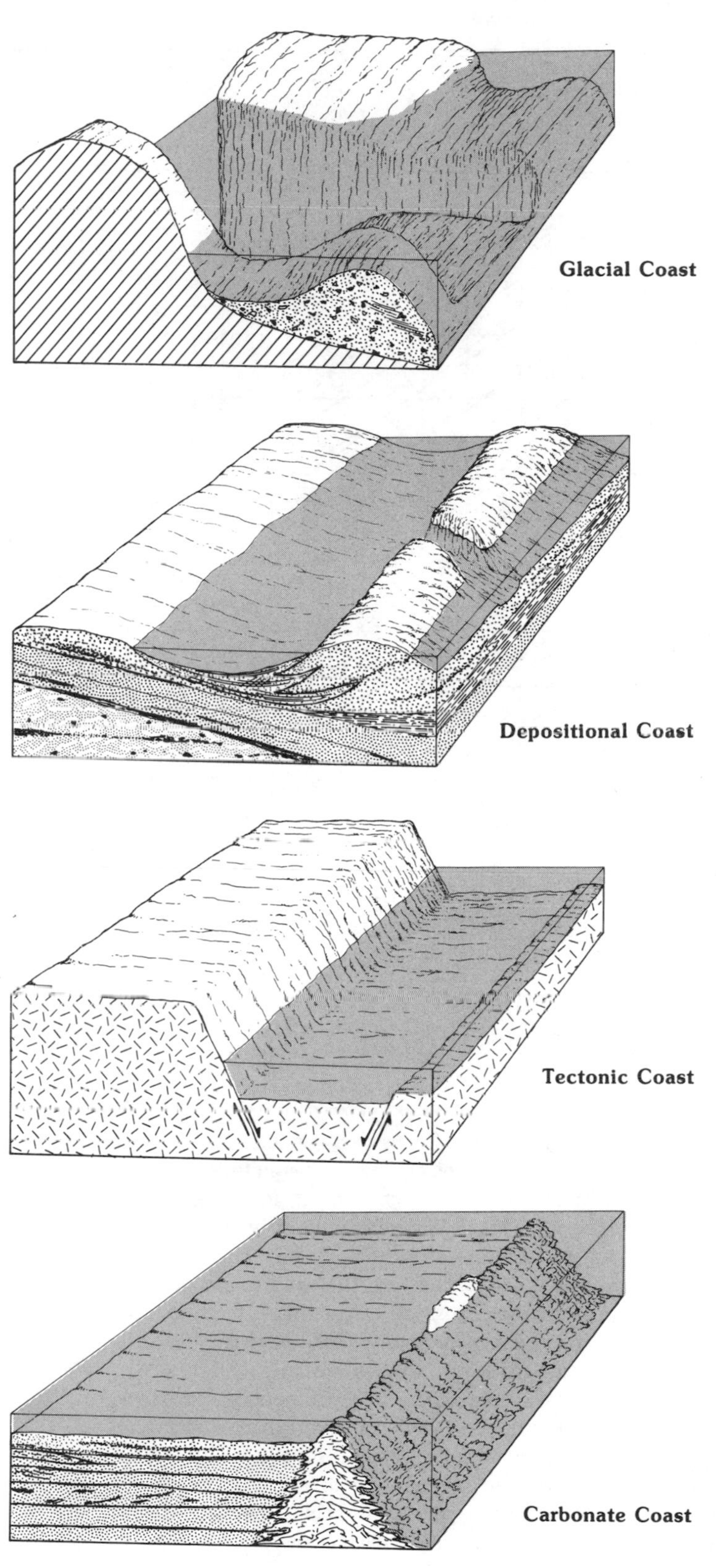

FIGURE 12–2 Block diagrams of coastal geomorphology.

CONCEPT 12–1

Wave Prediction

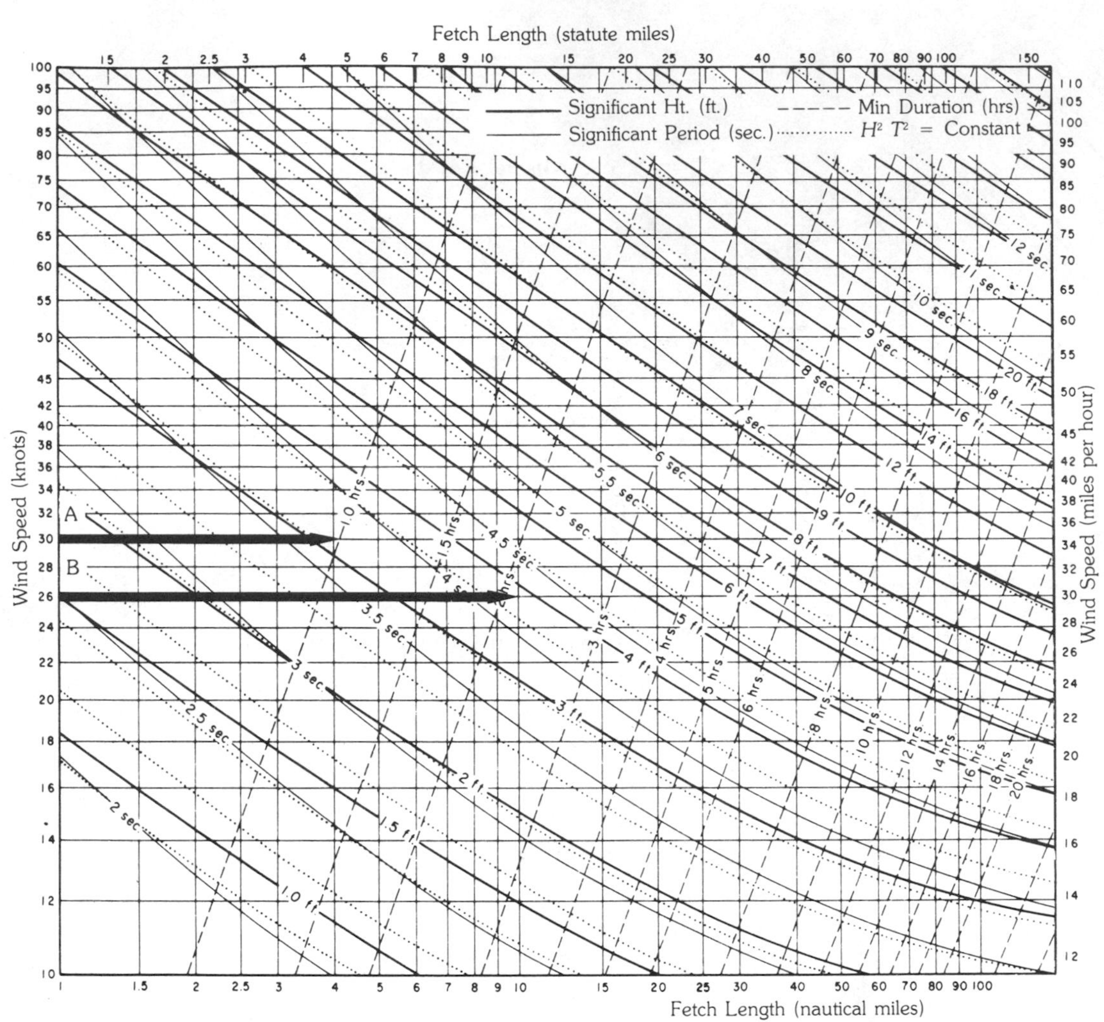

This deep-water wave prediction chart from the U.S. Army Corps of Engineers (1977) relates wind speed, fetch length, and duration to the predicted significant wave height in feet and the significant wave period in seconds. At any given wind speed, the height and period of the generated waves will be controlled by either the fetch length (distance of open water over which the wind is blowing) or the duration (amount of time the wind is blowing).

To find the predicted significant wave height and period, find the wind speed along the vertical axis. Once you locate the wind speed, move horizontally across the diagram until the diagonal duration line (dashed line) or the vertical fetch length line (solid line) is intersected. Significant wave height and significant period are read from the curved lines.

In example A, a 30-knot wind is blowing for 1 hour across a bay 10 nautical miles wide. Approaching the diagram from the left, move horizontally to 1-hour duration, which is intersected before the 10-nautical-mile fetch. You will read a significant height of 3.2 feet and significant period of 3.7 seconds. In this case the waves generated are duration limited. In example B, a 26-knot wind is blowing across the same bay for 5 hours. Again, approaching the diagram from the left (26 knots), move horizontally to the 10-nautical-mile line, which is intersected before the

5-hour duration line. This gives a reading of 3.9 feet for significant height and 4.2 seconds for significant period. The waves in this example are fetch limited.

Adapted from *Shore protection manual*, vol. 1, 1977. Washington D.C.: U.S. Army Corps of Engineers.

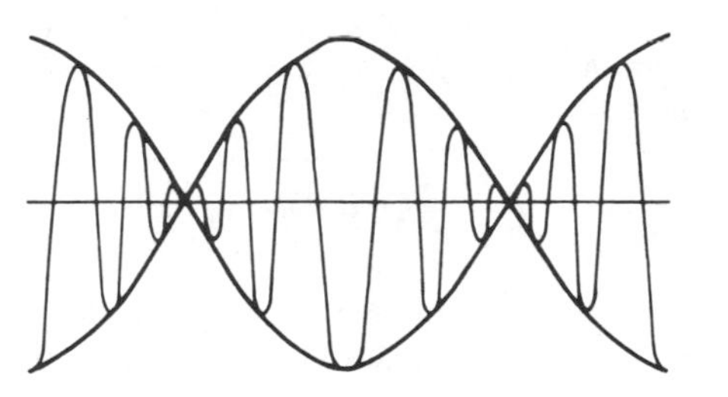

FIGURE 12–3 Surf beat, a long wavelength train composed of numerous smaller waves that add together.

wave forecasting curves for the U.S. Army Coastal Engineering Research Center (Concept 12–1).

WAVE MECHANICS

Waves in an open body of water are the observed result of the summation of numerous waves with various periods. Short-period waves have lower wave heights and a corresponding lower amount of wave energy than long-period waves. Waves reach the shore in a complex form that is difficult to analyze mathematically; however, wave analysis is important to the engineering geologist. The simplest approach is **airy wave** or **linear wave theory** (Figure 12–4), which assumes that the wave being analyzed is in the form of a *sine wave* having a height and period equal to the *significant wave*. Nonlinear **(strokes)** wave theories are more complex but do provide a closer match to true wave form.

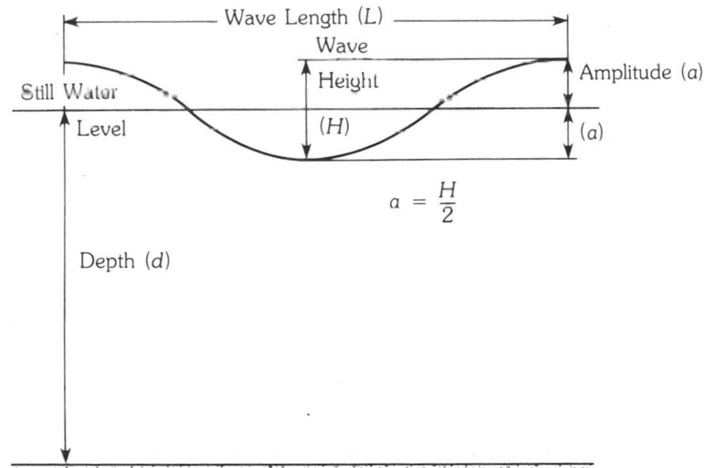

FIGURE 12–4 Characteristics of an elementary wave.

CONCEPT 12–2

Elementary Progressive Wave Equations

Elementary progressive waves are defined as simple sinusoidal waves having a wave length (L), a height (H), and a period (T). The equation that relates wave celerity (C), the velocity of propagation of the wave form, to wave length (L) and water depth (d) is

$$C = \sqrt{\frac{gL}{2\pi} \tan h\left(\frac{2\pi d}{L}\right)}$$

where g = acceleration due to gravity. Celerity is related to wave length by $C = L/T$; thus the relationship between celerity and period (T) is

$$C = \frac{gT}{2\pi} \tan h\left(\frac{2\pi d}{L}\right)$$

These three relationships are the basis for the elementary wave equations. With four unknowns (C, L, T, and d) in the equation and only one relationship, basic wave calculations require field measurements. Selected wave equations for shallow-water, transitional-water, and deep-water waves follow.

Shallow-water wave equations:

$$\frac{d}{L} \leq \frac{1}{25}$$

Wave celerity: $C = \frac{L}{T} = \sqrt{gd}$

Wave length: $L = CT = T\sqrt{gd}$

Transitional-water wave equations:

$$\frac{1}{25} < \frac{d}{L} < \frac{1}{2}$$

Wave celerity: $C = \frac{L}{T} = \frac{gT}{2\pi} \tan h\left(\frac{2\pi d}{L}\right)$

Wave length: $L = \frac{gT^2}{2\pi} \tan h\left(\frac{2\pi d}{L}\right)$

Deep-water wave èquations:

$$\frac{d}{L} \geq \frac{1}{2}$$

Wave celerity: $C = \frac{L}{T} = \frac{gT}{2\pi}$

Wave length: $L = CT = \frac{gT^2}{2\pi}$

Notice that when d/L is very small, $\tan h(2\pi d/L) = (2\pi d/L)$; and when d/L is very large, $\tan h(2\pi d/L) = 1$.

Airy wave parameters can be calculated from the basic equations derived in Concept 12–2. An analysis of the (tan h $2\pi d/L$) function shows that as the depth-to-length (d/L) ratio increases to one half or more, the hypertangent function approaches 1.0; and as the ratio decreases to 1/25 or less, the function approaches ($2\pi d/L$). These two limits, $d/L \geq ½$ and $d/L \leq$ 1/25, define **deep-water waves** and **shallow-water waves** and simplify the airy wave equation. When the d/L ratio lies between 1/25 and 1/2, the waves are defined as **transitional,** and their properties must be defined by the complete airy wave equation.

Wave energy is an important parameter in coastal process analysis. Wave energy includes the kinetic energy (E_k) used to elevate the water surface and potential energy (E_p), which develops when the water surface is elevated. Since kinetic energy is converted into potential energy as the water is elevated and is converted back to kinetic energy as the water is lowered, kinetic energy must be equal to potential energy ($E_k = E_p$) and **total wave energy** (E_t) is equal to the sum of the two ($E_t = E_k + E_p$) in airy wave theory. Total wave energy in a single airy wave per unit of width is:

$$E_t = \left(\frac{\rho g H^2 L}{8}\right)\left(\frac{1 - MH^2}{L^2}\right)$$

where

ρ = density of the fluid

g = acceleration due to gravity

M = wave energy coefficient ($\pi^2/[2 \tan h^2(2\pi d/L)]$

H = wave height

This equation simplifies to $E_t = 8LH^2$ for a deep-water oceanic wave. As a wave train moves into calm water, some of the wave energy is used to disturb the water surface. Thus the total amount of energy moving forward in a wave train (E_f) is less than the energy in a single wave by:

$$E_f = nE_t$$

where

$$n = \frac{1}{2}\left[\frac{1 + (4\pi d/L)}{\sin h(4\pi d/L)}\right]$$

In deep water, n approaches ½, and total wave energy moving forward in a wave train becomes $E_f = ½E_t$. As a wave train moves toward the shore, the wave form changes with the d/L ratio. The n-factor then comes into play, and wave energy is dissipated as the wave experiences bottom friction. The total wave energy in a wave having a period (T) reaching the shore (E_s) in any given time period (t) for any unit width of wave is:

$$E_s = E_f(t/T)$$

As waves approach the shore, in transition from deep water to shallow water, the properties of the wave change. The wave velocity (celerity, C) and length (L) both decrease, and wave height (H) increases. Three

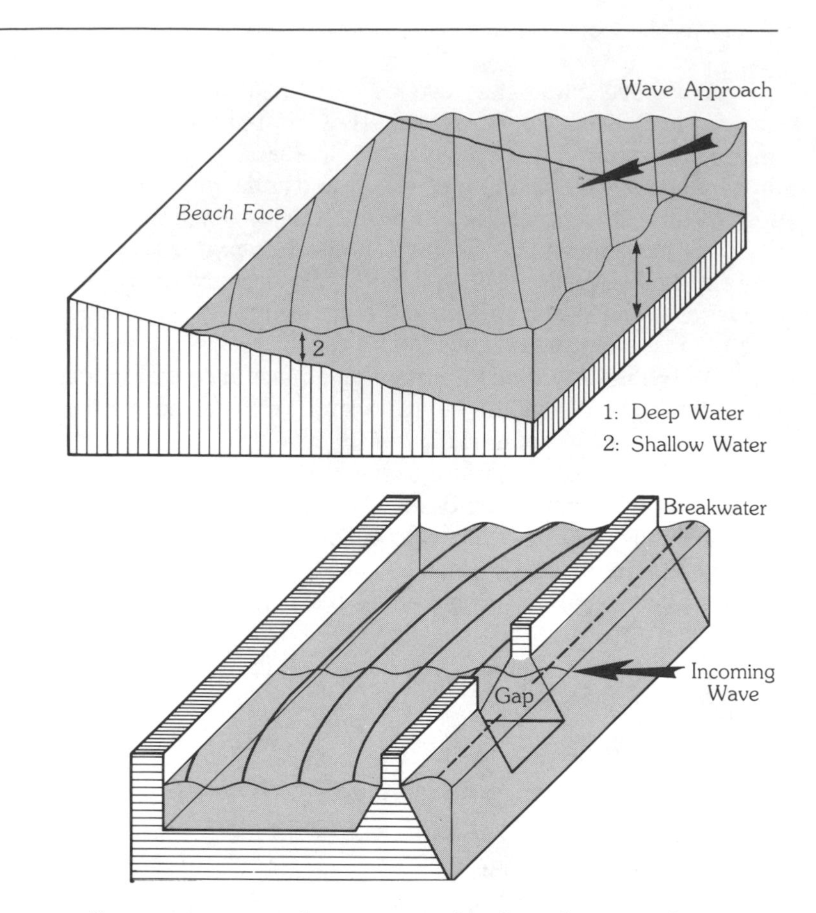

FIGURE 12–5 Wave refraction.

FIGURE 12–6 Wave diffraction.

processes affect waves as they approach the shore: **refraction, diffraction,** and **reflection.** Wave refraction is the process that causes the wave to change direction as it enters shallow water (Figure 12–5). Wave diffraction occurs as a wave enters a protected body of water through an inlet or gap (Figure 12–6). Waves are reflected by vertical to near-vertical walls, much as light is reflected from a smooth surface (Figure 12–7).

Of these three processes, wave refraction is the most common one to affect coastal engineering projects, but wave diffraction and reflection are

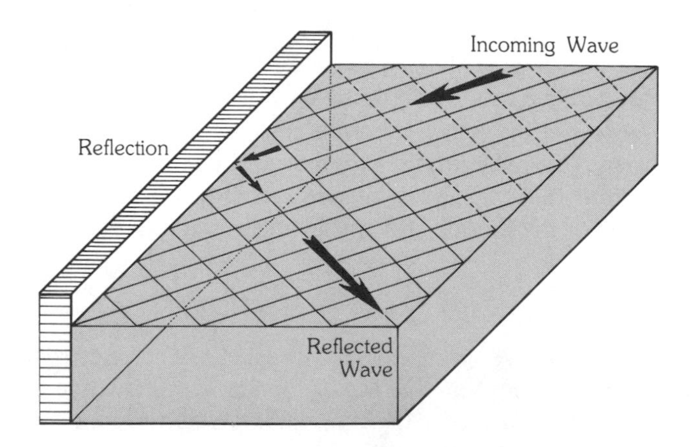

FIGURE 12–7 Wave reflection.

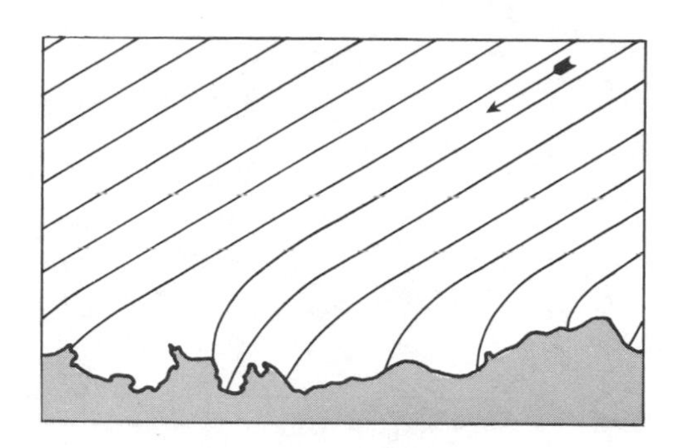

FIGURE 12–8 Wave ray diagram for part of the north coast of Molokai Island, Hawaii. Waves arrive from 060° and have a period of 20 seconds.

usually encountered in harbor design projects. The transfer of wave energy from the deep-water generation site to the shoreline is frequently analyzed using **wave diagrams** (Figure 12–8). A wave diagram starts in deep water where the **wave crest** and **wave rays** form an orthogonal network. The wave rays are then refracted into the area under analysis. In any wave diagram, the wave crest and wave ray lines must *always* intersect at 90°.

Since wave energy is conserved as the waves approach the shore, the amount of wave energy moving shoreward in one deep-water wave crest between any two wave rays that are (b_0) units apart must be:

$$E_f = (\tfrac{1}{2}E_t)b_0 = 4LH^2b_0$$

As the waves move into shallow water, the *n*-factor reenters the equation; however, in many cases, energy losses due to bottom friction can be assumed to be negligible. In this case, the amount of wave energy reaching the shoreline between the same two wave rays that are refracted to a spacing of b units wide is:

$$E_{f_{shore}} = \frac{b_0}{b}(\tfrac{1}{2}E_t) = \frac{b_0}{b}(4LH^2)$$

If the spacing between the wave rays (b) increases, the wave energy per unit of shoreline decreases, and if spacing decreases, the energy reaching a unit length of shoreline must increase (Figure 12–9).

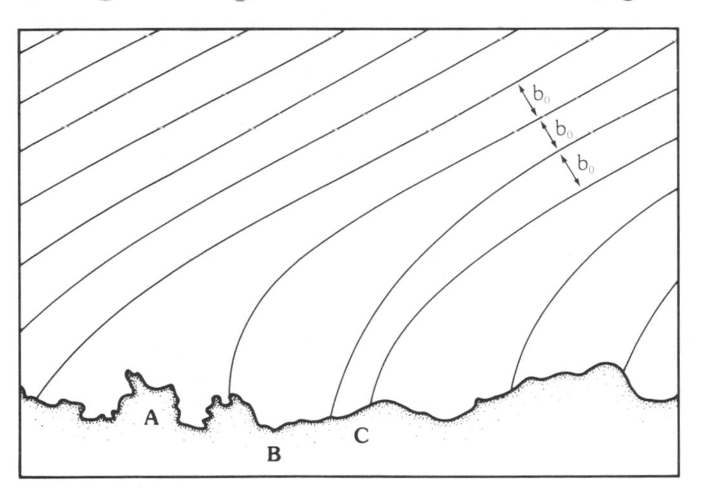

FIGURE 12–9 Wave diagram for part of the north shore of Molokai Island, Hawaii. Notice how the wave rays diverge from their spacing (b_0) in deep water. Wave energy arriving at the shore is related to the spacing between wave rays at the shore and in deep water. Area A has the lowest amount of incoming wave energy while area C has the highest amount.

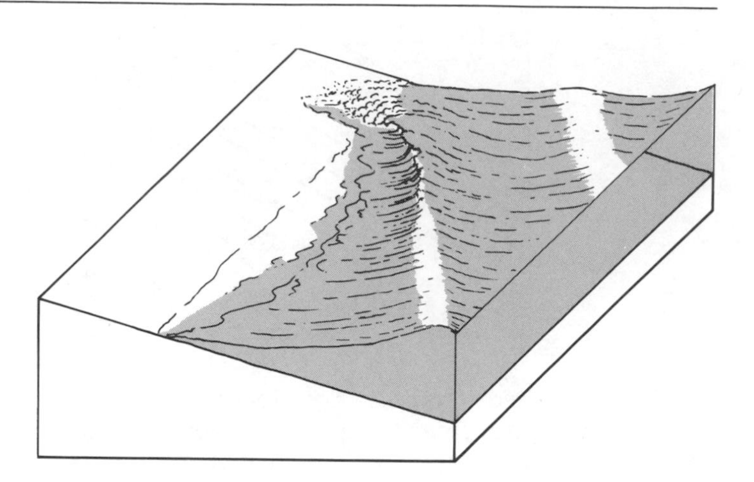

FIGURE 12–10 Plunging wave. Notice the curl of the crest.

The wave finally reaches a point where its height becomes unstable and it breaks. Breaking has been empirically related to depth by the following rules: The wave begins to peak when $d = 2H$, and the wave breaks when $d = 1.3H$.

Waves break in two basic patterns. The **plunging wave** is formed on steep, smooth bottoms where the crest is hurled ahead of the wave (Figure 12–10), and the **spilling wave** is formed on a smooth, flat bottom where the crest spills down the front of the wave (Figure 12–11). Most waves fall between these two extremes. Once a wave breaks, wave energy is dissipated as a **translationary wave.** It runs ashore and pushes water up the beach slope as **wave runup.**

The wave energy system begins with the transfer of wind energy through boundary shear to wave energy, which is eventually dissipated on shore by wave runup. Once this energy has been delivered to the shoreline, it is available to move sediments and alter the shoreline. Along coasts with sandy beaches, waves constantly erode, transport, and deposit sand particles. The net direction of sediment transport is the primary concern in

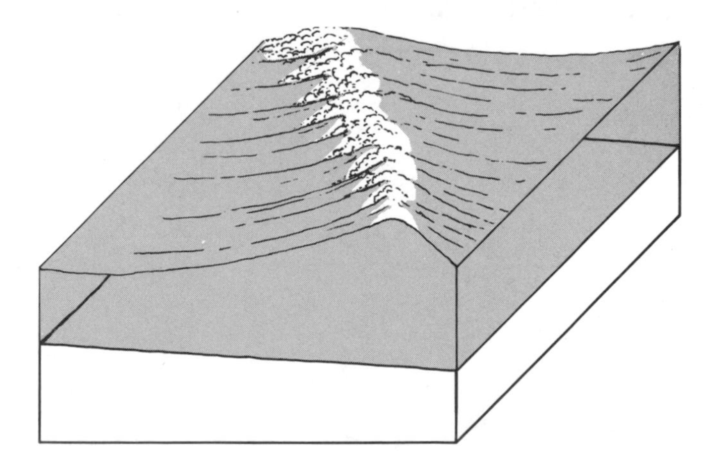

FIGURE 12–11 Spilling wave. Notice that the wave breaks by spilling down its face.

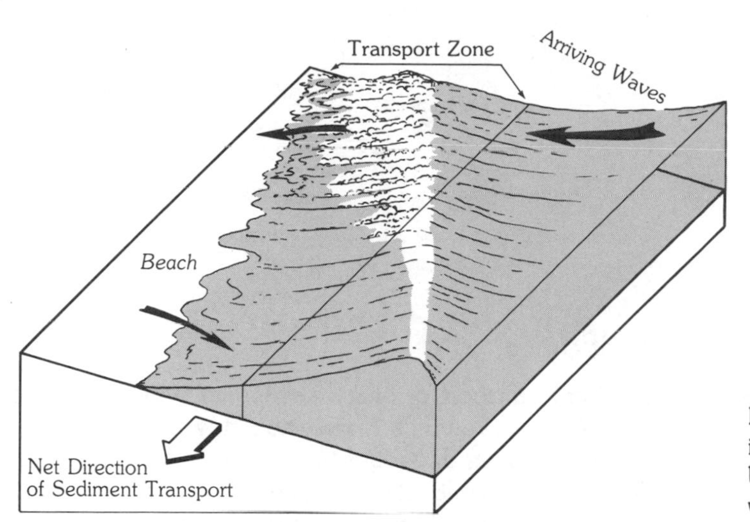

FIGURE 12–12 Block diagram showing the zone of littoral transport, bounded by the beach and the line of breaking waves.

these areas. Most sediment transport occurs in a narrow lane along the shore, bounded on one side by the line of breakers and on the other by the beach itself (Figure 12–12). Sediment in transport is called **littoral drift,** and the movement of the sediment is **littoral transport.**

The direction of littoral transport is related to the direction of the arrival of the translational wave. If the wave strikes the coast at an angle other than straight-on (90°), there is a **downcoast** component of wave energy (Figure 12–13). In addition to the direction of the approach of the waves, wave energy itself also influences shoreline erosion and deposition. High-energy waves, having a large **wave steepness** (*H*/*L*) will cause beach erosion, while waves having a small wave steepness and low energy will result in beach construction, or **aggradation.** Changes in wave energy from seasonal storms have resulted in the terms **winter beach,** where the beach is made narrow and steep by high-energy waves, and **summer beach,** where the beach is constructed wide and shallow by low-energy waves (Figure 12–14). These conditions do not occur along all beaches.

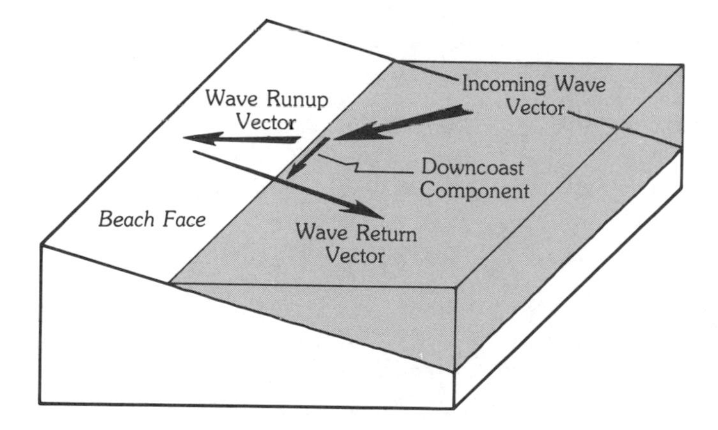

FIGURE 12–13 Vector diagram showing the development of a downcoast vector.

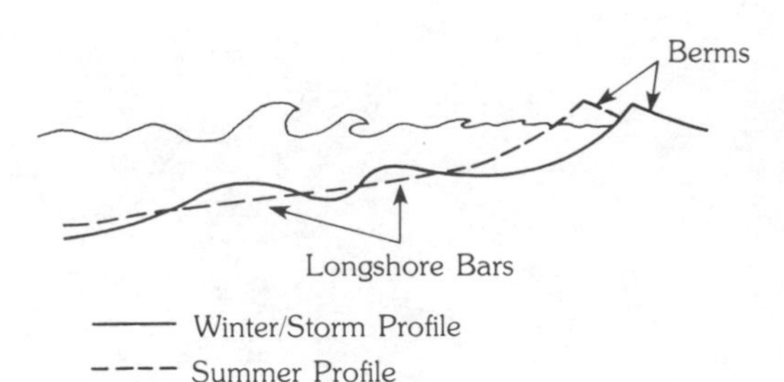

FIGURE 12–14 Cross section of a winter beach and a summer beach. The winter beach forms during periods of high wave energy, while the summer beach is a low-wave-energy profile.

WIND PROCESSES

Along some coasts, wind transport of beach sands can be a significant process. Many beaches are backed by a **foredune ridge,** or **dune line,** where windblown sand has been deposited (Figure 12–15). Dunes have received quite a bit of attention in recent years as coastal features that must be protected. The mechanics of dune formation and migration are complex and not fully understood, because wind erosion is a complex problem.

The wind erosion equation, however, can be used in the evaluation of dune processes. As the equation suggests, wind erosion and transport should be at a maximum in areas characterized by well-sorted, fine sand beaches in semiarid to arid climatic regions. During dry periods or after coastal grass fires, when the vegetative cover is under stress, wind processes may drastically alter a coastal area.

TIDES

Tides are the daily variation in the elevation of the still-water level in response to gravitational changes from the position of the sun-moon-earth system. Tides have a predictable periodicity that results from the rotation of

FIGURE 12–15 Foredune ridge along a sedimentary coastline.

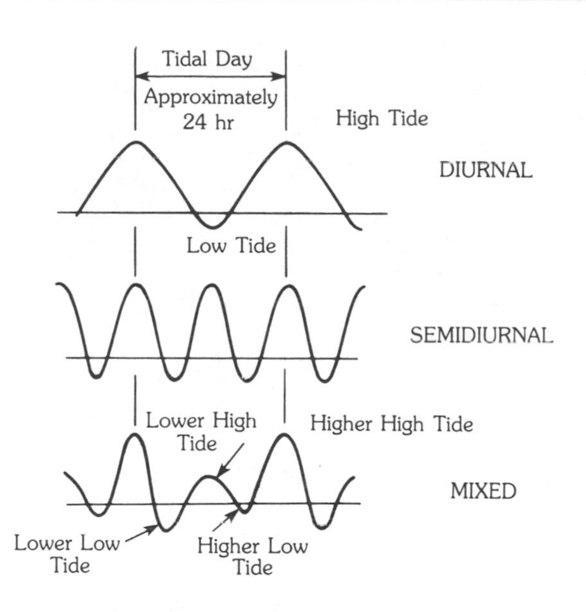

FIGURE 12–16 Representative tide records.

the earth combined with the shape of the basin, producing tides with unique wave forms (Figure 12–16). As the tide changes, the still-water depth changes, and there is a corresponding change in wave energy. During periods of high tide the water is deeper (*d* increases), and the nearshore wave processes increase.

Tidal currents develop in restricted channels such as inlets or narrow bays when the tides cause a difference in water level between the inland bay and the offshore (Figure 12–17). As the oceanic tide rises, water floods low-lying areas in what is termed **flood tide**. As the oceanic tide falls, water exposes the low-lying area in **ebb tide**. Tidal currents and stages are significant factors in sailing schedules and other harbor activities in certain areas of the world.

STORMS, HURRICANES, AND TYPHOONS

Storms are irregular, intermittent events that bring very high wave energy as the result of high wind activity of a low pressure cell. Storms involve four related processes: storm surge, high wave energy, extreme rainfall, and high winds. **Storm surge**, frequently called storm tide, is the elevation of the still-water level caused by the low pressure center of the storm and the physical transport of water by the very high wind shear along the water surface. Storm surge frequently affects the coastline long before the weather becomes noticeably stormy (Figure 12–18). As the storm approaches the shore, wind velocity and wave heights increase (Figure 12–19). When the storm moves inland, the already flooded coastal areas suffer additional flooding by storm rainfall. The beach and foredunes can be destroyed by erosion during a storm (Figure 12–20). Notice in Figure 12–20 that the storm waves erode the dunes rather than flood the land behind them. The dunes act as a first line of coastal defense during storms. In some storms the waves breach the foredune ridge, releasing the pressure that has been acting

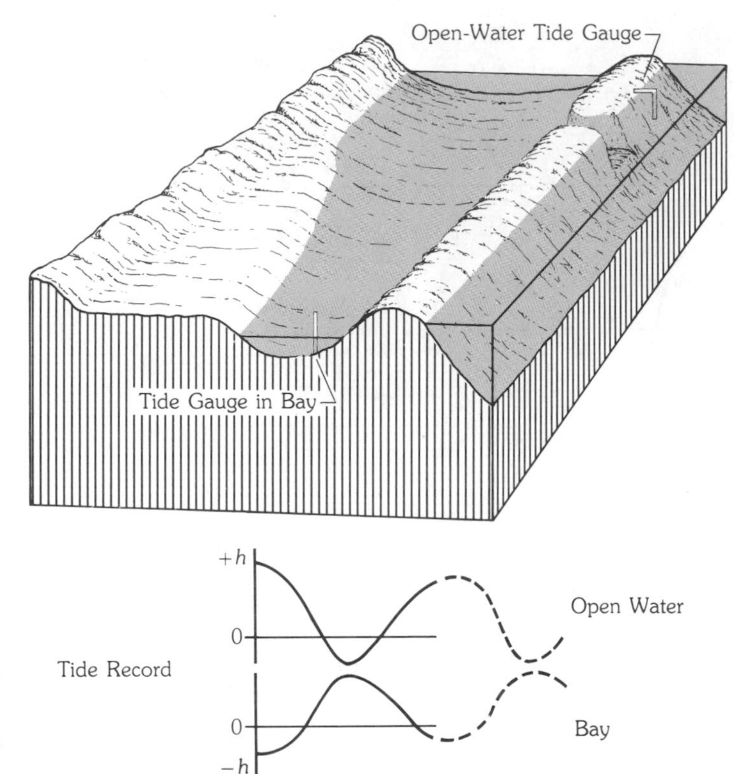

FIGURE 12–17 Mechanism for tidal currents. Notice that the tide record in the bay is out of phase with the open-water tide record. This difference produces the hydraulic gradient needed to cause tidal currents.

FIGURE 12–18 Reduction in beach width by storm surge. Notice that the beachfront cabins are surrounded by water.

FIGURE 12–19 Attack of storm waves on the beach and shorefront property during a hurricane. (Photo courtesy of U.S. Army Corps of Engineers.)

against the dunes, and rapidly flood the inland or back-island areas (Figure 12-21). Sediments deposited by this process form a **washover fan**.

COASTAL ENGINEERING GEOLOGY

The coastal zone, a narrow band of land and water, is a significant residential, recreational, commercial, and industrial area. It is the interface between the industrial and agricultural goods and commodities produced within a nation and the international marketplace. Marine shipping and fishing are primary coastal-zone industries. Limited space, high land values, and storm risks all make the role of engineering geology in the coastal zone important.

SHORELINE PROTECTION

Shoreline protection is the modification of the coast or coastal process in some manner to reduce erosion, wave energy, or the risk of loss. Two basic

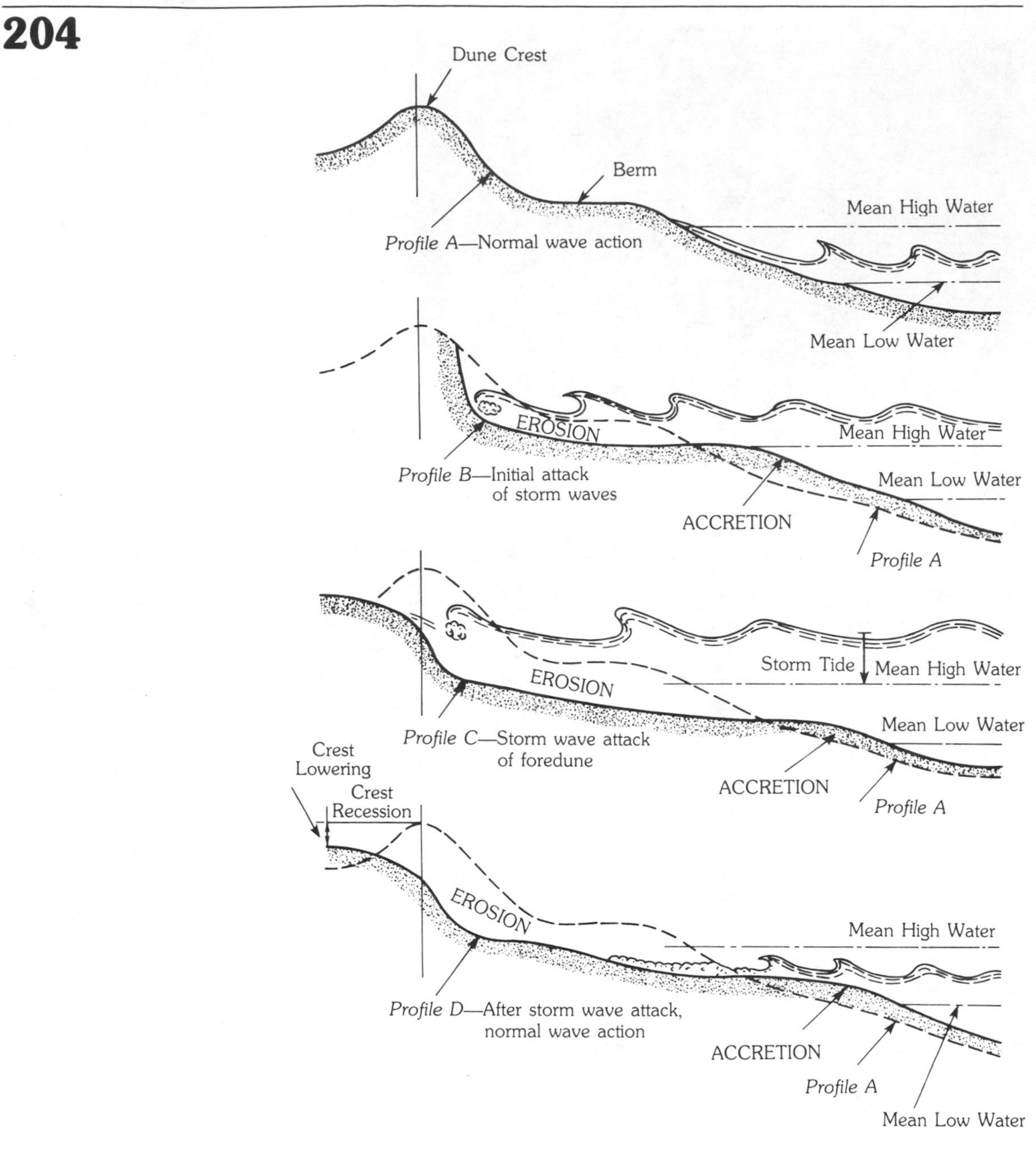

FIGURE 12–20 Diagram of storm wave attack of the beach and foredune of a depositional coast. (After U.S. Army Corps of Engineers.)

types of shoreline protection are used: **structural**, where some engineering structure is built, or **nonstructural**, where the shoreline processes are modified without building structures. The selection of the best type depends upon the land use to be protected.

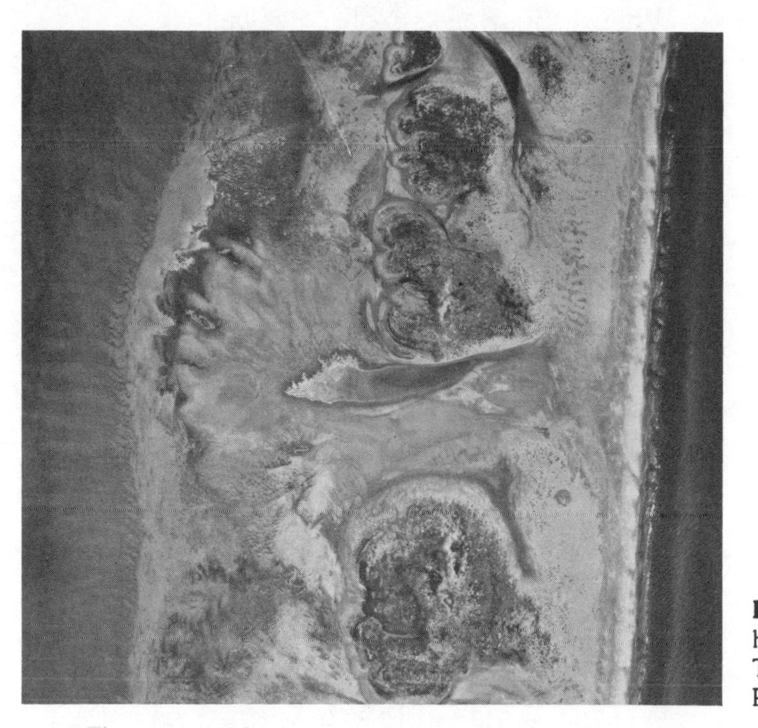

FIGURE 12–21 Aerial photograph of a hurricane washover. (Photo courtesy of Texas State Department of Highways and Public Transportation.)

The primary objective of structural methods of shoreline protection is to protect the erodible material by a structure that will dissipate the incoming wave energy. Structures can be placed in the **parallel** or **angular** position relative to the shoreline. Some structures, such as **breakwaters** and **jetties,** are designed to intercept incoming wave energy and provide an area of reduced energy (Figure 12–22). The selection of the type, size, and shape of any structural device must take into consideration the active geologic processes and wave energy that are expected at the site.

Nonstructural methods of shoreline protection are designed to modify the active geologic processes, using natural beach and dune materials. These methods have a lower initial cost but a significantly higher operating cost. **Beach nourishment** is the process of adding beach material at the

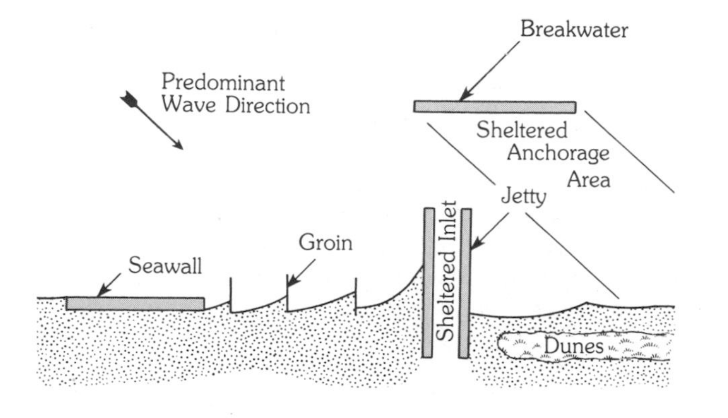

FIGURE 12–22 Common types of shoreline protection.

CASE IN POINT

When a Seawall Is Not a Seawall

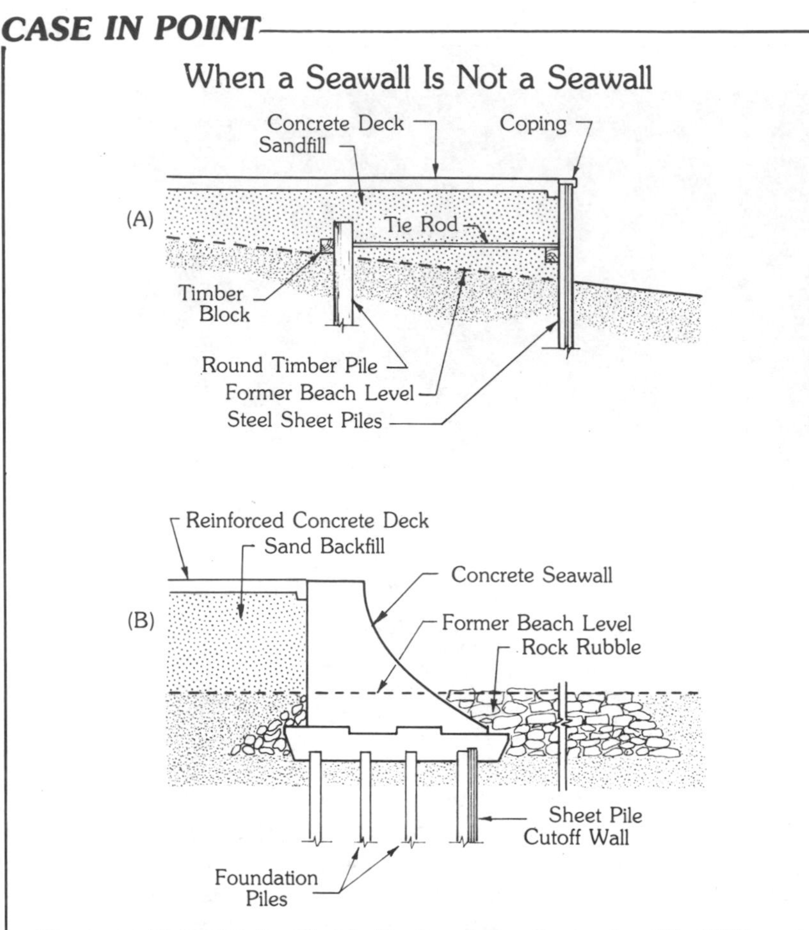

Hurricane Eloise hit the Florida Panhandle on September 23, 1975, and undermined and destroyed many "seawalls" that were intended to protect the high-value beachfront property. The loss of the "seawall" resulted in direct wave scour below foundations and corresponding failures. Why did these seawalls fail? They were not seawalls but retaining walls. Failure occurred because the structures were not designed for scour at their base or for wave-impact forces. Cross section A shows a typical coastal retaining wall that should be used along dredged channels and other environments of low wave energy but was used for beachfront protection. Notice that the only compressional member is the concrete deck. With flooding of the sand fill by storm surge and vibrationary loading of the wall face, the pore water pressure increases sharply with each wave impact. The combination of liquefaction of the confined saturated sand and horizontal forces from wave impact caused the only compressional member, the deck, to fail. Failure of the deck allowed excessive flexing of the face of the wall, which ultimately failed.

Compare the design of the retaining wall with a recurved seawall design (B) from the U.S. Army Corp of Engineers. The seawall is designed to redirect the impact of incoming waves and to absorb the wave forces. The rock rubble and sheet pile cutoff wall protect the foundation piles from scour. A seawall is designed to (1) absorb wave energy, and (2) resist foundation scour in order to protect the land behind it.

upcoast end of a littoral transport system and recovering it at the downcoast end. **Dune reconstruction** is the process of storing sand in the dunes for times of storm wave erosion and then rebuilding the dune ridge after the storm passes. **Building regulations** are another nonstructural method of reducing coastal losses.

INLETS

The natural formation of inlets (Figure 12–23) is the result of a variety of processes: rivers, storms, longshore drift, and other geologic phenomena. Any river entering the ocean forms a natural inlet. Although the flow regime of this type of inlet differs from that of a tidal inlet, basic characteristics may be the same.

Perhaps the most prominent agent of inlet formation along barrier island coasts is storm activity. Pierce (1970) noted that such inlets are most likely to occur across narrow islands where the adjacent lagoons are relatively deep. Inlets formed as the result of storms are usually ephemeral in nature and will normally close relatively rapidly, especially if the breach of the island occurs at a point where it is relatively wide.

Lateral deposition and growth of a spit or bar across the mouth of a bay can also form an inlet. The formation of a laterally growing bar is the result of transport of sedimentary material by longshore currents until it reaches the deeper, quieter water of the bay where it is deposited. This process of lateral deposition and growth across the bay continues until the inlet reaches a state

FIGURE 12–23 Tidal inlet.

of dynamic equilibrium with the physical and hydrologic parameters acting upon it.

Erosion by glacial activity during the Pleistocene formed extensive valleys. With the Holocene sea-level rise, many natural inlets were formed as the lower portion of the valleys were inundated by rising water. Inlets of this type are the **fjords** of Scandinavia and Alaska. Because fjords typically have rocky gorges, Bruum and Gerristsen (1960) note that normal relationships for inlets in alluvial material do not apply.

Figure 12–24 shows the general physiographic features of a typical alluvial inlet. The **offshore bar**, which is a feature along most sandy coasts, is bowed outward at the seaward end of the inlet as a result of the jetting action of the ebb tide. The **gorge** occurs at the point within the inlet where current velocity is at a maximum; thus it is the deepest portion. The **tidal delta**, or **bay shoal area**, is the result of deposition of material supplied to the bay by the flood tide. In areas where the ebb tide is the dominant tidal force, such as along the Pacific Coast of the United States, extensive offshore shoals may form at the expense of bay shoals.

Material for the shoal area in most cases is predominantly sediment supplied by littoral transport. During flood tide, some of the material supplied to the mouth of the inlet is drawn in and discharged into the bay. For several reasons the amount of material supplied to the bay does not equal that carried back out during the ebb tide, even if the two tidal currents are of the same magnitude. As material is jetted into the relatively quiet waters of the bay, deposition occurs, and the average grain size decreases with the decreasing energy regime of the flood current away from the inlet channel. During ebb tide, the distribution of the current energy regime is generally the same as that of the flood tide, with flow velocities increasing toward the inlet channel. Since the velocity required to erode the finer material is higher than that required to transport it, some of the material will be left behind. Bay shoals tend to be extensive in areas of low tidal current energy and fine-grained sediments. The other major contribution to this process is wave energy, where the energy over the bay shoals is less than the energy acting along the open beach. Thus, more energy is available to erode beachfront sediments than is available to erode the shoals.

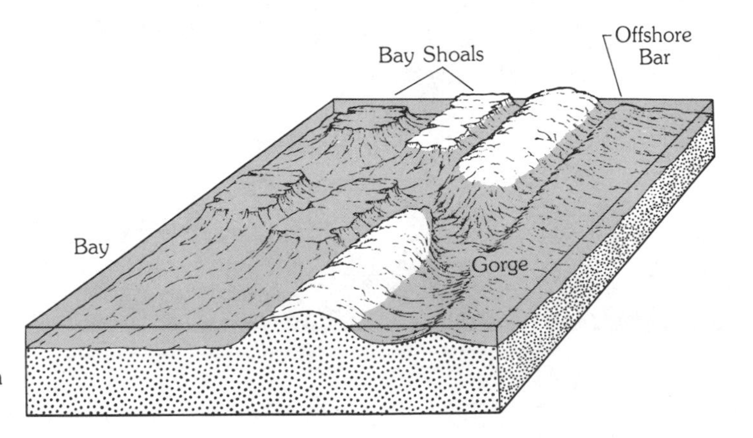

FIGURE 12–24 Block diagram of a tidal inlet.

Coastal inlets are a natural trap for littoral drift. Although much of the material is captured by the inlet and stored in the shoal areas, some of it bypasses the inlet by way of the offshore bar, particularly if the bar is in relatively shallow water. Littoral drift not only supplies material for shoaling, but it is also responsible for inlet migration. In areas with high net littoral drift rates, lateral deposition along the updrift side of the inlet mouth causes the maximum velocity streamline to be forced against the opposite side, causing erosion and migration of the inlet. In turn the channel is lengthened overall. As the channel becomes longer, increased frictional resistance causes a loss in the ability of the tidal currents to erode and transport material, resulting in shoaling and eventual closure of the inlet.

Inlets are often stabilized by jetties to provide for safe, navigable entrances to the sea. These structures serve four basic purposes: halting migration, restricting the movement of littoral material into the inlet, maintaining a safe channel depth as free from shoaling as possible, and providing protection from storms. The most important factor to consider in

CASE IN POINT

An Inlet That's Not?

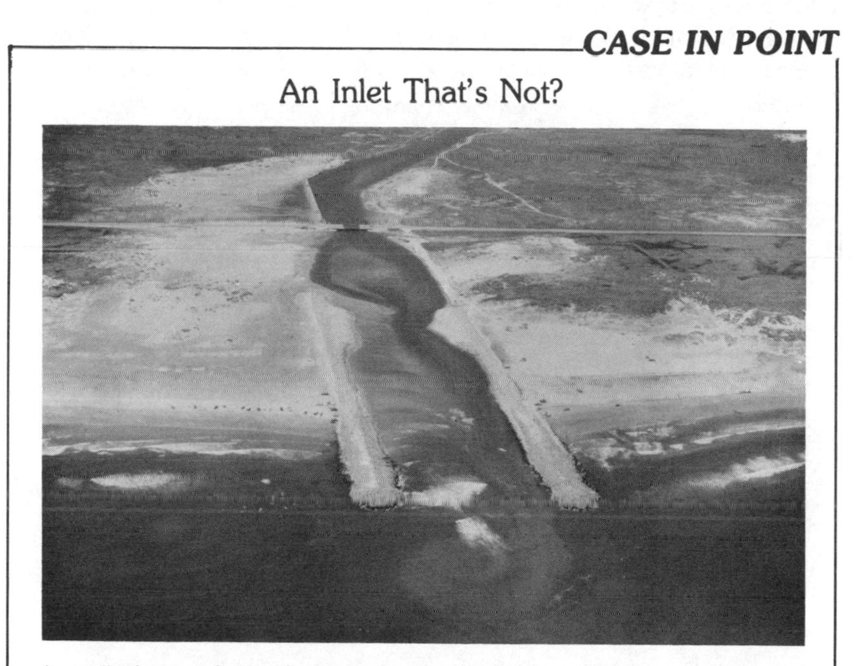

A small inlet was dredged across a barrier island to provide for an exchange of water between the open ocean and the lagoon in order to improve the fishing in the area. Because of land problems, the inlet had to make a turn of about 40° one third of the way across the island. In addition, a narrow highway bridge was constructed just inland from the jetties. Problems started soon after the inlet was completed. The outside of the turn eroded, requiring erosion control measures, and the jettied portion of the inlet began to silt up. If this inlet is to remain an open pass, maintenance dredging will now be required—at a cost that was not considered when the inlet was originally designed.

designing a jetty system is the channel configuration that the structures create. It must be at an optimum so that confined flow through the channel will maintain the desired depth. If it is too wide, the resulting velocity will be too low, thus causing shoaling. At the opposite extreme, if it is too narrow, erosive velocities, which could undermine jetty foundations, result.

The need to create access to the sea and/or to improve bay or lagoonal circulation in some areas has led to the creation of artificial inlets. As with jettied natural inlets, the problem of the proper channel configuration is important. Concept 12–3 shows the relation of flow velocity to inlet stability. Failure to account for this relationship has led to expensive coastal property damage from excessive inlet erosion or to inlet failure from inadequate hydraulic flushing of the sediments.

SEDIMENT BYPASSING

Jetties, which serve to stabilize an inlet, or breakwaters, which protect an anchorage, often create problems, especially in areas of high littoral drift rates. Material that once was carried downshore by the dominant longshore current becomes impounded by the updrift jetty or behind the breakwater. Since the material can no longer bypass the area, undernourishment of the

CONCEPT 12–3

Hydrodynamics of Alluvial Inlets

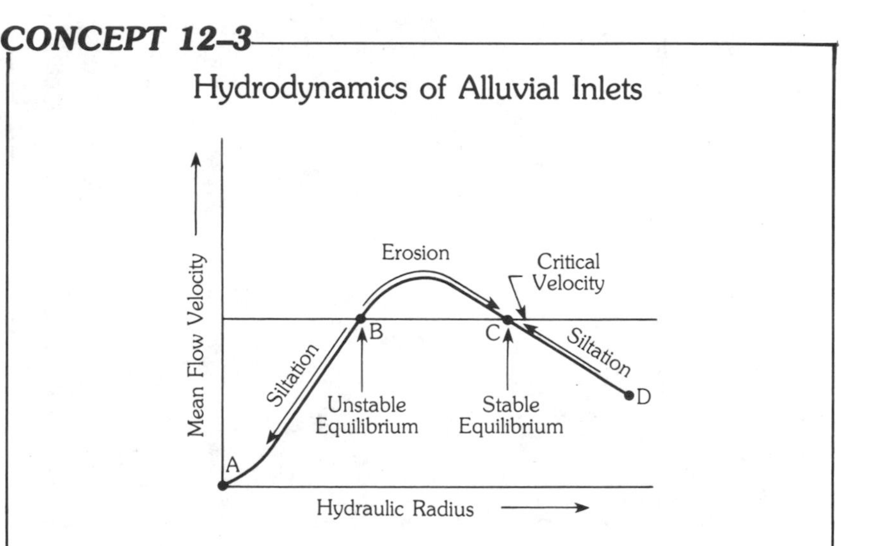

Sediment transport and erosion of an alluvial inlet are related to the mean flow velocity and inlet shape, or hydraulic radius. In small inlets, high boundary shear reduces the mean flow velocity and leads to sediment deposition and inlet siltation (A–B). In very large inlets, the flow velocity decreases because the increased cross-sectional area also leads to siltation (D–C). A critical velocity, related to the sediment grain size, is required to keep sediments moving through the inlet. Any flow velocity above this critical level leads to inlet erosion. An inlet will remain open if the hydraulic radius allows a flow velocity above the critical velocity (B–C). An unstabilized inlet will erode or silt up to maintain a hydraulic radius at C. If a larger or smaller inlet is desired, artificial inlet control measures will be required.

Sediment Bypassing Channel Islands Harbor, California

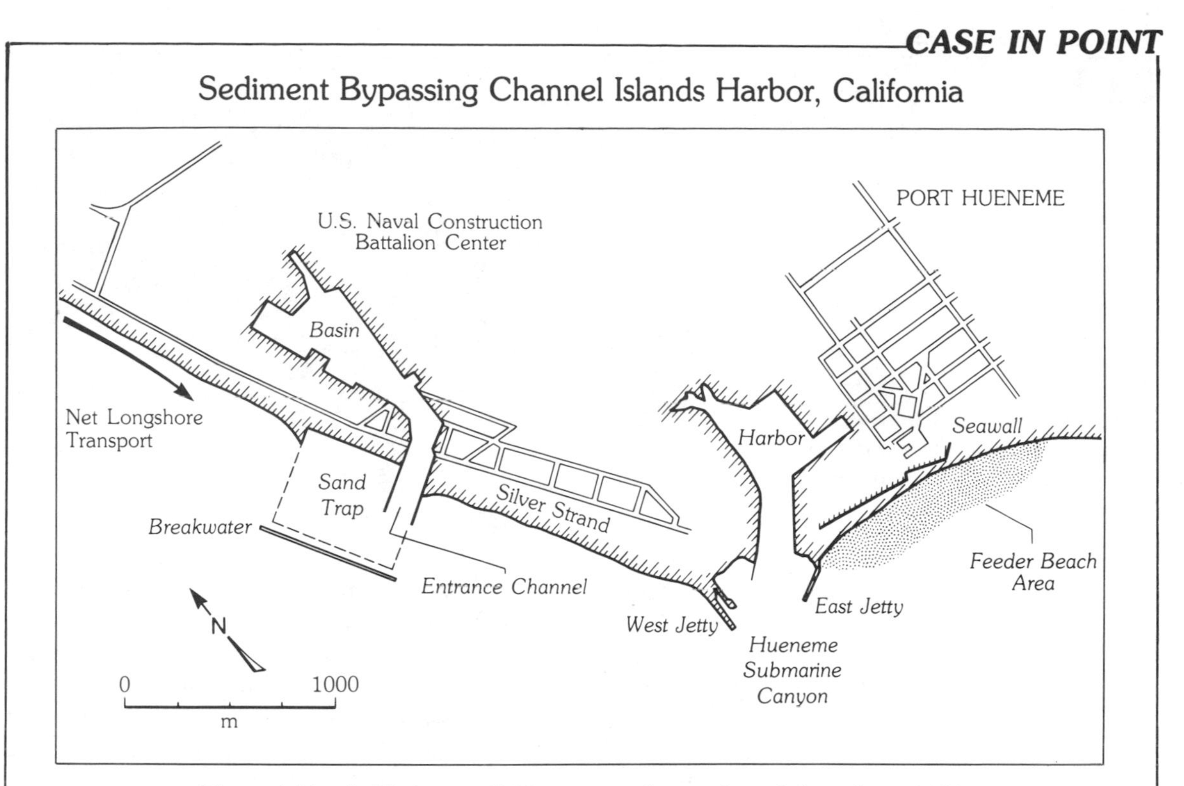

Channel Islands Harbor in California was designed to shelter about 1100 private small craft, and was constructed about a mile northwest of the entrance channel to Port Hueneme. The design objectives of the littoral barrier were to trap nearly all of the southward-moving littoral drift, to prevent losses of drift into the Hueneme Canyon, to prevent shoaling of the harbor entrance, and to protect a floating dredge from waves. The sand-bypassing operations transferred dredged sand across both the Channel Islands Harbor entrance and the Port Hueneme entrance to the eroded shore downdrift (southeast) of Port Hueneme.

The project consisted of an offshore breakwater and two entrance jetties. The breakwater, 2300 feet long and located at the 30-foot contour, is a rubble-mound structure with a crest elevation 14 feet above mean lower low water. Its location and orientation enable it to trap almost all of the downcoast littoral drift. The breakwater provides protection from waves for the dredge and for the small craft entering the harbor. The rubble-mound entrance jetties have a crest elevation of 14 feet above mean lower low water, and extend to about the 14-foot isobath. They prevent shoaling of the entrance channel, which has a project depth of 20 feet.

A floating dredge has cleaned the trap periodically since 1960. In 1960–61, dredging of the sand trap, the entrance channel, and the first phase of harbor development provided about 6 million cubic yards of sand. In 1963, 2 million cubic yards were dredged; in 1965, 3 million cubic yards were transferred. The Port Hueneme operation transferred 2 million cubic yards in 1953. This total of 13 million cubic yards had stabilized the eroded downdrift shores by 1965. Since 1965, bypassing has continued at intervals of about 2 years.

Reprinted from *Shore protection manual,* vol. 1, 1977. Washington, D.C.: U.S. Army Corps of Engineers.

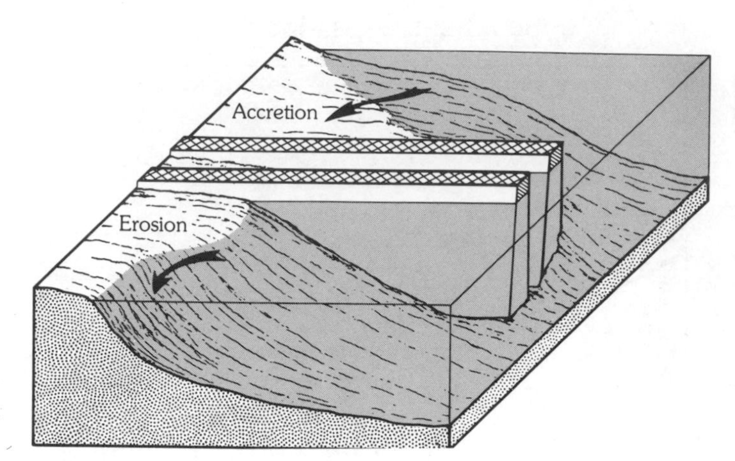

FIGURE 12–25 Block diagram of the influence of jetties on shoreline processes. Notice the accretion of sediment on the updrift side of the jetties and erosion on the downdrift shore.

downdrift shoreline occurs, and beach erosion ensues (Figure 12–25). The only apparent solution to this problem is a method of artificially bypassing the littoral drift to the downdrift side of the site.

Because physical conditions along a coastal zone can vary greatly, the artificial bypass system to be employed must be individually tailored to the particular inlet involved. Table 12–2 gives some of the more important factors to be considered in designing a bypass system. This list of factors emphasizes the complex nature of coastal sedimentary processes. The engineering geologist and the design engineer must construct an artificial system that is matched to the variable conditions at each particular site.

DREDGING

With the continuing development of coastal zones for industrial, commercial, and residential use, these areas have had to be modified to supply enough space for development and safe passages for ships. A great deal of

TABLE 12–2 Factors affecting sediment bypass systems

Coastal geomorphology
Subsurface conditions
Water-level variations
Winds and waves
Allowable wave runup and overtopping
Shoreline and offshore depth changes
Direction, rate, character, and quantity of littoral drift to be bypassed
Effects of nearby inlets or other structures
Effects of prior constructive works
Zone of collection and placement of littoral drift
Availability of materials for construction and maintenance
Effective time of operation of the bypass system
Comparative costs

Source: Watts, 1962.

modification has occurred in coastal estuaries to provide sites for industries that need waterways for bulk hauling, and also for residential and recreational developments. To make these locations attractive, dredging has been used to modify the estuaries by creating channels for access and landfill for area expansion. Dredging operations have become so extensive in some estuaries that it is the dominant operating geologic process (D. D. Smith, 1975).

Dredging may be simply defined as a reworking of the sediments to reshape the estuary. Methods of dredging are **hydraulic** and **mechanical.** Disposal of dredged material can be either **confined, open-water**, or **onshore**. Confined disposal is the discharge of dredged sediments behind dikes or levees. The sediment settles out of the water, the water is drained off, and the disposal site eventually fills (Figure 12–26). Open-water disposal is the discharge of dredged sediments in open water without any confinement, usually forming subaqueous mounds which, as disposal continues, may or may not become subaerial (Figure 12–27). Onshore disposal sites are dikes or open areas on existing land areas. These sites are selected to enhance the value of the land by filling or to confine polluted sediments (Figure 12–28). About two-thirds of all dredged material in the United States is dumped in open water, with the remainder being deposited in confined areas or onshore (Boyd et al., 1972).

Dredging in estuaries changes the natural configuration and affects the natural processes and conditions: currents, waves, bathymetry, and the surrounding geomorphic features. Change will occur in the circulation patterns, flushing, salinity, dissolved oxygen, temperature distribution, sediment transport, and areas of sediment erosion and deposition within the estuary. Confined disposal eliminates many of the environmental and physical changes caused by the alteration of the estuary once the disposal site has been selected. Open-water disposal is a great deal more complex, since each site represents a new alteration of the geomorphology. Table

FIGURE 12–26 Confined dredge disposal site.

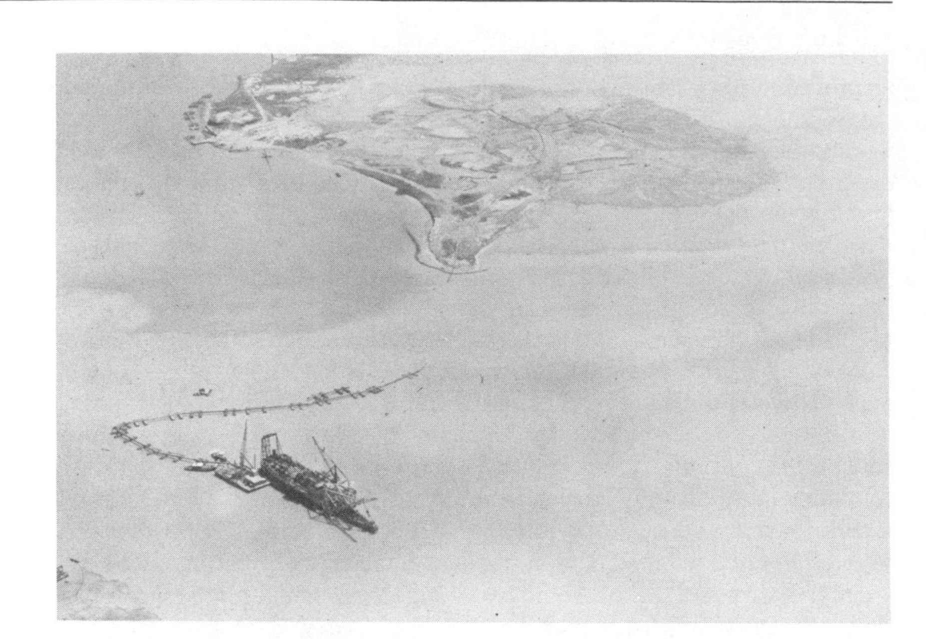

FIGURE 12–27 Unconfined disposal of dredge material.

FIGURE 12–28 Disposal of dredged material on land to develop construction sites.

12–3 lists a number of physical factors that affect the erosion and redeposition of dredged-material islands. Complex interactions between these factors further complicate the selection of an open-water disposal site.

The engineering geologist must be able to evaluate the interaction between a proposed disposal site and the physical factors that will act to erode and redeposit the dredged sediment.

TABLE 12–3 Physical factors in dredged-material islands

Wind	Climate
Waves	Basin physiography
Tides	Island design
a. Astronomic	People
b. Wind	a. Ship wake
Currents	b. Subsidence
a. Wind	Biology
b. Fluvial	a. Fauna
c. Tidal	b. Flora
Dredge material	
a. Sand	
b. Clay	
1. Cohesive	
2. Clay balls	
c. Silt	

Source: Mathewson and McHam, 1977.

Slope Processes

thirteen

Gravity produces a component of stress, directed downward, on the materials in every natural or artificial slope. Slope processes are downslope movements of solid earth materials in response to a stress field produced by the geometry of a slope and the force of gravity. These processes have been called **mass wasting** or **mass movements** by geomorphologists, or more popularly, **landslides**. These terms can be confusing if they are not fully defined each time they are used. Varnes (1972) has suggested that **slope movement** be used as an all-inclusive term for any downslope movement of earth materials.

Slope movement classification systems can be based on many features or characteristics, such as type of material, geometry of the failure, mechanics of deformation, type of movement, rate of movement, or a combination. Each of these features or characteristics can be related to the prediction, prevention, and control of slope movements. Varnes (1972) developed a classification system based on the type of movement, with the type of material being used as a secondary classification tool. This system has the advantage that the features used to classify the slope movement are preserved, thereby making classification of past events possible. In addition, classification by type of movement also establishes the characteristics of the mechanism of deformation, which is important in the analysis of slope stability.

CLASSIFICATION OF SLOPE MOVEMENTS

Slope movements can be divided into five types: **falls**, **topples**, **slides**, **spreads**, and **flows**. Within each class, however, the rate of movement, moisture conditions, rock type, rock substance, and rock-mass characteristics may vary. Figure 13–1 shows a generalized classification of slope movements.

TYPE OF MOVEMENT	TYPE OF MATERIAL		
	Rock	Soil	
		Coarse Grain	Fine Grain
Fall	Rock Fall	Debris Fall	Earth Fall
Topple	Rock Topple	Debris Topple	Earth Topple
Rotational Slide	Rock Slump	Debris Slump	Earth Slump
Translational	Rock Block Slide	Debris Block Slide	Earth Block Slide
Slide	Rock Slide	Debris Slide	Earth Slide
Spread	Rock Spread	Debris Spread	Earth Spread
Flow	Rock Flow	Debris Flow	Earth Flow
Complex	Combination of Any of the Above		

FIGURE 13–1 Classification of slope processes. (After D. J. Varnes, in *Transportation Research Board Special Report 176*, ed. M. Clark, 1978.)

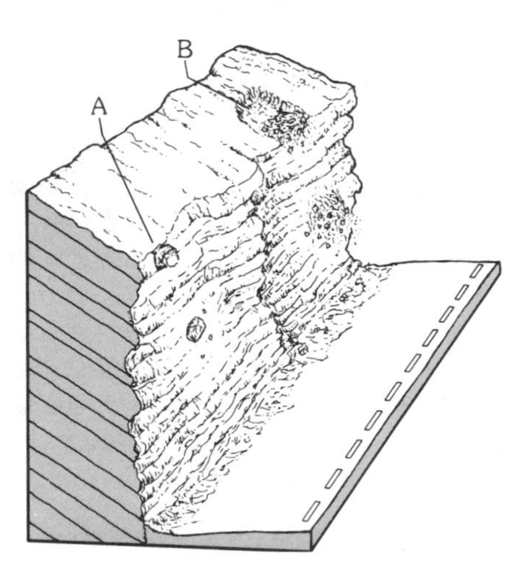

FIGURE 13–2 Falls: (A) rock fall and (B) debris fall.

FALLS

Falls are slope movements on steep slopes where a discrete mass of each material, regardless of size, is detached and moves downslope by traveling through the air, bouncing, or rolling (Figure 13–2). Little or no sense of shear or sliding precedes the failure. The rate of movement may be rapid to extremely rapid, depending on the path of downslope movement. These types of slope movements have been called **raveling** failures.

More detailed classifications of falls are based on the type of material. **Rock falls** involve bedrock, while **debris falls** involve coarse-grained fragments, and **earth falls** involve fine-grained aggregates of material. The

CASE IN POINT

Protecting the Tracks

Two railroad lines follow the Thompson and Fraser rivers through the Rocky Mountains of British Columbia, Canada. In order to meet grade (slope) requirements for the tracks, the railroad was cut into the valley walls. Rock slides and falls from the slopes above have to be directed away from the tracks to keep the route open and safe. The solution was to construct covers over the tracks to direct the falling material into the river. In some cases, short tunnels were driven to maintain the mountain slope and thereby avoid oversteepening it.

termination of a fall, where the mass lands, depends upon the type of material and the geometry of the slope (Figure 13–3). In some cases, a rock may travel far from its original site by rolling or bouncing down the lower slopes. Debris-fall and earth-fall masses tend to terminate near their source because they frequently disaggregate on impact.

TOPPLES

Topples occur when a tensile failure in the material, ice or plant wedging, or some other cause of instability causes rotating about a point (Figure 13–4).

FIGURE 13–3 Rocks filling the drainage ditch are from the slope above and move downslope as falls.

FIGURE 13–4 Topple failure.

These types of movements usually occur on steep slopes and may terminate as a fall or slide, depending on the geometry of the slope below the point of rotation. Topples can occur in any cohesive material and may range in size from very small to extremely large. The size of the mass that topples is controlled by the substance and mass characteristics of the rock.

SLIDES

Slides are movements in which the sense of deformation is pure shear, and the moving mass slides along a discrete failure surface or in a definable shear zone. Slides may occur in any earth material. The character of the moving mass and the shape of the failure surface are frequently used to refine the classification of the slide. If the slide mass moved as a single unit, the slide is a **block,** or **intact,** slide; if it moved as numerous independent units, it is a **broken,** or **disrupted,** slide. The shape of the failure surface is a more definitive refinement of the classification because the shape reflects the

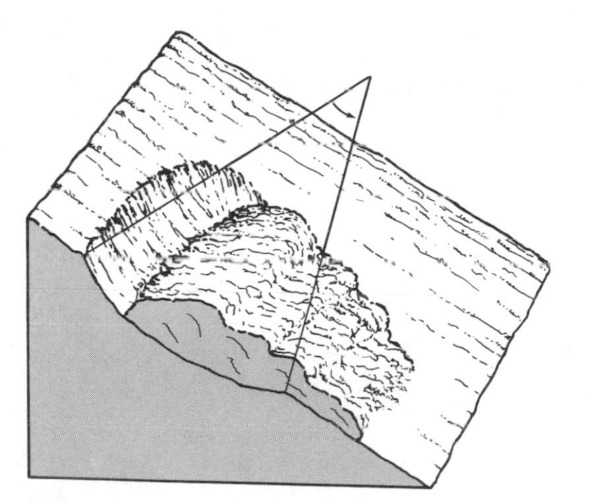

Rotational Slide

Translational Slide

FIGURE 13–5 Classification of slides.

geology and mechanics of the failure. Slide movement is either **rotational** or **translational** (Figure 13–5). Rotational slides occur in earth materials where the strength of the rock substance is nearly equal to the strength along discontinuities in the rock mass. The failure surface is generally uncontrolled, although some rotational slide surfaces are controlled by the geology of the site and the mechanical properties of the geologic units. Failure of the slope occurs along a circular or spoon-shaped surface, because the maximum shear stress develops along such a surface.

Rotational slides commonly appear as slumps of earth material along slopes, river channels, roadcuts, and fills. The original land surface usually rotates into the slope, tensional or extension fractures frequently form in the upper portion of the slide mass, and the lower portion may be broken and disrupted. Sites of old slides may have sag ponds in the upper portion (Figure 13–6).

The influence of a rotational slide tends to be limited to the immediate vicinity of the slide because the driving stress decreases as the mass rotates. There are, of course, exceptions. Rotational slides may overload the slope below or remove support from the slope above and cause a series of failures, or the slide material may liquefy and flow downslope.

Translational slides occur along **planar**, or gently undulating, failure surfaces. These types of slides usually occur in earth materials where the

FIGURE 13–6 Scarp formed by the rotational failure of glacial clays.

strength of the rock substance is greater than the shear strength along discontinuities in the rock mass. The material appears to slide along a discontinuity rather than rotate. Translational slides are frequently called **planar slides**.

Because translational slides require some planar failure surface, the geology and geometry of discontinuities in the rock mass have a significant control on slope stability. Translational slides can occur in any earth material

CASE IN POINT

Engineering Geology of Multiple Landsliding Along I-45 Road Cut Near Centerville, Texas

Multiple landslides in the highway backslopes of Interstate 45 near Centerville, Texas, appear to be the result of (1) the stratigraphy within the backslopes, (2) the shear strengths of the earth materials, and (3) alteration of the natural ground surface during construction.

Geologically, the roadcut is primarily within the Eocene Queen City formation which, at this locality, is a fluvial deposit consisting of natural levee clays, crevasse splay sands, and floodplain-marsh organic clays. The downdip, impermeable floodplain clays retard the movement of groundwater, creating perched water tables and artesian conditions within the natural levee clays and crevasse sands updip. Overlying the Queen City formation on both sides of the highway is the Weches formation, which is noted for its instability in slopes. The positions of the two landslides opposite each other in the north end of the cut and within the levee clays and sands are a direct result of the stratigraphic and hydrologic conditions.

In engineering terms, the highway backslopes are composed of overconsolidated, low- to high-plastic clays and silts, and poorly graded sands. Stability analysis results, using shear strength parameters determined from laboratory testing, indicate initial slope failure occurred as a result of the combination of high pore water pressures and oversteepening of the slopes. Subsequent failures have occurred in response to residual shear strengths and increases in the pore water pressures during periods of heavy rainfall.

FIGURE 13–7 Translational slide. Notice the failure surface (dashed line) and adverse structure (arrows).

that has a discontinuity or failure surface and a sufficient state of stress. These types of slides have occurred as blocks of soil sliding on frozen ground, debris sliding on a bedrock surface, slabs of rock sliding on weaker rocks, and masses of rock sliding on fault plains (Figure 13–7).

The influence of a translational slide frequently extends beyond the immediate failure site because movement of the slide mass does not reduce the driving stress. In fact, a translational slide mass can develop a significant

FIGURE 13–8 Energy developed in a sliding block caused the failure of the safety wall that is more than 1.8 m high.

amount of kinetic energy if it slides down a long slope (Figure 13–8). The termination of a translational slide can take many forms. For example, it can end as a large block, or numerous blocks that roll and bounce downslope, or a fluid mass if the material liquefies. A translational slide can also trigger slides either above or below the site of the original failure.

SPREADS

In spreads the sense of movement is nearly horizontal, and the earth material fails both by shear along a failure surface or in a failure zone, and by tension, or extension, along a nearly vertical surface (Figure 13–9). These types of movements require that some underlying geologic unit fails and

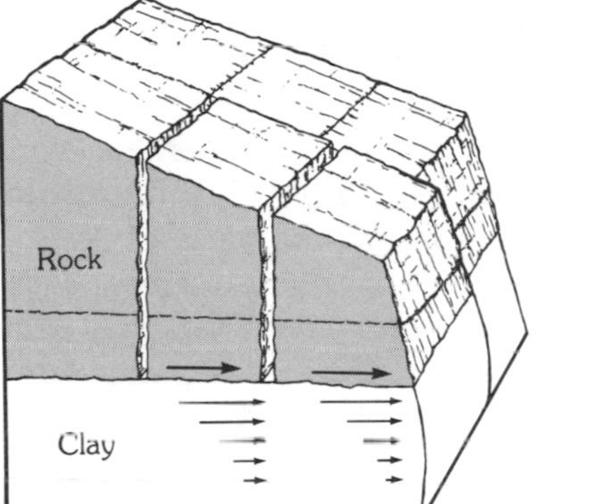

FIGURE 13–9 Spread failure.

FIGURE 13–10 Spread failure in which a plastic clay material is moving into the roadcut and undercutting the overlying limestone.

moves outward, carrying the overlying material. The stability of the slope is controlled by the geologic units at the site and the loading conditions (Figure 13–10). Another type of spread occurs near ridges where bedrock is exposed. The rock fails in tension with little, if any, horizontal deformation of the rock mass. Possibly this type of failure is caused by strain relief mechanisms that exceed the tensile strength of the rock.

The termination of spread movements may be limited to the immediate area of the failure or may be far away. Extensive effects of spreads can result when a failure occurs in flat-lying rocks exposed on upper slopes and the failure mass rolls or slides downslope, or when the failure is caused by the liquefaction of supporting geologic units.

FLOWS

Flows are slope movements in which the mechanical properties of the slope material behave as a **viscous fluid, plastic body,** or a true **fluid**. Flows may deform at any rate, fast or slow, or in any moisture condition, dry to saturated. Such slope movements as **creep, rock glaciers, debris avalanche, mudflow, sand flow, dry flow, block flow, turbidity currents,** and **block streams** are types of flows.

Slow-moving flow failures are frequently classed as creep-type failures; however, the term **creep** has a variety of definitions. As we use the term here, it means deformation that is continuous when subjected to a constant stress; in other words, the stress-strain curve is horizontal. Although this definition is specific, the actual deformation of an earth material probably does *not* fit the true mechanics of creep because the stress conditions and material properties of a moving mass are not constant. There are numerous creep mechanisms that reflect changing conditions (Figure 13–11).

Creep movements can be initiated by melting of the active layer in Arctic regions—**solifluction**—and by ice formation in broken rock masses—**rock glaciers**, which will be discussed in Chapter 15. Wetting and drying cycles in an expansive soil can produce a form of downhill creep (Figure 13–12). True mechanical creep of some rocks may also account for

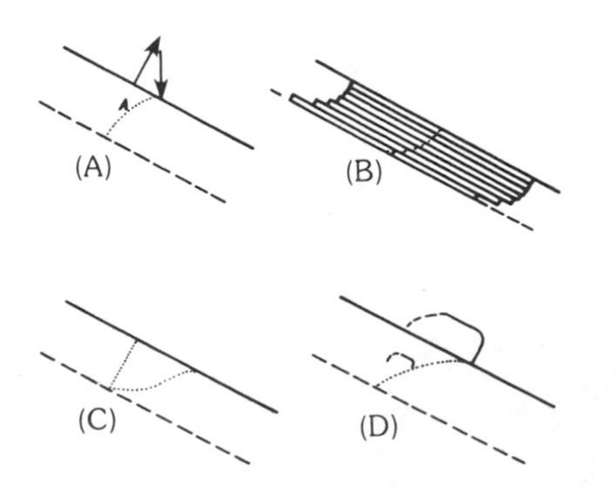

FIGURE 13–11 Creep deformation paths: (A) frost creep; (B) viscous flow; (C) combination of A and B; and (D) creep in expansive soils.

FIGURE 13–12 Damage caused by downhill creep movement (arrow points in the direction of movement).

FIGURE 13–13 Downhill creep-induced folds in layered rocks. (Photo by L. D. Dyke.)

CASE IN POINT

The Hope Slide

On January 5, 1965, about 130 million metric tons of rock swept across Highway 3 west of Hope, British Columbia, Canada. The failure surface is in a volcanic greenstone that dips toward the valley at 45° to 60° and strikes parallel to the valley. The regional geologic structure, however, dips at a steeper angle; thus, the failure surface cuts across structure and is not along a major surface. The failure surface appears to be formed by a combination of local structural and rock-type features. It is believed that the slide was triggered by one or both of two small earthquakes of magnitude 3.2 and 3.1, which were centered about 1.5 km from the slide. The rock slide produced a mudflow that was driven up the opposite slope and as far as 5 km down the valley.

slow downslope movements. Weathering of steeply inclined, thinly bedded sedimentary and foliated or bedded metamorphic rocks on moderate slopes may produce a form of creep movement as the general strength of the rock is reduced (Figure 13–13). Slow-moving flows are generally restricted to plastic deformation mechanisms.

Rapidly moving flows occur in materials that behave as viscous fluids, while very rapid flows behave as a low-viscosity fluid. Dry materials flow as **sand flows** and as **block**, or **rock, flows**, in which the moving mass is composed of discrete particles, ranging from sand size to large blocks of rock (Figure 13–14). **Earth flows** occur in silt- and clay-size materials and range greatly in their rate of downslope movement (Figure 13-15). The rate depends on the moisture conditions in the material and on the slope angle.

FIGURE 13–14 Sand flows down the face of a dune. (The pocket comb shows scale.)

FIGURE 13–15 Earth flow in soil. (Photo courtesy of U.S. Department of Agriculture, Soil Conservation Service.)

Very rapidly moving flows are **debris avalanches, mudflows**, and **turbidity currents**. Since the word *avalanche* means a slope movement of ice and snow, care should be taken to use the complete term, *debris avalanche,* when referring to this type of flow. Classification as a "very rapid debris flow" is a less confusing term. These types of flows contain a significant quantity of coarse material, which may range in size up to blocks weighing many tons. Mudflows are composed of sand-, silt-, and clay-sized materials (Figure 13–16). Liquefaction of quick clays can also produce mudflows on the sides of submarine canyons or on the continental slope.

These currents carry shallow marine sediments into the oceanic basins and produce sea-floor features that are very similar to features produced by rivers.

The influence of flows varies as much as the types of flow failures. Slow-moving flows may move at imperceptible rates and may not be

FIGURE 13–16 Mudflow.

FIGURE 13–17 Secondary evidence for very slow downslope movements.

FIGURE 13–18 Very rapid flow remains. A failure above (designated by arrow) rapidly deposited the material at the base of the slope.

discovered unless some secondary feature such as **pistol butt** tree growth, tilted poles, or deformed retaining walls shows up (Figure 13–17). Their fluid properties may cause rapid and very rapid flows to travel extensive distances from their site of origin (Figure 13–18).

AVALANCHES

Although avalanches usually involve the downslope movement of ice and snow with very little rock or soil, they are significant in mountainous terrain (Figure 13–19). Some avalanches can incorporate large quantities of rock and debris or even trigger additional slope movements in materials downslope of the initial site of the failure. Avalanche conditions develop when wind-transported snow and ice accumulate as a frozen mass that becomes unstable when the thaw begins or when loading continues. Many major **avalanche runways** appear to be scars of past debris flows or rock slides, because the avalanche removes the vegetation and leaves a cleared area as it moves downslope (Figure 13–20). The engineering geologist should verify whether or not such scars were left by an avalanche or by some other type of slope movement.

FIGURE 13–19 Avalanche. (Photo by L. D. Dyke.)

FIGURE 13–20 Avalanche runway. The source of the snow and ice is at the crest of the mountain (designated by arrow), where prevailing winds deposit it.

MECHANICS OF SLOPE MOVEMENTS

The engineering geologist must understand the basic mechanics of slope movements in order to evaluate the stability of a natural or proposed slope. The previous discussion of slope movements has concentrated on the types and the materials involved. The engineering mechanics of slope movements reclassify slope processes into three basic types: **rock, soil,** and **fluid**. Rock slope analyses are based on the mechanical properties and geometry of the rock discontinuities, or **structure**. Soil slope analyses are based on the mechanical properties of the slope material. Movements that involve flow are analyzed using fluid mechanics.

ROCK SLOPES

Failures in rock slopes include falls, topples, and translational and spread types of movement that are controlled by the geometry of the structure or discontinuities in the rock mass. Rock slope analyses must, therefore, be based on an understanding of rock structure. As Figure 3–7 shows, rock discontinuities are generally planar features. These features, which form a plane, can be represented in three-dimensional space by two mutually perpendicular lines. One line is horizontal and oriented in the compass direction—**strike**—and the other lies in the plane of the plane being represented. The angle from a horizontal plane to the line lying in the plane being represented is the **dip** (Figure 13–21). Each discontinuity is therefore represented by a bearing and an angle—strike and dip.

Where the structure is simple and an obvious set of rock discontinuities is easily seen, the strike and dip can be measured at the site and immediately

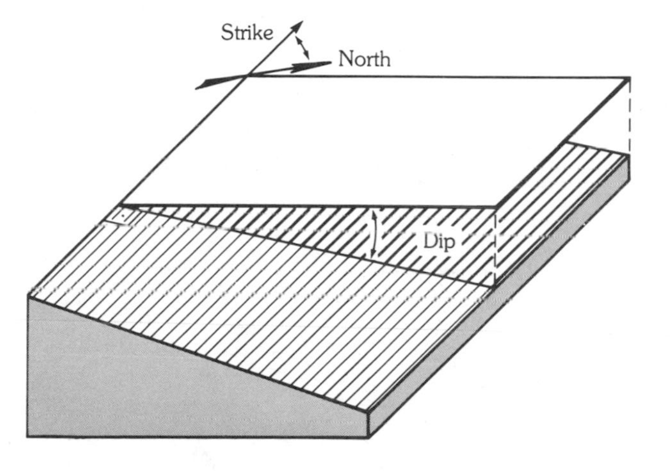

FIGURE 13–21 Relationship between strike and dip. The orientation of the plane in space is defined by the compass bearing of the plane with the horizontal (strike) and the angle between the horizontal and the plane (dip).

FIGURE 13–22 Simple structure (discontinuity) geometry that controls the stability of the slope. The arrow identifies an incipient failure.

incorporated into the stability analysis (Figure 13–22). However, most rock units have been subjected to a complex stress history, and there is no obvious potential failure surface (Figure 13–23). At complex sites, a statistical analysis of the structure is necessary to identify the orientation and density of the discontinuities. A graphical solution, using a spherical projection, is prepared to identify the significant structures (see Concept 13–1). The one most commonly used is an equal-area projection, which is

CONCEPT 13–1

Stereographic Projection

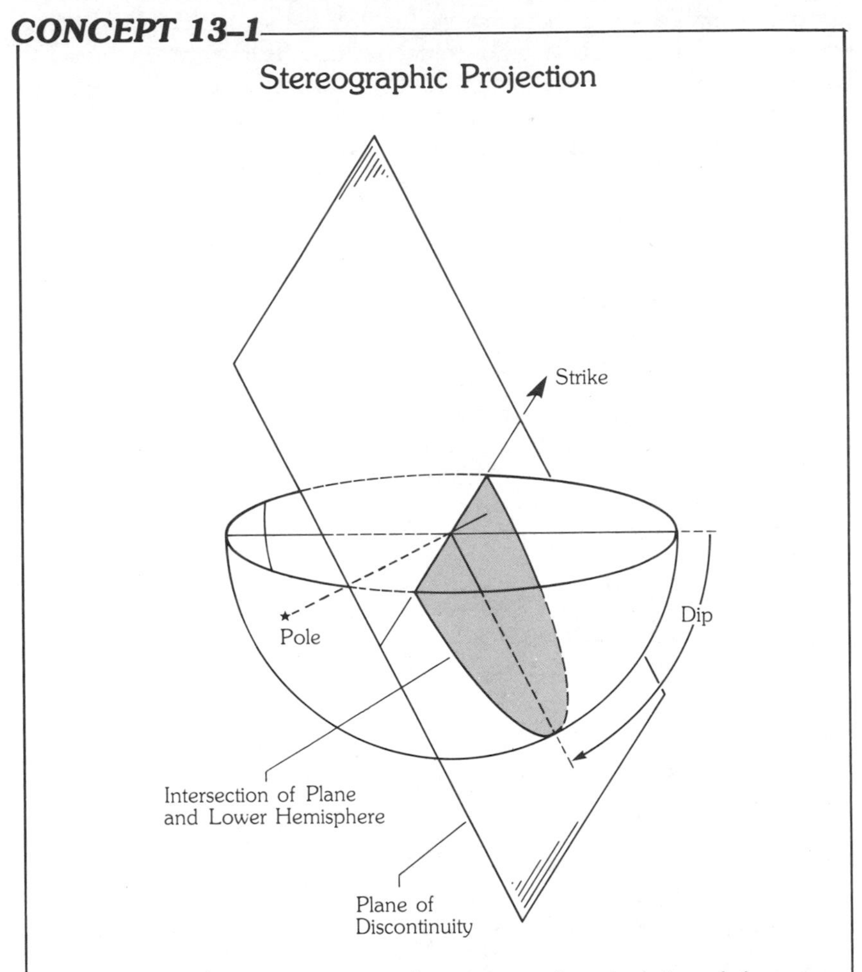

The stereographic projection is used to represent the orientation of planes in space on a flat surface. The plane is projected through the center of an oriented hemisphere. The strike of the plane forms a line across the equatorial surface of the hemisphere, while the intersection of the plane and the sphere forms a curved line. The dip of the plane is the angular distance from the equator to the curve. The pole of the plane intersects the sphere at a unique point. This projection makes it possible to plot many strike and dip measurements and to identify whether a common structural trend exists.

FIGURE 13–23 Complex rock structure where no obvious failure surface is visible.

also known as a Lambert projection or Schmidt net. (This projection is used on many maps to show the spherical shape of the earth on a flat page.) The advantage of an equal-area projection is that the density of each structural set can be contoured and the significant discontinuities defined. Computer programs are also available to identify significant discontinuities. Once they are defined, the basic geometry of the potential failure surfaces is established. Figure 13–24 shows common failure geometries of rock slopes.

The stability of a rock slope is evaluated by treating the problem as a block sliding along the significant discontinuities (see Concept 13–2). A major problem in the analysis is the determination of the shear and tensile strength along each discontinuity. Laboratory samples of the discontinuities are severely limited in size, and therefore laboratory test results may not be representative of actual field conditions. For example, laboratory testing cannot measure the large-scale variations in fracture shape or fillings. The final stability analysis must be based on a thorough geological investigation, careful laboratory testing procedures, and experience.

SOIL SLOPES

The approach to a stability analysis of soil slopes differs from rock slope stability, because the potential failure surface is not usually defined by

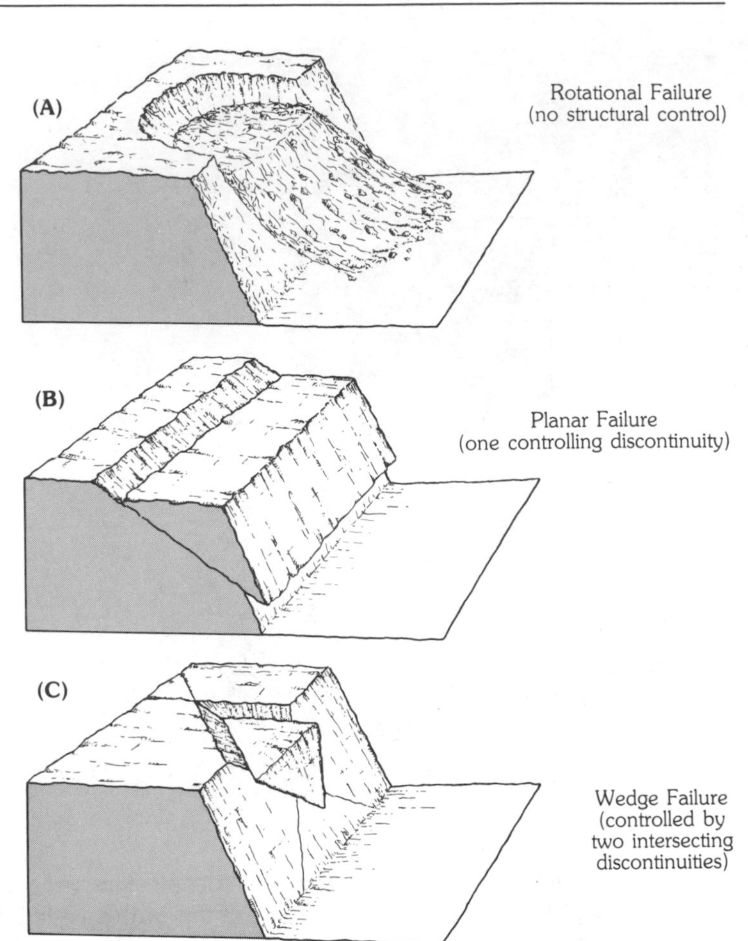

FIGURE 13–24 Effect of structure on planar slope failures.

CONCEPT 13–2

Stability of Translational Slides

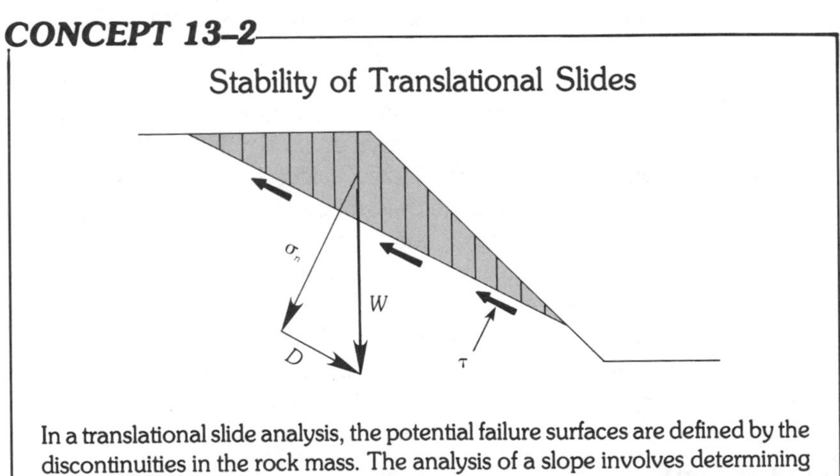

In a translational slide analysis, the potential failure surfaces are defined by the discontinuities in the rock mass. The analysis of a slope involves determining the shearing resistance (τ) along the failure surface and the downslope driving force (D) due to the weight of the slide mass. If the driving force is greater than the resisting force, the slope is unstable.

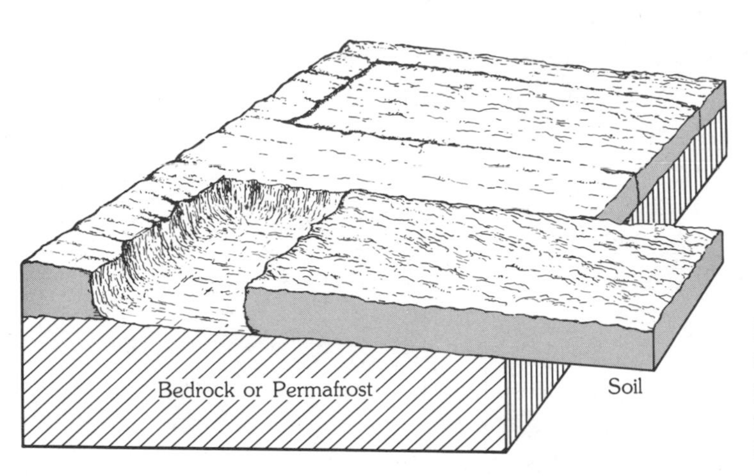

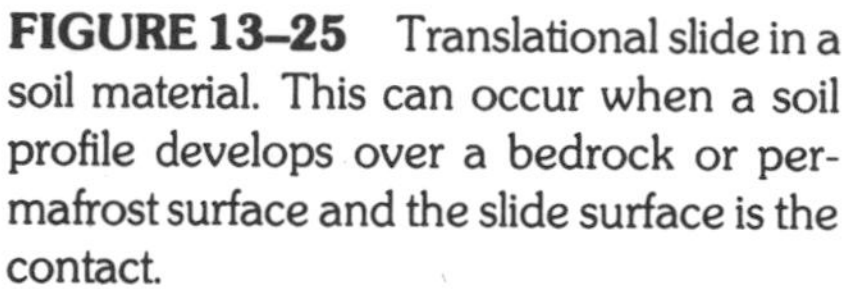

FIGURE 13–25 Translational slide in a soil material. This can occur when a soil profile develops over a bedrock or permafrost surface and the slide surface is the contact.

discontinuities. The failure surface must therefore be located during the analysis procedure. Geologic and stratigraphic information may be available to guide the analysis; however, the analysis should not completely ignore the general solution.

Soil analyses most commonly examine rotational slope movements; however, some translational slides also fall into this class (Figure 13–25). Since the failure surface is unknown, the analysis procedure assumes a failure surface and calculates the ratio of resisting force, which is developed by the shear strength of the material, to the driving force, which is produced by the weight of the material and the geometry of the assumed failure surface (Concept 13–3). This ratio is the **factor of safety**. The analysis is then repeated until the *lowest* factor of safety is determined. If the factor of safety is less than 1.0, the slope is unstable and will probably fail.

FLUID SLOPES

Stability analyses of flows are complex processes. Slow-moving flows can often be treated as true creep processes in which the rate of deformation is assumed to be constant. The problem is that the actual state of stress at the site is usually unknown. Laboratory creep tests are only approximations of field conditions. The analysis can be greatly improved if in situ measurements of slope movements can be made.

Analyses of rapid flow movements, in contrast, involve two distinct components. An analysis of the stability of potential failure sites must first be made. These analyses require extensive geologic investigations just to identify potential failure sites and then realistic assumptions about the triggering events and conditions. Some attempts to evaluate the runout of rapid flow movements must also be made in order to predict the extent of such a flow. In some cases it is possible to predict the runout of rapid flow movements by treating the problem with a fluid dynamics solution.

CONCEPT 13–3

Stability of Rotational Slides

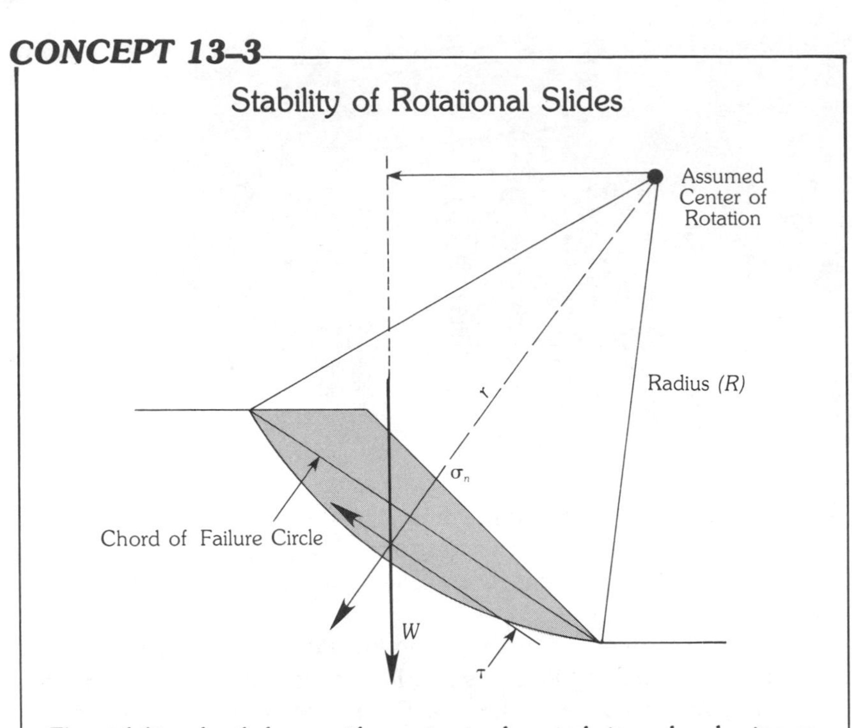

The stability of soil slopes with no structural controls is analyzed using an assumed failure surface that takes the form of a circle. Once the failure surface is selected, the driving force (weight) and resisting force (shear strength) are determined. If rotational forces causing failure are greater than rotational forces resisting it, the slope is unstable. If the resisting forces exceed the driving forces, the slope is assumed to be stable. A slope stability analysis will assume a variety of failure surfaces in order to identify the failure condition having the lowest ratio of resisting forces to driving forces (factor of safety).

FACTORS CONTRIBUTING TO SLOPE MOVEMENTS

Slope movements are affected by two basic factors: the material in the slope—its geology, structure, and groundwater—and external processes—climate, topography, slope geometry, other geologic processes, and human activity. Thus far, our discussion has dealt with the characteristics of the slope materials and their influence on slope movements, but the external processes continually act on the slope and modify the conditions. A slope that is presently stable may be rendered unstable by the external processes or by changes in the slope material. The factors that contribute to slope movements can be divided into factors that decrease the shearing resistance in the slope materials and factors that increase the shearing stress.

FACTORS THAT DECREASE RESISTANCE

PORE WATER PRESSURE This is one of the most common causes of instability, because an increase in pore water pressure reduces the effective normal stress acting across a given potential failure surface (see Concept 2–3). Precipitation is the most common source of water. Permeable or fissured soils are most susceptible to rain-caused pore water increases; however, rapid drawdown of reservoirs can cause excess hydrostatic head if the water table does not lower at the same rate as the reservoir water level.

Loosely packed, saturated soils may densify if subjected to cyclic loading with a consequent increase in pore water pressure. If the pore pressure increases to equal the overburden pressure, the shear strength of the soil decreases to that of water, and liquefaction takes place. Much laboratory and field evidence exists for the liquefaction of sands and silts (Seed, 1973), but little is known about the behavior of clays.

MATERIAL CHANGES Changes in the water content may cause a decrease in shear strength due to chemical and/or physical processes. The presence of water may lead to the weathering of clays as bonds between particles are weakened. Changes in the fabric of sands and silts to produce denser packing may cause liquefaction if enclosed pore water cannot immediately escape. Shrinkage caused by drying of plastic clays causes fissuring, which allows water greater freedom of access and also weakens a soil.

Very slow flow movement (creep in a soil horizon) may take place continuously or be controlled by rainfall and freeze-thaw cycles. Creep represents a steady-state response of the soil to gravitational loading, which may change the clay fabric and cause enough decrease in shear strength to allow failure along a discrete surface or zone. Movement along this zone produces the disturbance that progressively weakens the material, thereby reducing the factor of safety against recurrent movement.

STRUCTURE Unfavorable, fracture, or fault orientation and density may develop as a rock material responds to unloading, strain relief, or as ice wedging, chemical weathering, and vegetative wedging occur. Since the stability of a rock slope is controlled by the discontinuities in the material, any activity that increases the structure in the rock will decrease shearing resistance. Wedging by ice or plant roots may produce a form of shear

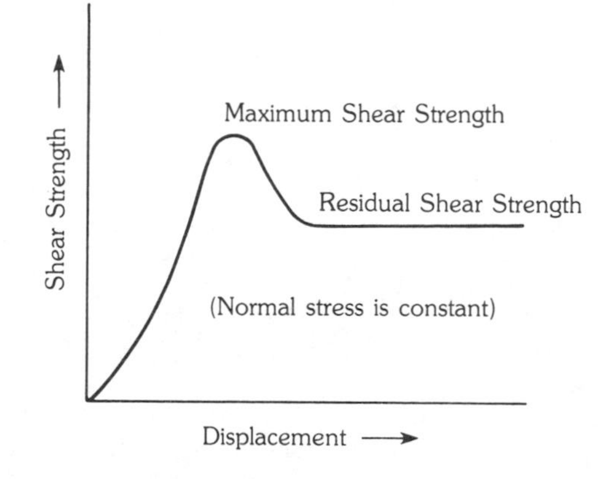

FIGURE 13–26 Shear strength versus displacement curve, showing the reduction in shear strength after the material has been displaced beyond the point of maximum shear strength. This type of deformation can occur in slopes that are creeping.

displacement that can decrease the shear strength along a potential failure plane to the **residual strength** (Figure 13–26).

FACTORS THAT INCREASE SHEAR STRESS

REMOVAL OF LATERAL OR UNDERLYING SUPPORT This process is usually the result of running water or waves that undercut the slope and remove material that is acting to oppose failure. Washing out of granular material, or *piping*, may take place if the slope has high groundwater flows because seepage erosion at the site of groundwater discharge has the same effect as undercutting (Figure 13–27). Artificial cuts are generally formed rapidly and may remove the slope support. Similarly, lowering the water level in a lake or reservoir is a potential factor because a rapid decrease in water level removes support.

INCREASED LOADING Placing a load or surcharge at the top of a slope will increase the shear stress along a potential failure surface. Such loading can be caused by natural processes, such as accumulations of water or snow or material transported downhill from a failure on the upper part of a slope, or by pressures from human activities, such as ore stockpiles, mine tailings, waste dumps, road fills, and buildings. Absorption into the slope material of rain or runoff can also cause internal loading, thereby increasing the unit weight of the material.

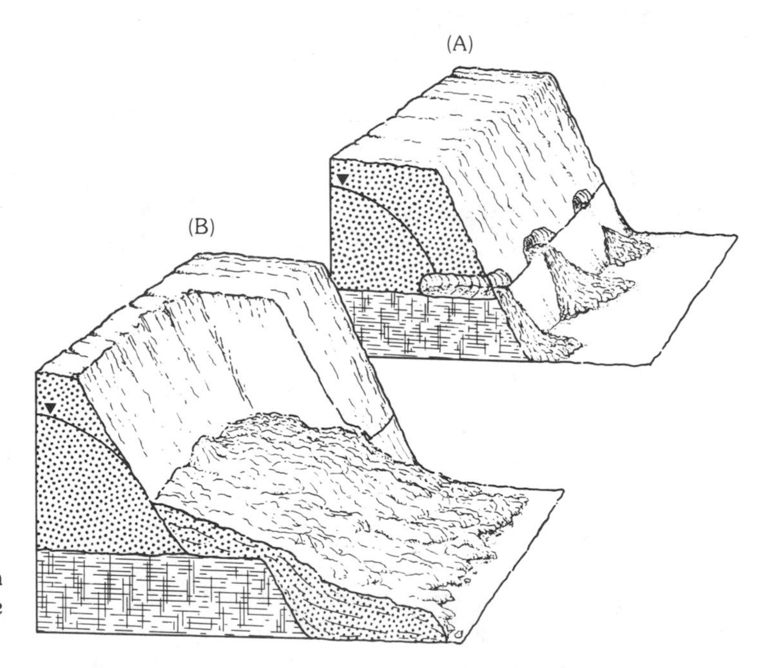

FIGURE 13–27 Piping failure of a slope caused by groundwater seepage pressures.

FIGURE 13–28 Failure of a vegetative mat. (Photo by E. A. Keller.)

TRANSITORY EARTH STRESS Vibratory loading from earthquakes (the most effective type) or construction activities can increase the shear stress along a potential failure surface. The materials in the slope are in a state of equilibrium with the existing stationary or nearly stationary stress field. Vibrating loads, however, produce a spectrum of oscillations, which can set up cyclic loading. Sands and unconsolidated granular or blocky materials may lose some cohesion, quick clays may liquefy, pore water pressures may fluctuate widely, and fracture fillings may fail.

VEGETATION Little detailed work on the influence of vegetation on the stability of the slopes has been carried out. Increases in slope movements appear to be caused by the removal of trees because the tree root systems provide some reinforcement and remove groundwater, and because the trees absorb much of the raindrop impact, reducing the rate of water delivery. The addition of vegetation to slopes has also caused slope movement because the vegetative mass increases the weight of the slope material (Figure 13–28).

PREVENTION OF SLOPE MOVEMENTS

The primary concern regarding slopes in many engineering projects is to maintain existing or constructed slopes. Two basic methods are available to achieve this objective. The factors that contribute to slope movements may be modified or altered to reduce their influence, or slope-stabilizing techniques can be used to increase the shear resistance or hold back the slope.

Slope drainage is the most commonly applied engineering technique to modify the factors that contribute to slope movements. Other techniques include sealing the slope to prevent the infiltration of water, diverting surface runoff, planting woody vegetation, and sealing open fractures.

There is a wide variety of techniques to achieve slope stabilization through construction design or slope reinforcement. If adequate land is

CASE IN POINT

Stable Rock Slope

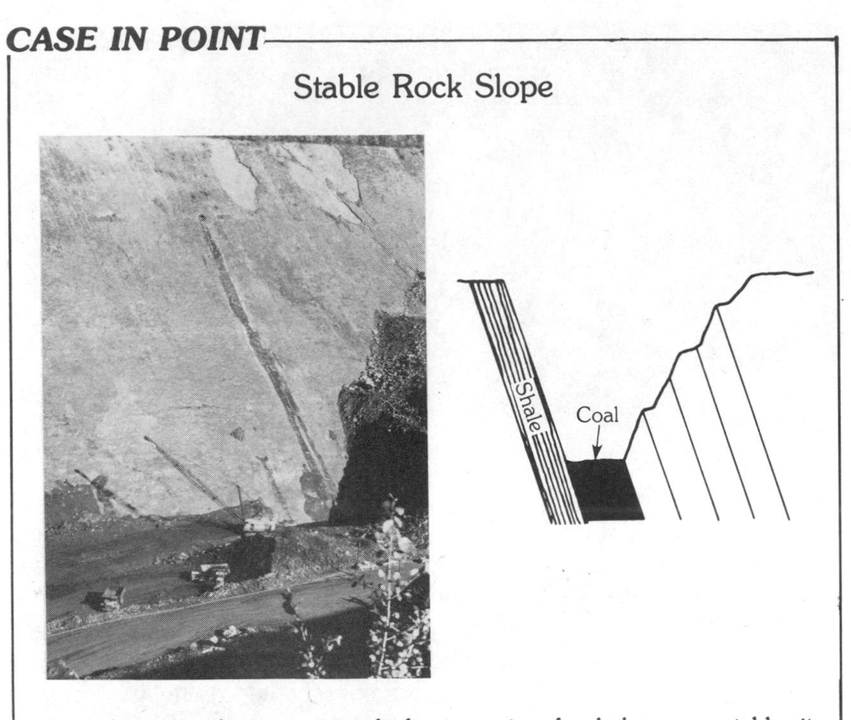

An anthracite coal mine in a steeply dipping series of rocks has a very stable pit slope on one side of the mine, because the pit slope is a bedding plane between the coal and the underclay (shale). The opposite pit slope, however, is not as stable because the mine must undercut the bedding planes. The tensile strength across a bedding plane is weak and fails; thus this slope is much flatter.

FIGURE 13–29 Retaining walls used to support the slope above. Notice that there are three types of walls shown.

FIGURE 13–30 Rock bolts (designated by arrow) and chain link fence used to support a rock slope.

FIGURE 13–31 Benches (designated by arrows) used to control falls in a roadcut.

available, the grade of the slope, or slope angle, can be reduced during construction. **Buttresses** and **retaining walls** are the most common slope-stabilizing structures used in soil materials (Figure 13–29). Rock slopes can be stabilized with **rock bolts**, buttresses, or retaining walls (Figure 13–30). Falls on rock slopes are usually controlled by **trapping** the falling material with **benches** (Figure 13–31), or by sealing the exposed rock face. Careful design of the blasting program during the excavation of rock cuts can significantly reduce the risk of falls by preventing **overbreak**. Excessive blasting will fracture the rock beyond the limits of the excavation, thereby increasing the density of the discontinuities and decreasing the stability of the slope.

Tectonic Processes

fourteen

The earth has undergone and is still undergoing great regional to continental-scale deformations. Geologists, geophysicists, and tectonophysicists are still investigating the driving forces behind these processes. Driving forces that have been identified in the tectonic history of the earth are related to thermal processes, the rotational stresses in the earth, fluid dynamics, and gravitational forces, to list a few. The forcing mechanisms may be of academic interest to the engineering geologist; however, the responses of earth materials as seen in their deformation, discontinuities, and orientations are of prime importance.

Although the worldwide study of earth tectonics is at such a scale that any engineering project is effectively invisible, it is important for the engineering geologist to understand global tectonics. The most active tectonic processes—earthquakes and volcanic eruptions—can be described in terms of **plate tectonics**, the basic mechanism of global tectonics.

The basic assumption of plate tectonic theory is that the earth's crust is broken into a series of plates. Some, such as the Pacific plate, are composed entirely of oceanic crust, and others, such as the North American plate, include both continental and oceanic crust. These plates float on the viscous upper mantle, just as plates of ice float on a frozen river. The plates move apart at oceanic ridges, where new oceanic crust is formed. These areas are generally free of the active tectonic processes that affect engineering projects. Where the plates come together, however, the crust is forced downward below an advancing plate. Deep marine trenches and major mountain ranges are associated landforms in these areas.

Stresses generated by the downward movement **(subduction)** of the plate produce most of the active tectonic processes affecting engineering projects. The most tectonically active plate boundary occurs where the Pacific plate and other smaller oceanic plates in the Pacific Ocean meet the Australian, Eurasian, North American, and South American plates. These plate boundaries correspond with the approximate boundaries of the Pacific Ocean and mark the "ring of fire." The ring of fire incorporates the volcanic and earthquake activity of the Pacific coasts of North and South America,

Japan, the islands of the western Pacific Ocean, and New Zealand. Engineering geologists must be aware that global tectonic processes may be a significant factor when they evaluate the tectonic history and risks associated with a proposed engineering project.

The mechanical properties of the earth materials, under the stress conditions that existed during their deformation, may be significantly different from their properties at or near the earth's surface. The geometry of the rock unit as it was deformed by tectonic processes is in most cases significantly different from the geometry of the body as it is influenced by an engineering project. The mechanical behavior of any earth material is influenced to a significant degree by temperature, loading conditions, strain rate, and the geometry of the body. Variations of any of the factors within an earth-scale can produce widely differing mechanical behaviors.

Temperature plays a significant role in the deformation of a rock. As temperature increases, the rock becomes more ductile. As temperature approaches the melting point, the material may behave like a viscous fluid and flow (Figure 14–1). The **loading conditions** also affect the deformation behavior; as confining pressure (σ_3) increases, rocks become more ductile (Figure 14–2). **Strain rate**, the rate of deformation, also influences mechanical behavior. As strain rate increases, the rock becomes more brittle (Figure 14–3). The **shape** or geometry of the body being deformed influences the mechanisms of deformation (Figure 14–4). Interpreting the characteristics of an earth material after it has been

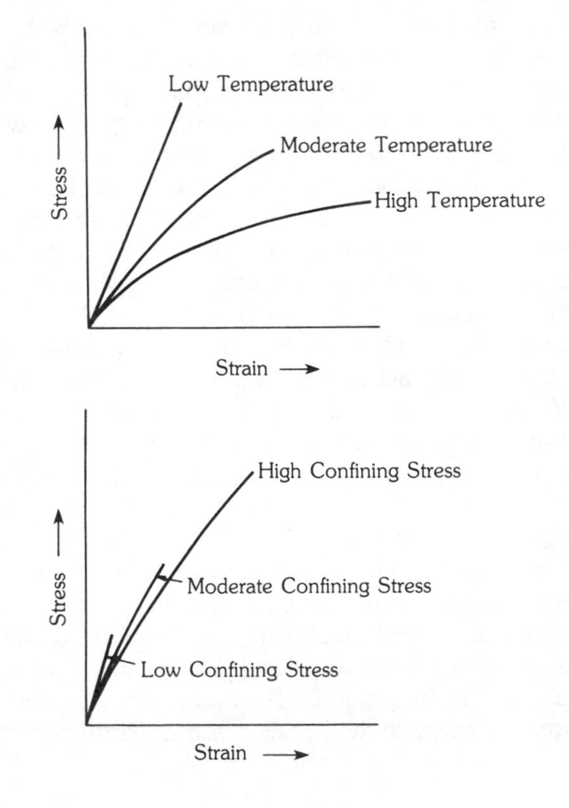

FIGURE 14–1 Influence of temperature on the deformational characteristics of a rock.

FIGURE 14–2 Influence of confining stress on the deformational characteristics of a rock.

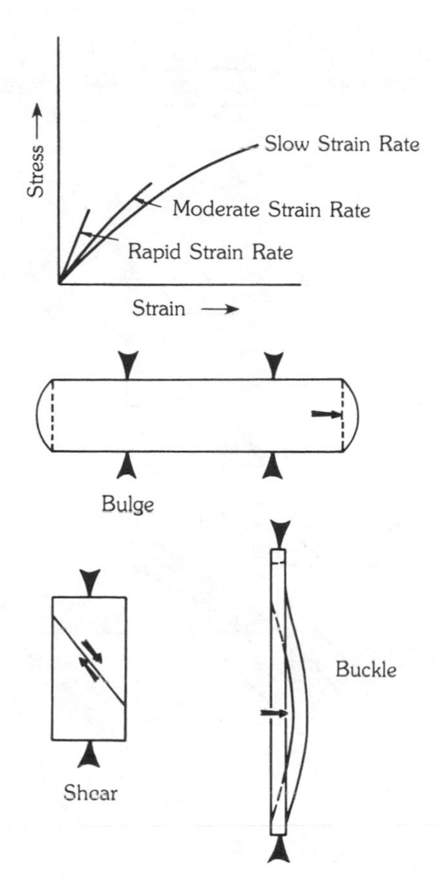

FIGURE 14–3 Influence of strain rate on the deformational characteristics of a rock.

FIGURE 14–4 Effect of specimen shape on the deformation of the specimen.

tectonically deformed and predicting its behavior in an engineering project is the engineering geologist's responsibility.

FRACTURES AND FOLDS

Any earth material, given the right temperature, loading, strain rate, and geometry can either **fold** or **fracture**. Two primary mechanisms, **horizontal compression** and **lateral deflection** produce folds (Figure 14–5). Fractures can be produced by **shear** or **tensile** stresses. Folds and fractures are classified by their descriptive aspects, regardless of the process that produced them, folds by their orientation in space (Figure 14-6) and fractures by their sense of movement (Figure 14–7). **Faults**, as used in this text, are fractures along which movement has taken place. Fractures are frequently called joints; however, this term is confusing and is not recommended.

In most tectonically deformed terrain, the rocks are both folded and fractured. The relative scale of these features may be different, for example, fractured folds or folds that result from fracture movement (Figure 14–8). The conditions that existed during initial deformation and by changes in the stress field in the time since deformation control the severity and characteristics of the secondary features. Rocks that formed under high

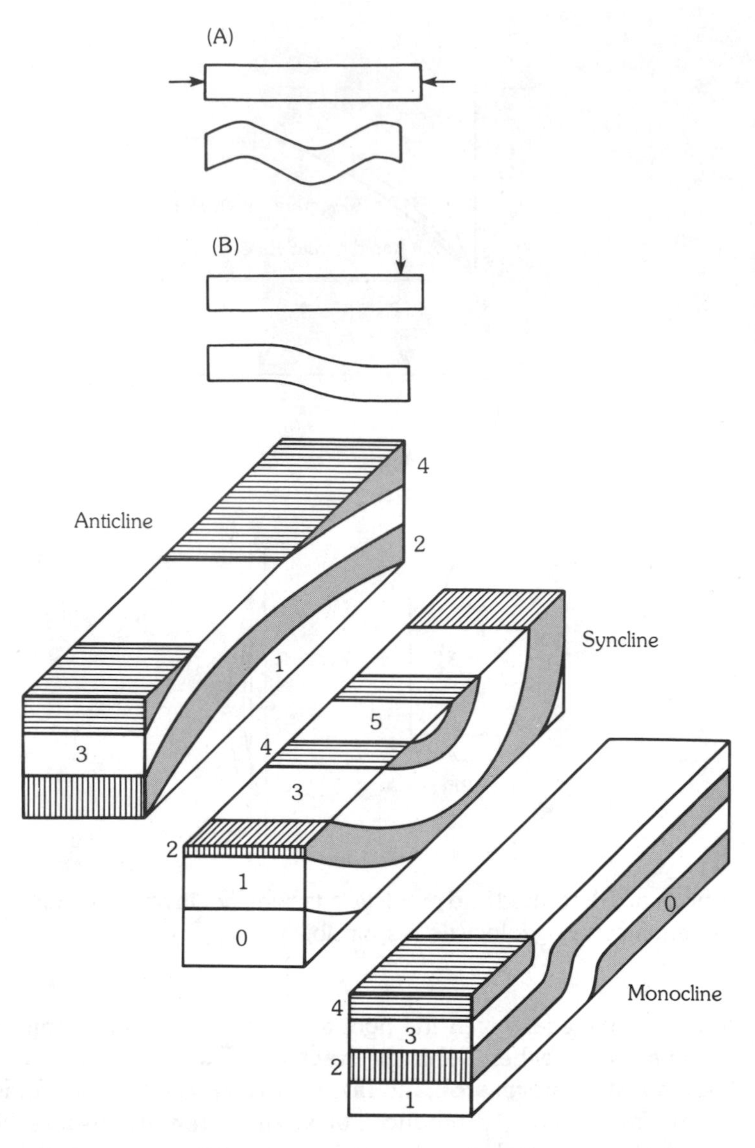

FIGURE 14–5 Effect of loading conditions on folding. In part (A) end loading produces a buckle, while in part (B) perpendicular loading produces a flexure. Both loading conditions produce a fold.

FIGURE 14–6 Classification of folds.

confining stresses, such as a granite, will have some remnant of this past stress preserved in the rock. This stress, plus any regional tectonic stress field, is known as the **in situ state of stress.**

Measurement of the in situ state of stress is difficult. The entire region is the forcing member (σ_1), and the engineering project site is the site where this stress is released (σ_3). Since measuring the in situ stresses directly is impossible, the state of stress is related to a measurement of **strain relief**. If a specimen is loaded to some unknown stress and the strain is measured and set equal to zero, the specimen will strain in response to a change in the stress (Figure 14–9). A strain gauge is cemented to a rock while it is still in place and read (strain = 0). The rock is then cut out of the surrounding rock, reducing the in situ stress to near zero and the new strain is read (strain = strain relief).

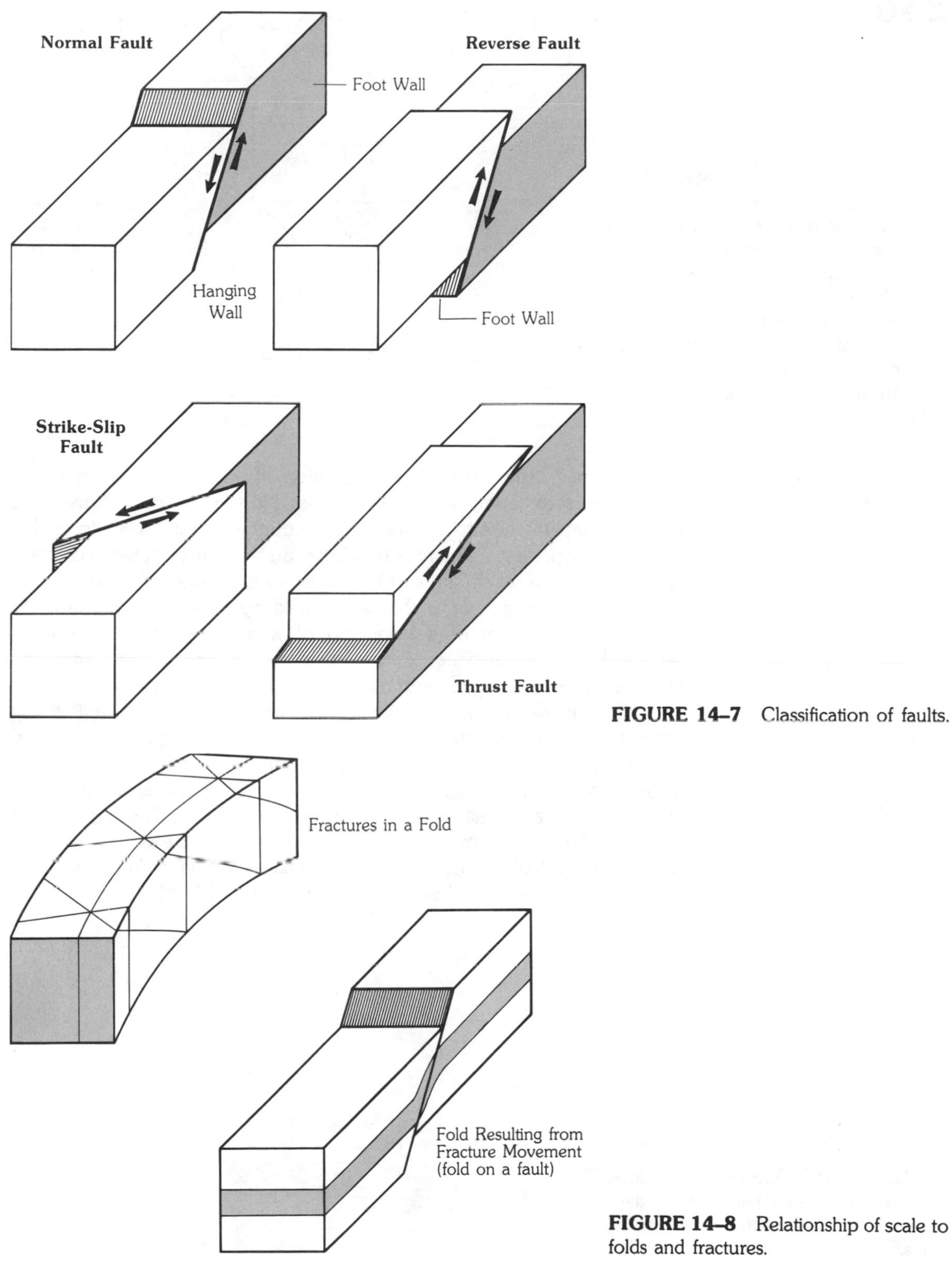

FIGURE 14–7 Classification of faults.

FIGURE 14–8 Relationship of scale to folds and fractures.

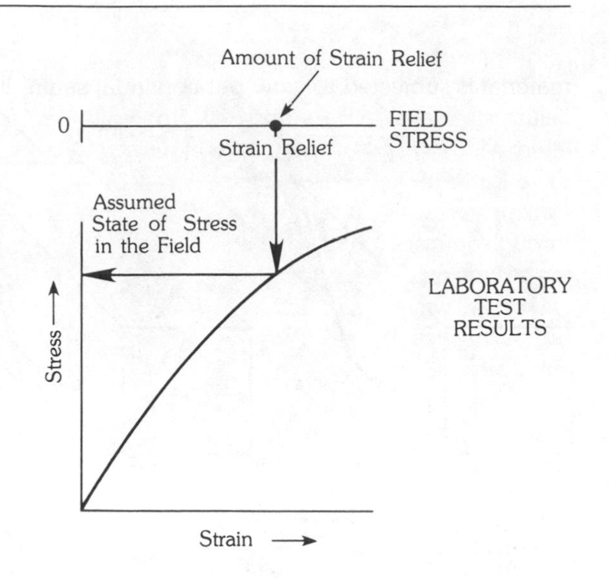

FIGURE 14–9 Determination of the state of stress in the field by strain-relief methods. The strain is measured as the rock responds to the removal of the field stress. The state of stress is then determined by comparing the field strain to the laboratory-determined strain in a stress-strain diagram. The state of stress is assumed to be equal to the stress needed to strain the rock to the level that exists in the field.

The amount of strain relief is an indication of the in situ state of stress. Strain-relief measurements can also be made in the field using **bore hole extensometers**, which measure the amount of strain relief toward a constructed opening. Since rock behavior can be defined in terms of strain or stress, stress-relief studies can be related to the expected behavior of a rock mass as it responds to changes caused by an engineering project.

The mechanism of deformation of a rock also plays a role in establishing the behavior of the rock as an engineering material. Deformation of any given earth material is related to the conditions at the time of deformation and the geologic characteristics of the unit being deformed. Geologic features such as rock type, mineralogy, grain or crystal size, and depositional features are all important. Rock type, mineralogy, and grain or crystal size influence the behavior of the material at any given temperature, load, strain rate, and geometric condition. The depositional features influence the mechanism of deformation. For example, a single massive bed will follow a deformational mechanism different from that of the same thickness of thinly bedded material (Figure 14–10).

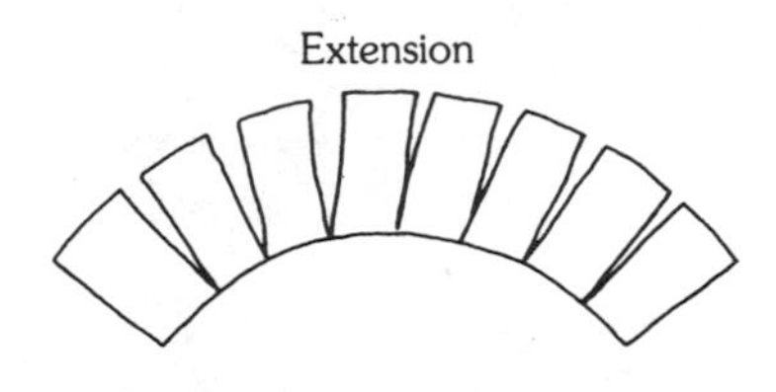

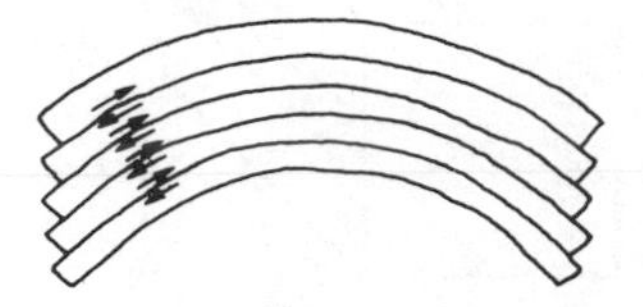

FIGURE 14–10 Influence of geologic discontinuities on the behavior of a thick rock unit when subjected to a bending force.

EARTHQUAKES

Earthquakes are the result of mechanical failure of rock materials. As the material is subjected to slow but continual strain, it will ultimately reach a failure state and fail. As this built-up strain is released during the failure, energy is released and felt at the surface as an earthquake.

Earthquake energy is transmitted in the form of wave energy: **compressional**, **shear**, and **surface**. A compressional wave travels through any medium at a higher velocity than the others and has been called the **primary**, or P-wave, since it arrives at a recording station first. Shear waves, (S-waves) can only travel in a medium that has shear strength. Surface waves (L-waves) travel along the land-air or water-air interface and are dissipated rapidly.

The amount of energy released by any earthquake is measured by two basic methods: **intensity** and **magnitude**. The intensity of an earthquake is based upon the amount of surface damage and divided into twelve levels in the Modified Mercalli Intensity Scale of 1931 (Table 14–1). This determination of earthquake intensity is subjective, as it depends upon the perception of the observer, and is therefore difficult to use as a basis for earthquake-resistant structure design. Earthquake magnitude is a quantitative measure of earthquake energy expressed as a log function and based upon the characteristics of the earthquake waves recorded at a distant **seismograph** (Concept 14–1).

CONCEPT 14–1

Determination of Earthquake Magnitude

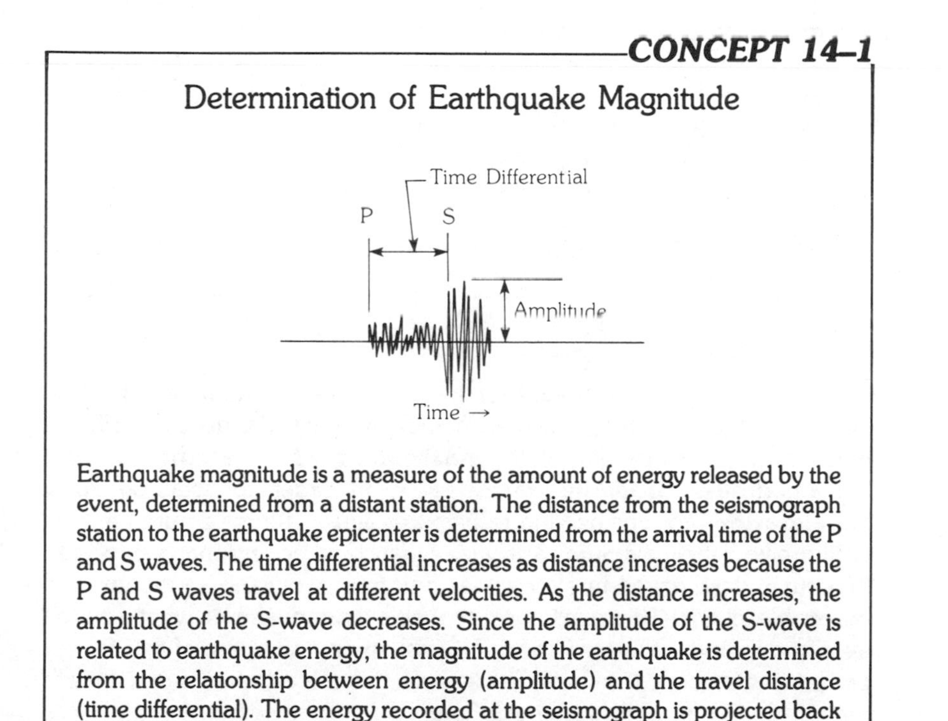

Earthquake magnitude is a measure of the amount of energy released by the event, determined from a distant station. The distance from the seismograph station to the earthquake epicenter is determined from the arrival time of the P and S waves. The time differential increases as distance increases because the P and S waves travel at different velocities. As the distance increases, the amplitude of the S-wave decreases. Since the amplitude of the S-wave is related to earthquake energy, the magnitude of the earthquake is determined from the relationship between energy (amplitude) and the travel distance (time differential). The energy recorded at the seismograph is projected back to the site of the earthquake based on travel distance.

TABLE 14–1 Earthquake magnitude and intensity

	Intensity	Description	Richter Scale Magnitude	Approximate Radius of Perceptibility
I	Instrumental	Detected only by seismography		
II	Feeble	Noticed only by sensitive people	3.5 to 4.2	2–15 mi.
III	Slight	Like the vibrations due to a passing lorry (truck); felt by people at rest, especially on upper floors		
IV	Moderate	Felt by people while walking; rocking of loose objects; including standing vehicles	4.3 to 4.8	15–30 mi.
V	Rather strong	Felt generally; most sleepers are awakened and bells ring		
VI	Strong	Trees sway and all suspended objects swing; damage by overturning and falling of loose objects	4.9 to 5.4	30–70 mi.
VII	Very strong	General public alarm; walls crack; plaster falls	5.5 to 6.1	70–125 mi.
VIII	Destructive	Car drivers seriously disturbed; masonry fissured; chimneys fall; poorly constructed building damaged	6.2 to 6.9	125–250 mi.
IX	Ruinous	Some houses collapse where ground begins to crack; pipes break open		
X	Disastrous	Ground cracks badly; many buildings destroyed and railway lines bent; landslides on steep slopes	7.0 to 7.3	250–450 mi.
XI	Very disastrous	Few buildings remain standing; bridges destroyed; all services (railway, pipes, and cables) out of action; great landslides and floods	7.4 to 8.1	250–450 mi.
XII	Catastrophic	Total destruction; objects thrown into air; ground rises and falls in waves	8.1 +	250–450 mi.

The Richter Scale of earthquake magnitude is determined from an event that has occurred. But the engineering geologist needs to predict future earthquake magnitudes, the **maximum credible earthquake** and the **maximum probable earthquake**, in order to establish **rock accelerations** for engineering resistant designs. The maximum credible earthquake is the maximum earthquake that could be reasonably expected to occur under the presently known tectonic framework and is a rational and believable event. The maximum probable earthquake is what can be expected in any 100-year period. The risk in any one year of a 100-year earthquake is about 1 percent. Rock acceleration is a measure of the earth motion during an earthquake and is used for engineering design.

MAXIMUM CREDIBLE EARTHQUAKE

The maximum credible earthquake is determined with little regard to the probability of occurrence, or recurrence interval, since this event is defined as the greatest earthquake that could be expected. The determination is based on the existing knowledge of the geology and seismology of the area. The maximum credible earthquake is determined from the following geologic and seismic factors:

1. The seismic history of the vicinity being evaluated.
2. The seismic history of the geologic province.
3. The length of the significant fault or faults which can affect the site within 100 km.
4. The type or types of faults.
5. The tectonic and/or structural history of the area.
6. The tectonic and/or structural pattern or regional setting.

The seismic history of the site being evaluated is based on the past record of earthquake activity. A generalized seismic history map of the United States, showing seismic risks, is shown in Figure 14–11. Note that parts of South Carolina, California, and New York all have a high

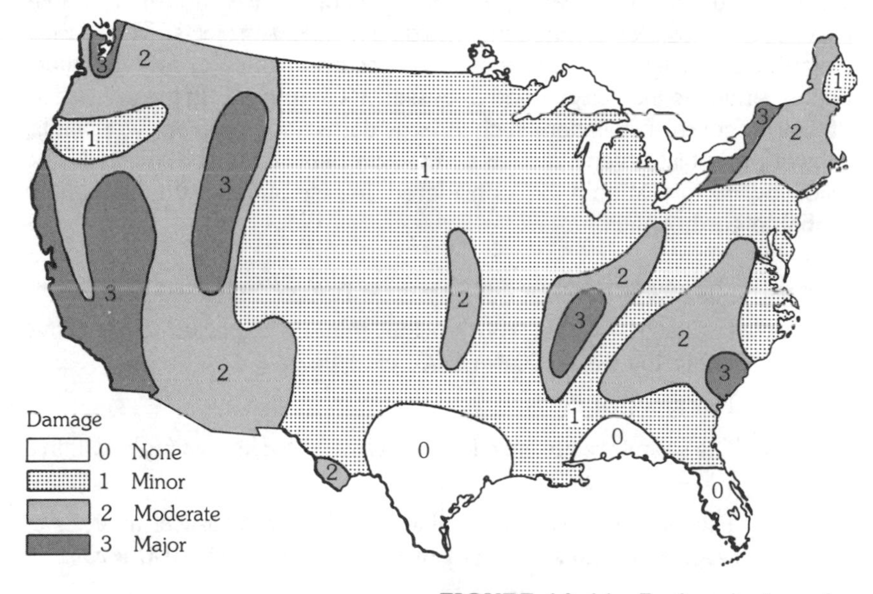

FIGURE 14–11 Earthquake hazard zones in the United States. The zones define hazards based on damage; therefore, this is an intensity map and not a probability map. (After National Oceanic and Atmospheric Administration.)

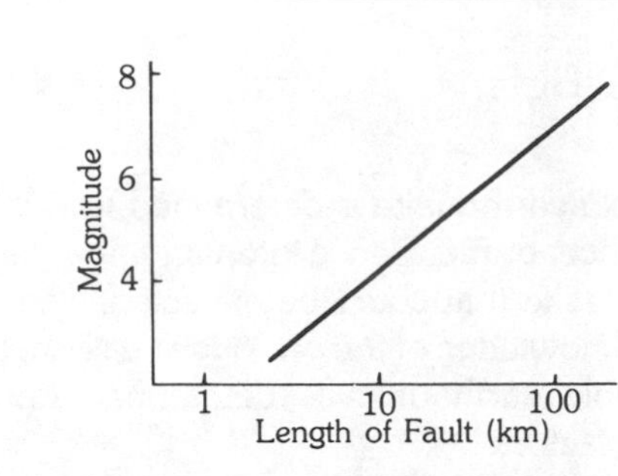

FIGURE 14–12 Relationship between length of a fault and magnitude of the earthquake.

earthquake risk, but that California has much more frequent earthquakes than the other sites.

The relationship between the length of a fault and earthquake magnitude has been determined from empirical studies of past earthquakes (Figure 14–12). A similar relationship has also been developed to relate the maximum surface displacement to the magnitude of the earthquake. Expected earthquake magnitudes for each fault are calculated and combined with the other tectonic factors to determine the maximum credible earthquake expected to occur at the site being investigated. This determination establishes an upper limit of earthquake magnitude.

MAXIMUM PROBABLE EARTHQUAKE

The maximum probable earthquake is a functional-basis earthquake and is used as the basis of engineering design for many projects. This design magnitude usually has a probability of occurring once in any 100-year period, similar to the design flood discussed in Chapter 11. In the evaluation of sites for critical installations such as large dams and nuclear power plants, a lower probability event, such as a 500-year period, may be selected. The following factors should be considered in establishing the maximum probable earthquake:

1. The regional seismicity based on past seismic history.
2. The fault(s) within 100 km that may be expected to be active within the next 100 (or 500, etc.) years.
3. The types of faults.
4. The seismic recurrence for the area and for known faults within a 100-km radius.
5. The mathematical probability based on a statistical analysis of seismic activity associated with the faults within 100 km.

Note that the factors used to determine the maximum probable earthquake include the effect of time—seismic recurrence—while time is not considered in the determination of the maximum credible earthquake. Regional seismicity and the expected 100-year earthquake magnitude usually form the basis of this determination.

EARTHQUAKE ENGINEERING

Although engineering geologists will not be directly involved in the structural design of an engineering project, their determination of the expected rock accelerations caused by the design earthquake will have a direct bearing on the design. The performance of earth materials during earthquake accelerations is a significant consideration.

The manner in which earth materials respond when loaded by cyclic stresses caused by earthquake accelerations varies widely. The amplitude of horizontal and vertical accelerations, or shaking, is directly related to the mechanical properties of the earth materials. Massive, high-strength rocks are more dense than loose, low-strength rocks; therefore, more energy (stress) will be required to deform the massive rock to the same amplitude (strain) as the low-strength rock. Both the rock substance and rock-mass characteristics must be evaluated when determining surface accelerations (Figure 14–13).

Saturated, low-density sands must be considered as potential sites for liquefaction if they are on a slope, or compaction if they are confined. **Mud boils**, **mud eruptions**, or **mud geysers** have all been reported during earthquakes. Mud boils, eruptions, and geysers appear at the surface as excess pore water escapes (Figure 14–14). The collapse of a saturated mass of loose sand or silt during cyclic loading can cause flows. The abrupt reduction in porosity as these units collapse causes a sharp increase in free pore water and a resulting flow if a slope exists.

Vibratory loading during an earthquake can have a significant impact on the shear strength of earth materials. Rapid horizontal acceleration

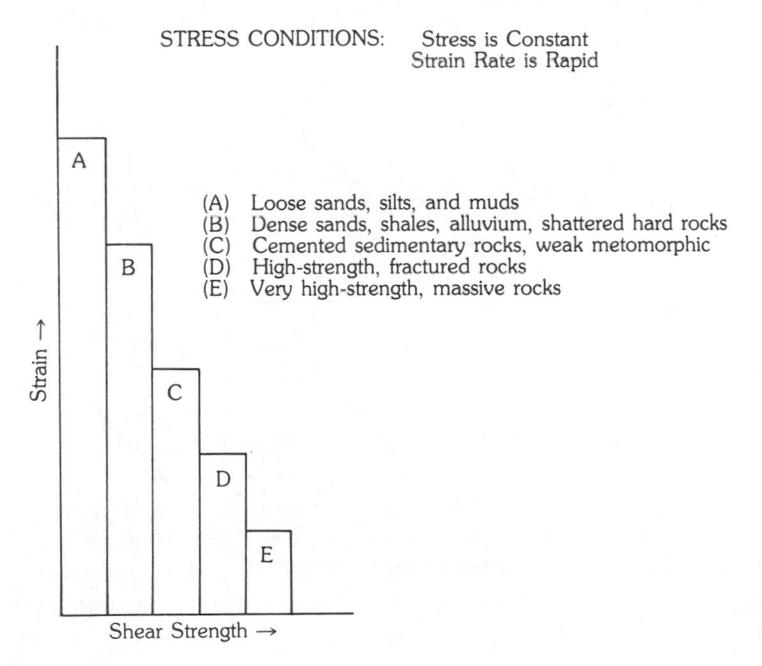

FIGURE 14–13 Amount of strain expected at the same stress level for various rock and soil types.

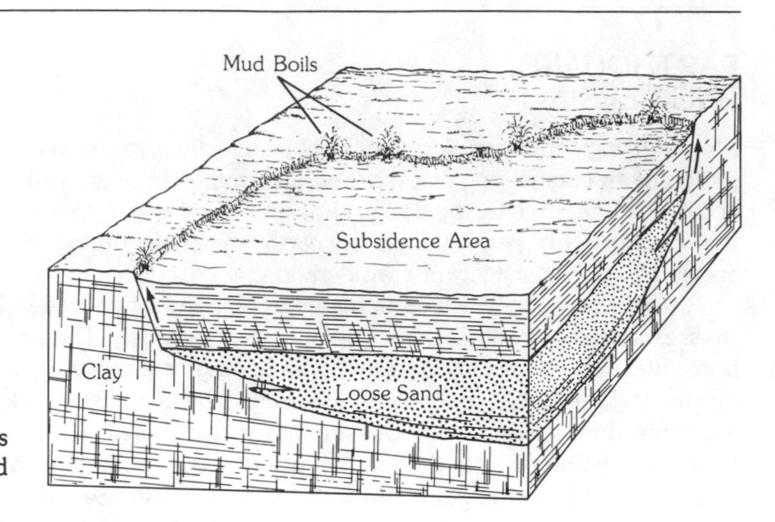

FIGURE 14–14 Surface characteristics caused by the densification of a confined loose sand at depth.

parallel to a significant discontinuity in a rock slope can destroy or greatly reduce the cohesion (C) along the discontinuity. Vertical accelerations can reduce the normal stress (σ_n) across the discontinuity. A combined decrease in both cohesion and normal stress can reduce the shear strength to a level that produces a slope movement.

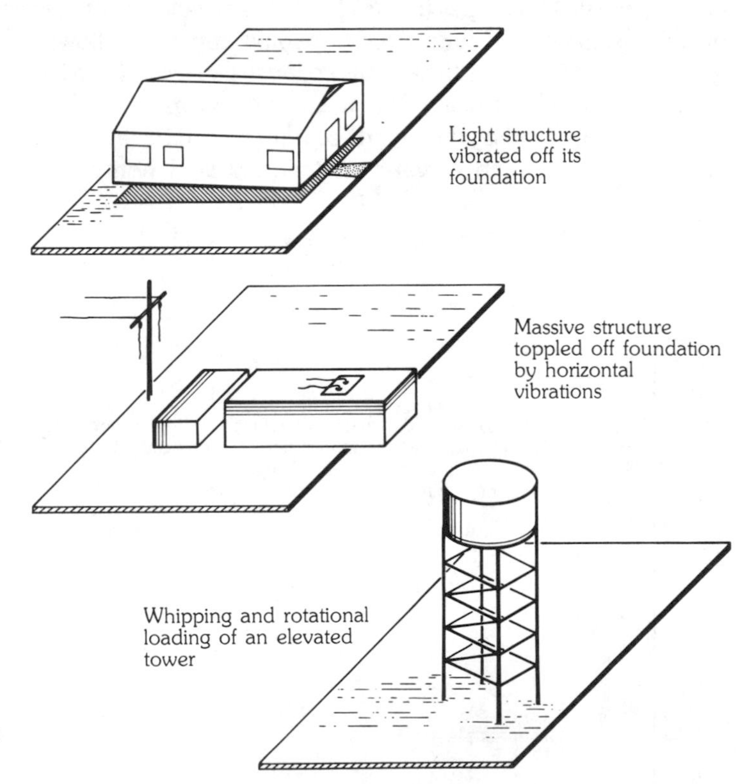

FIGURE 14–15 Structural loading on various structures during earthquakes.

The geologic rock substance and rock-mass characteristics of the foundation material for an engineering structure have a significant influence on the magnitude of the accelerations that the structure will experience. Earth accelerations transmitted into a structure will depend upon the coupling between the structure and the earth. If the coupling is very inefficient because its shear strength is low, very little building vibration will occur; however, the building may be relocated as it "walks" or twists on the surface. In contrast, an efficient coupling may transmit all accelerations into the building. Very heavy, massive structures, which have a high inertia, will attempt to remain stationary while light structures with a low inertia will follow the earth vibrations. Towers and other elevated masses will tend to "whip" under cyclic loads and may topple if they are deflected beyond their stable limits. Figure 14–15 shows earthquake loading systems diagrammatically.

FAULT OR TECTONIC CREEP

Movements along faults are not necessarily the result of an earthquake. Movement of a fault to release the strain as it develops under some regional stress is mechanically a creep type of movement. As long as there is no buildup of strain along the fault, strain can be gradually released, and no earthquake energy develops.

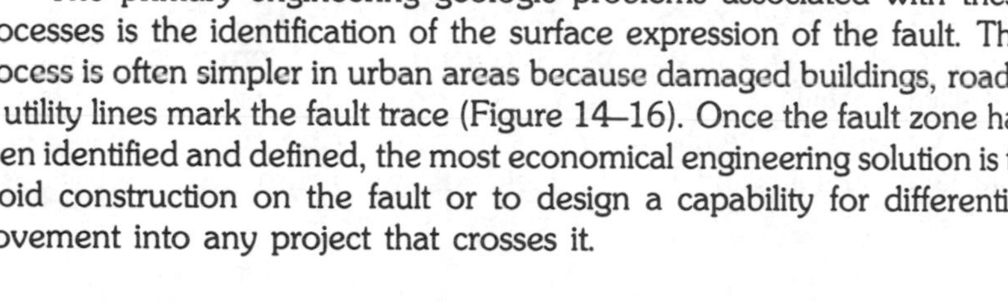

The primary engineering geologic problems associated with these processes is the identification of the surface expression of the fault. This process is often simpler in urban areas because damaged buildings, roads, or utility lines mark the fault trace (Figure 14–16). Once the fault zone has been identified and defined, the most economical engineering solution is to avoid construction on the fault or to design a capability for differential movement into any project that crosses it.

FIGURE 14–16 Structural damage caused by fault or tectonic creep.

VOLCANOES

Active volcanic processes are large-scale geologic features that can be easily identified. The determination of the level of risk from these features is, however, a more complex problem. For example, Mt. St. Helens, Washington, was considered to be active, but the magnitude of the eruption in the spring of 1980 was not expected. Because eruptions are irregular and intermittent, volcanoes are defined by their state or level of activity as **active**, **dormant**, or **extinct**. An active volcano has erupted during historic time (Mt. St. Helens), a dormant volcano shows geothermal indications of heat (Mt. Mazama, Crater Lake, Oregon), and an extinct volcano shows no measurable geothermal anomaly (Diamond Head, Hawaii). Much of the Cascade Range of California, Oregon, and Washington is dotted with active and dormant volcanoes, which represent a potential threat to the region. Of concern to the engineering geologist are the secondary processes that occur during and after an eruption.

Volcanoes are classified by their geomorphology and the characteristics of their eruption as **shield**, **composite**, and **dome** volcanoes. Shield volcanoes have broad, gentle slopes (**flow slopes**) and fluid magma (Figure 14–17). Composite volcanoes are steep-sided cones composed of loose **pyroclastics** (cinders, ash, volcanic bombs), and flow basalts formed by the eruption of "dry" magma (Figure 14–18). The slope of the cone is controlled by the angle of internal friction (ϕ) of the pyroclastics. Dome volcanoes are composed of a mixture of pyroclastics and felsitic lavas and are often formed by explosive eruptions (Figure 14–19). The geochemistry of the rocks in these three types also varies from a ferromagnesium-rich, low-silica, basaltic lava in a shield volcano to a high-silica, rhyolitic lava in dome volcanoes. The characteristics of the eruption range from generally quiet lava fountains and flows in shield volcanoes, through ash eruptions in

FIGURE 14–17 Shield volcano in Hawaii. Notice the flat flow slopes. The top photograph shows the entire volcano, and the bottom photograph shows the cauldron.

FIGURE 14–18 Cone volcano.

FIGURE 14–19 Volcanic dome, Mt. St. Helens, Washington, shown prior to the 1980 eruption. (Photo courtesy of U.S. Geological Survey.)

composite volcanoes, to explosive eruptions that can move at rates as high as 100 km/hr.

Volcanic flows fall into two general classes: **fluid flows** and **ash flows**. Fluid flows are usually called **lava flows**. Regardless of the type of flow, they follow the basic physical rules of a fluid. That is, they move downslope under the influence of gravity and therefore follow topographic lows. As a flow cools, it becomes more viscous and may build levees along its sides. The change in the viscosity of lava flows is readily apparent in the character of the flow surface. **Pahoehoe** is a smooth, ropey surface that forms while the flow mass is hot. Farther from the source this smooth surface changes to **aa**, which is very coarse, blocky, and angular (Figure 14–20).

Attempts to control a lava flow by directing it away from developed areas or by stopping its advance have had limited success. Extensive international support and a large supply of cooling water succeeded in redirecting the lava flow away from the harbor on Heimaey Island, Iceland. In most cases, the cost of attempting to control the flow will exceed the

FIGURE 14–20 Characteristics of lava. Pahoehoe (top) is massive while aa (bottom) is broken and rough. These two rocks have significantly different engineering properties.

benefits. The engineering geologist should consider the possibility of a volcanic eruption and the probable path of a flow, should one develop, when siting critical installations in active volcanic areas.

In addition to considerations of the risk and flow path of lavas or ash, the engineering geologist must consider the secondary processes, mudflows, other slope movements, and flooding due to snow melt. The heat generated during an eruption can rapidly melt the snow pack on a volcanic mountain. This sudden supply of water frequently causes massive mudflows that may do more damage than the lava flow.

The assessment of the risk of these secondary processes is based first on the risk of a volcanic eruption and then on the physical characteristics of the site. Are the conditions favorable for mudflows or other slope movements? Is a source of water available? Each risk analysis is, therefore, unique to the volcanic region being evaluated.

CASE IN POINT

Lahars (Mudflows) Volcanic Hazards, Mt. St. Helens

Earthquake swarms up to magnitude 4.2, beginning on March 20 and continuing through the first eruption of Mt. St. Helens Volcano on March 27, 1980, seem to have been the most credible predictive signs of an impending eruption. The original crater, about 75 m in diameter and 45 m deep, expanded to about 520 m in diameter and 260 m deep within days of the original eruption. Tephra from the first 5.2-km elevation plume was carried eastward for a distance probably not much greater than 8 to 16 km. *Tephra* is fragmental volcanic material blown from the source vent and transported by air currents. The resulting deposits are generally called *ash fall.*

Nearly continuous earthquakes, which increased to 4.8 in magnitude during the next 1½ months, created harmonic tremors and signaled the rising of gases—and, most likely, magma—within Mt. St. Helens. The first eruptive phase culiminated in a phreatic eruption on May 18, 1980, that had an explosive force equal to 10 megatons and drove greater than 1 km^3 of pumaceous tephra to an elevation of 19.2 km above the volcano. The finest-sized tephra were viewed as a haze in South Dakota on May 20. Eventually ash reached the U.S. Atlantic coast (at altitudes about 7 km), about five days after the May 18 eruption. The explosion formed a crater approximately 3.2 km long, 1.6 km wide, and greater than 1.6 km deep. The volcano's original peak elevation of 2.95 km was decreased to about 2.56 km at the crater's southwest rim. The crater area may actually be called a *caldera* because ash venting continues within and because of the length of the magma dome-emplacement phase of activity.

Mass movements on Mt. St. Helens began prior to the initial tephra eruption of March 27 with snow avalanches, generally triggered by earthquakes, sliding to near timberline elevation (about 1340 m). Beginning about March 27 and continuing through the May 18 tephra explosion, mountainslope mudflows occurred, especially on the northern slopes. Some of the mudflows resulted from an increase in slope gradient, or oversteepening of the northern flank of the volcano, near a topographic expression locally called "The Bulge." In the area of "The Bulge," internal volcanic forces were causing a pressure ridge near the mountain summit which moved outward at an approximate rate of 1.2 m per day preceding the May 18 eruption.

A hazardous geologic occurrence resulting from the May 18, 1980, eruption was the *lahar* (mudflow) that flowed down the westward-draining valley of the Toutle River and into the Cowlitz and Columbia rivers (see map). The lahar formed as hot, dry pyroclastic debris generated by the initial base surge from the source vent, located on the northern flank of the mountain, mixed with river water and glacier snowmelt. The terms *lahar* and *debris earthflow* can probably be used synonomously to describe the flow in the lower sections of the Toutle River. During the eruption, mudflows also occurred in the southeast-flowing drainage of the Muddy River and extended to near the Lewis River. The average grain size of the ash fall close to the mountain was probably that of gravel with abundant cobbles and boulders of pumice.

The lahar included runoff and snowmelt of the Toutle River drainage system, mixed with the Mt. St. Helens tephra. The tephra's density was

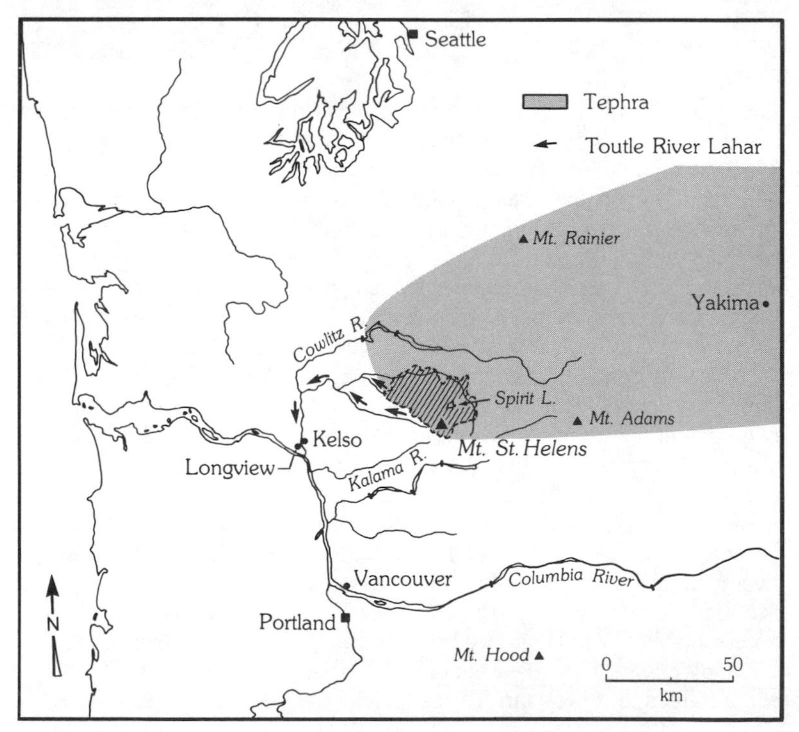

approximately 1450 to 1550 kg/cm^3. The mudflow in the Toutle River valley was steaming as the flow's frontal lobe passed underneath a highway bridge, about 20 km from the source vent. Approximately 390 km^2 of timber (some trees older than 100 years) were blown flat by the May 18 base surge blast of tephra and hot volcanic gases. Many trees, stripped of all their limbs, were carried along the upper portions of the lahar as it advanced downstream. In places, the trees appeared to be floating on the mudflow in laminar fashion.

Existing fluvial deposits within the Toutle valley were burrowed into and undercut in places by the advancing lahar. However, the lahar was generally deposited to a thickness of 75 m above existing stream deposit material. Silt- to cobble-sized lahar material was deposited in the Cowlitz River near Castle Rock, Washington, at the mouth of the Toutle River; in the Columbia River near Longview, Washington, at the mouth of the Cowlitz; and probably near the mouth of the Columbia at the Pacific Ocean (see map). The total river distance traveled by the ash of the May 18 lahar was at least 200 km.

Following the May 18 eruption, approximately 10.7 million m^3 of the lahar were reported to have been deposited near the mouth of the Cowlitz River and in the Columbia River near Longview, Washington. With deposition, water depth in the Columbia River rapidly decreased to about 4.3 m, isolating approximately 70 ships due to a 7-m ship draft limitation. The shipping channel is normally 183 m wide and 12.2 m deep. At least three shipping ports were temporarily abandoned, also due to ship drafts more shallow than the depth required for navigation. The deposit within the river appears to be in the medium sand-size range and is composed of juvenile ash, blown-down trees, pumaceous lapilli-breccia, and intermixed fluvial mud. The Columbia River may have absorbed as much as 38.2 million m^3 of the mudflow. About 8.9 river kilometers of dredging were necessary to open the Columbia River completely for ship navigation. Dredged material was placed into existing holes on the bottom of the Columbia. The U.S. Army Corps of Engineers was responsible for dredging the material, at a reported total estimated cost of at least $200 million. The Cowlitz River Salmon Hatchery, one of the largest in the world, had to be closed indefinitely, necessitating the truck hauling of about 8 million stranded yearling salmon to other hatcheries along the Cowlitz River system. Adult salmon did not take their normal migration routes up the Cowlitz River, but instead diverted to the Kalama River.

Total regional damage from the catastrophic May 18 blast has been estimated to be over $2 billion, including lost timber and crops and cleanup downwind from the volcano. Costs to the Portland, Oregon, commercial port trading was estimated to be $1 to 2 million per month until a safe 12-m channel depth could be reached by dredging. The flood carrying capacity of the Columbia River was also reduced by the lahar deposits. Water filtration plants at several towns along the Cowlitz River, including Toutle and Castle Rock, were essentially destroyed by the lahar.

The eruptive activity of Mt. St. Helens has provided dynamic illustrations of volcanic hazards, which have been documented in detail for the first time in the continental United States. Of the thirty dead and approximately thirty-five persons missing due to the May 18 eruption and subsequent lahar within the Toutle River valley, most were killed by the heat and velocity of the pyroclastic direct blast. Major volcanic hazards observed in the Mt. St. Helens eruptive activity include heat and suffocating ash from the pyroclastic flow and the volumes of tephra emitted; Columbia River navigation impairment; potable water intake clogging; timber industry losses; ashfall deposit damage to residences; severe auto and truck transportation problems; fish and wildlife destruction; and recreation area lossses.

Written by Richard C. Kent of Kent Associates, Geologic Consultants, Lake Oswego, Oregon. Photo courtesy of U.S. Forest Service.

Ice Processes

fifteen

A significant portion of the earth's surface is affected in some manner by the simple process of water freezing. Recent development of energy reserves in the Arctic regions of North America has brought modern technology into a generally "new" environment. The engineering geologist is involved in the evaluation of a wide range of problems caused by ice and ice processes. These problems range from glacial processes to frost.

Active glacial processes are limited to polar regions and to high mountains. The geomorphic results of past glaciation, however, are visible throughout much of Canada, the northern United States, and the eastern and western mountains of the United States. Glaciers occur in two basic forms: **continental** and **alpine.** Continental glaciers, on Greenland and Antarctica, cover large areas. Alpine glaciers are presently confined to high-altitude mountain valleys at middle to higher latitudes.

Most engineering geologists will not become involved in the analysis and evaluation of active glacial processes unless they work on engineering projects in the Antarctic or Greenland. In these areas, the only available earth material for a foundation is glacial ice and snow. Problems such as ice movements, settlement, and squeezing due to loading must be evaluated.

Near-active alpine glaciers in the mountainous northern latitudes and the mechanical properties and geological significance of glacially produced sediments and seasonal floods must be considered. In the higher latitude oceans the effects of **sea ice** and **icebergs** can be significant factors in site selection.

FROZEN GROUND

The most significant ice-related geological processes are associated with frozen ground. Frozen ground is any soil or rock where the temperature is at or below 0°C through any two-year period. Frozen ground extends over a wide area of the northern latitudes (Figure 15–1). In **seasonally frozen**

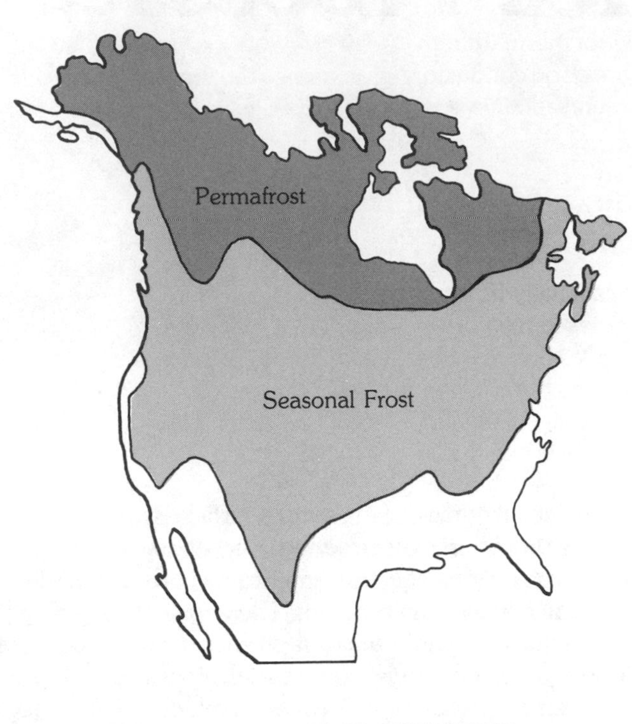

FIGURE 15–1 Frost and permafrost regions of North America.

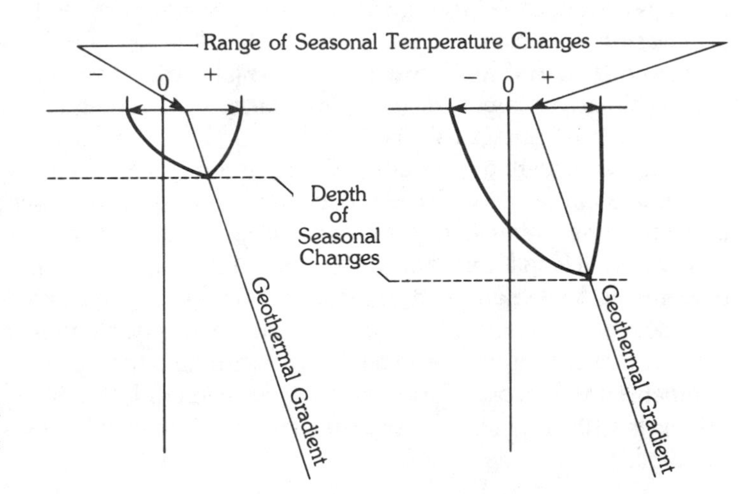

FIGURE 15–2 Schematic temperature-depth profiles showing the influence of the surface conductivity.

ground the ground is at or below freezing for at least part of the year for two years. Permanently frozen ground is **permafrost.** *Notice that the classification of frozen ground does not require that ice or water be involved in the freezing.*

The freezing cycle in the ground is a thermal balance between the heat flow from the center of the earth and the rate at which heat is removed by the air at the land-air interface. The thermal characteristics of the surface have a significant impact on the depth of the seasonal freezing. Surfaces that are poor thermal conductors will retard freezing and thawing, while surfaces that are good conductors will enhance the thermal processes. Figure 15–2 shows representative temperature profiles in frozen soil areas.

SEASONALLY FROZEN GROUND

Seasonally frozen ground includes the materials that undergo freeze-thaw cycles in middle-latitude or colder regions and the upper, **active layer,** in Arctic climates. Materials in the temperate regions freeze from the surface down, but thaw both from the surface and from below. In permafrost regions, the material freezes from the surface downward and the permafrost upward, but thaws only from the surface downward.

Freezing of dry ground will have little effect on the properties of the earth materials, because no ice can form. In wet ground, however, the formation and subsequent melting of ice has a significant influence on the mechanical properties of the ground.

Ice or **frost** problems increase with additional freeze-thaw cycles each year. When the pore water is frozen, the ground has a high shear strength. After the pore water has melted but before it can drain away, however, fine-grained materials may be very weak or even behave as a fluid. In areas of numerous freeze-thaw cycles, the repeated change in the material properties can cause serious deterioration of highways and shallow foundations. Spring melt periods are more hazardous to roadways and shallow foundations because the frozen ground below the melted zone is effectively impermeable, thereby prohibiting the drainage of meltwater.

In the ice-related process of **frost heave**, portions of a frozen layer swell, sometimes by as much as 30 percent of the thickness of the layer. Frost heaving most frequently occurs in silt-sized soil materials or between fractures in large rocks. As the pore water begins to freeze, additional water is supplied to the ice by capillary rise in the pores or through fracture permeability. The final result is a series of **ice lenses** scattered through the material. The ideal geologic conditions for ice lens formation are a supply of water, a fine-grained material with high absorption potential, and a high permeability to provide easy water migration. All three conditions fit silty (ML) soils, fine silty sands, and some clayey sands very well (Figure 15–3).

Since seasonal freeze-thaw cycles cannot be controlled, conditions that cause frost heave must be prevented or the freeze-thaw zone must be avoided. Frost heave is prevented by selection of materials that are not susceptible to the process and by designs that either keep or drain water away from the site (Figure 15–4). These techniques are applied to roadway, railroad, airport, and other projects where avoiding the frost zone is impossible or uneconomical. Projects with a limited aerial extent, such as buildings, frequently avoid the frost zone by placing foundations below the depth of freeze, and, if necessary, insulating the structure from the frost zone by adding fill between the foundation and the frost-susceptible materials.

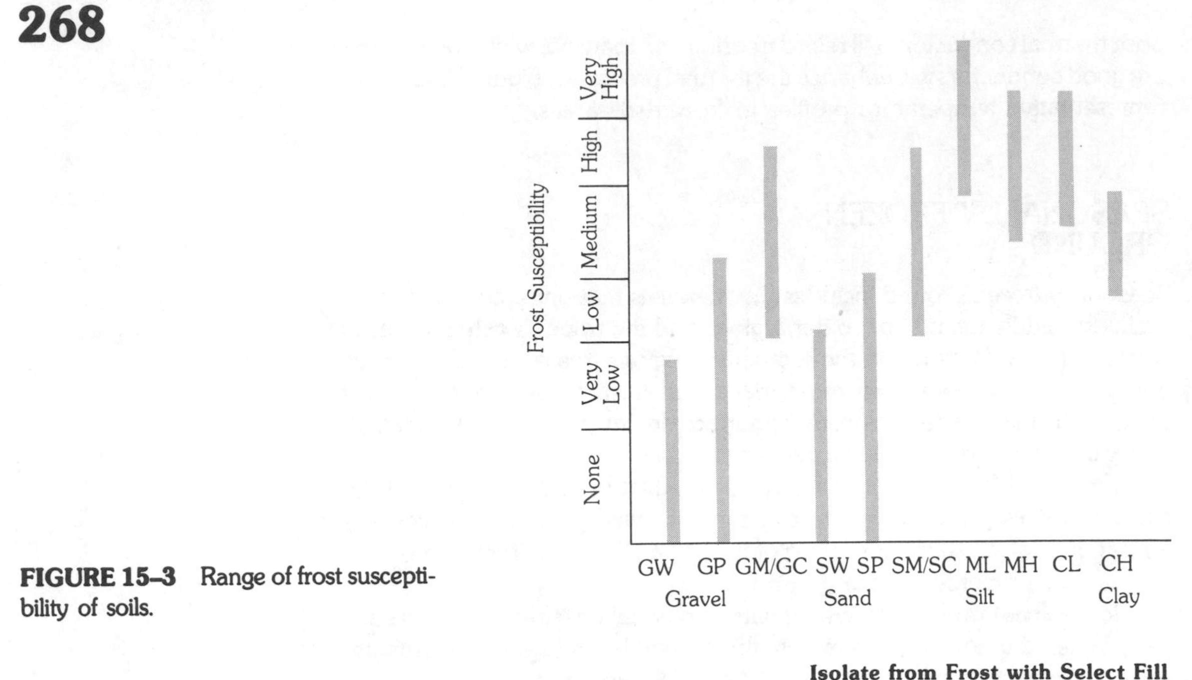

FIGURE 15–3 Range of frost susceptibility of soils.

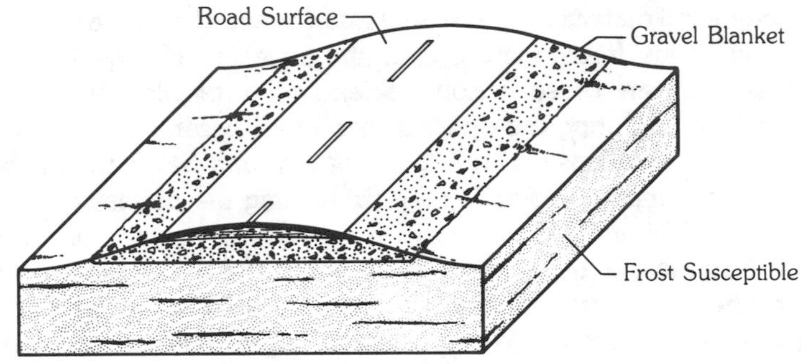

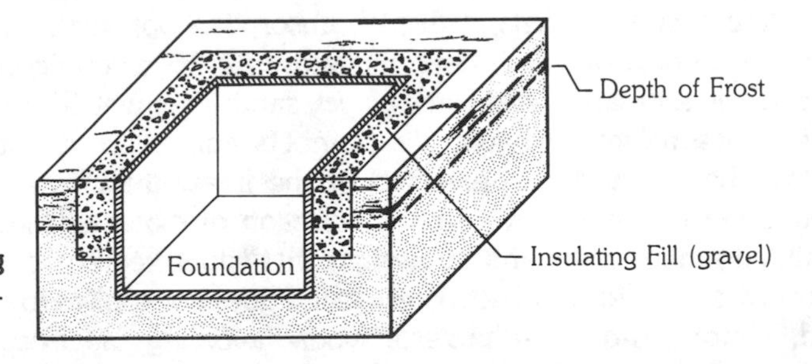

FIGURE 15–4 Common engineering solutions to construction in frost-susceptible areas.

PERMAFROST

Permafrost is permanently frozen ground that has been at or below freezing continuously for at least two years. The characteristics of permafrost are related to climatic (temperature) variations (Figure 15–5). Near the warmer

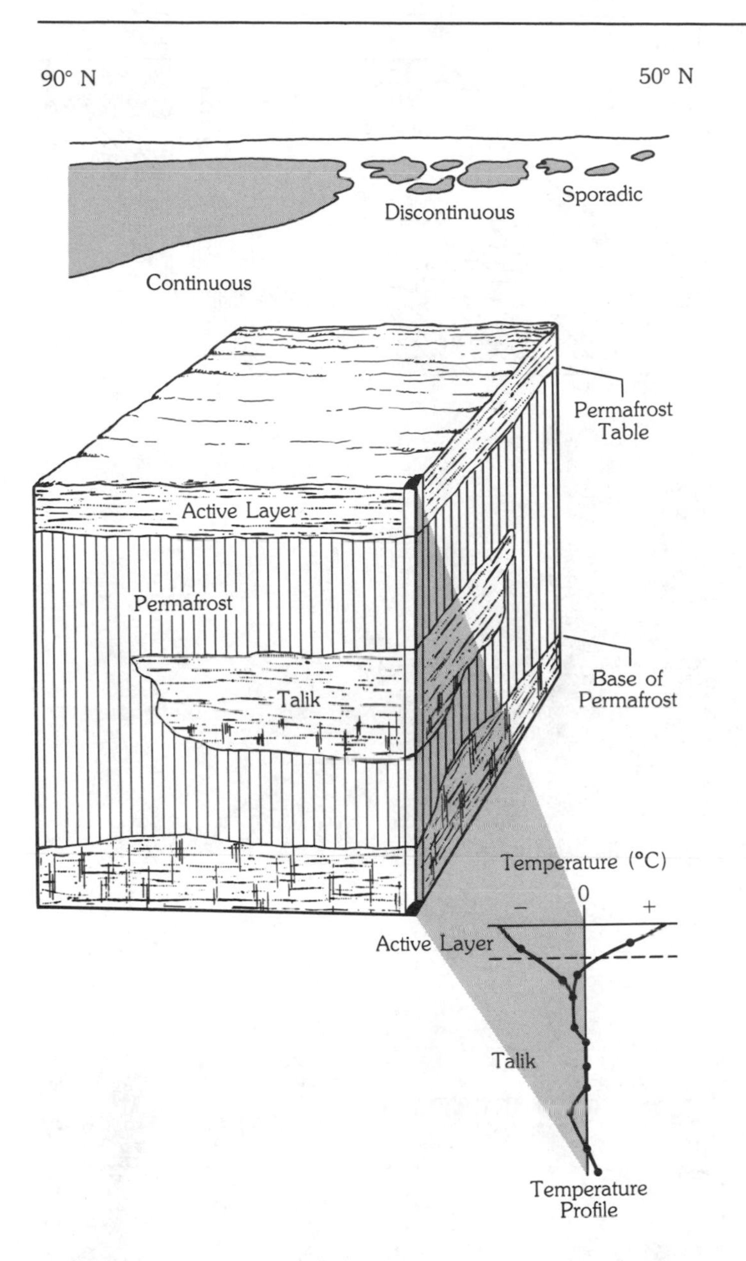

FIGURE 15–5 Cross section of permafrost terrain from south to north.

FIGURE 15–6 Layers in permafrost terrain.

southern limits of the region, the permafrost is **sporadic** and occurs in thin, isolated patches. Farther north, the permafrost is **discontinuous** with alternating layers of frozen and thawed ground. In areas of **continuous permafrost** all the ground is frozen. A vertical profile through a continuous permafrost area shows three primary layers: **active layer, permafrost layer,** and **talik** (Figure 15–6). The active layer undergoes seasonal freeze-thaw. Unfrozen ground is known as talik. The boundary between the active layer and the permafrost is the **permafrost table.** In areas of discontinuous and sporadic permafrost, the vertical section is more complex. A **frost table** may develop within the active layer during an

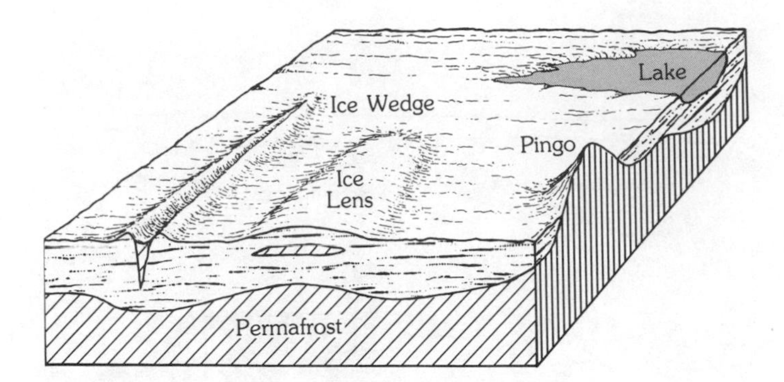

FIGURE 15–7 Common geomorphic features in permafrost terrain.

FIGURE 15–8 Ice lens exposed in the scarp of a rotational landslide. The ice lens (designated by arrow) is about 20 cm thick.

FIGURE 15–9 Pingo, or ice hill, projecting above the flat terrain. (Photo by P. Gretener, University of Calgary.)

FIGURE 15–10 Patterned ground.

FIGURE 15–11 Rock heave in the Arctic. These shales were originally dipping at about 15° until they were disturbed by frost-heave processes.

unusually cold summer. This temporarily frozen zone, however, will melt during an average year.

Numerous ice features that can be significant in their size and engineering problems develop in permafrost regions (Figure 15–7). **Ice wedges** and **ice lenses** that can exceed 500 km in magnitude can form if sufficient water is available (Figure 15–8). Ice hills, or **pingos,** reaching heights in excess of 100 m have been found. In fact, pingos form the only relief in much of the Mackenzie Delta, Yukon Territory, Canada (Figure 15–9). Lakes and rivers can have a profound impact on permafrost conditions. Below the lake bed and around the shore the presence of warmer water depresses the permafrost table.

The active layer is similar to the frost zone in nonpermafrost areas except that the permafrost table always prohibits downward drainage of meltwater. The seasonal freeze-thaw and resulting soil-moisture changes produce a highly active zone. Differences in thermal conductivity between rock and soil result in the selective heaving of the rock and the formation of patterned ground (Figure 15–10). In fractured or bedded rocks, the freeze-thaw cycles tend to turn the rocks and separate them along their bedding plane (Figure 15–11). Fine-grained soils are frequently unstable, even on flat slopes of 2°, so that there is seasonal downslope creep, or **solifluction** (Figure 15–12). Thin mudflows and slope failures occur when the saturated soil mass fails along or near the permafrost table (Figure 15–13).

During the freezing process, the active layer remains active. As the layer freezes from above and below, the saturated, thawed zone between the freezing points can undergo volume expansion as it freezes. On slopes,

FIGURE 15–12 Solifluction lobes. (Photo by L.D. Dyke.)

FIGURE 15–13 Mudflows on a 5° slope (designated by arrows). The mudflows occur in the active layer and are less than a meter thick.

meltwater migrates downslope and is trapped in this thawed zone, leading to significant frost heave and surface deformation. This process frequently leads to a **drunken forest** where trees point in all directions.

ENGINEERING GEOLOGY IN PERMAFROST REGIONS

Permafrost engineering must recognize the hydrologic conditions, soil and rock characteristics, and the thermal processes of Arctic regions. Many engineering projects are seasonal, but the preferred season for operations in some Arctic areas is late winter and early spring, while the active layer is still frozen. Thus working conditions may be severe—dark nearly twenty-four hours a day with a high temperature of −20°C.

Permafrost conditions can be divided into two basic types: **dry permafrost** and **saturated-ground permafrost.** In dry-permafrost areas, the ground contains little if any ice, and soil moisture-related problems do not occur. These sites generally require no special engineering attention. Areas of saturated-ground permafrost contain significant quantities of ice that, if melted, will lead to appreciable changes in the behavior of the earth materials. Engineering practices in these areas must consider the characteristics of the material in both frozen and thawed state. Saturated gravels have significantly different properties from saturated silts and clays.

At sites where a change in the permafrost table will cause significant changes in the behavior of the ground, the engineer has two basic choices: remove the undesirable material or protect the permafrost table. Removal of unsuitable material can be economically accomplished only in areas of thin, sporadic permafrost where suitable, coarse-grained fill is available. Permafrost protection must consider the hydrology, active-layer characteristics, and the permafrost. Hydrology is important because the freezing and thawing cycles of the soil water can produce significant frost heave and

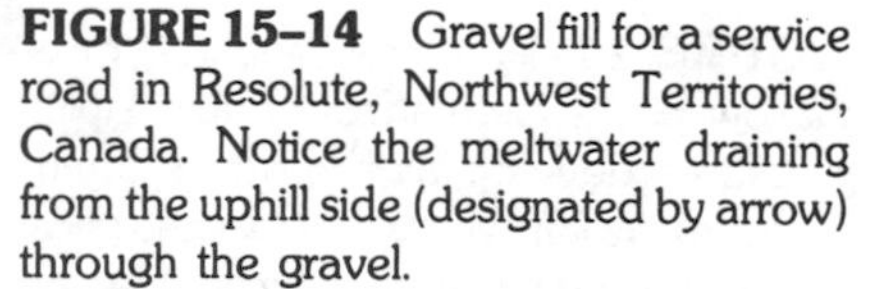

FIGURE 15–14 Gravel fill for a service road in Resolute, Northwest Territories, Canada. Notice the meltwater draining from the uphill side (designated by arrow) through the gravel.

differential settlement problems. The active layer must either be allowed to remain active or be permanently frozen. Differential thaw of the permafrost around piles or below footings can cause a loss of support.

The permafrost can be protected by the use of an insulating fill or piles. For fill a blanket of coarse sand or gravel, several meters thick, is placed directly on the ground surface (Figure 15–14). Pile construction uses a wood or other non-heat-conducting post to support the structure above the surface of the ground. The space between the ground and the floor of the structure allows the thermal processes to continue in the active layer (Figure 15–15).

Each of these common construction methods has advantages and disadvantages. The fill method, commonly used for transportation routes, can form a dam that retards surface runoff, and frost heave and the risk of flooding become problems (Figure 15–16). In addition, differential

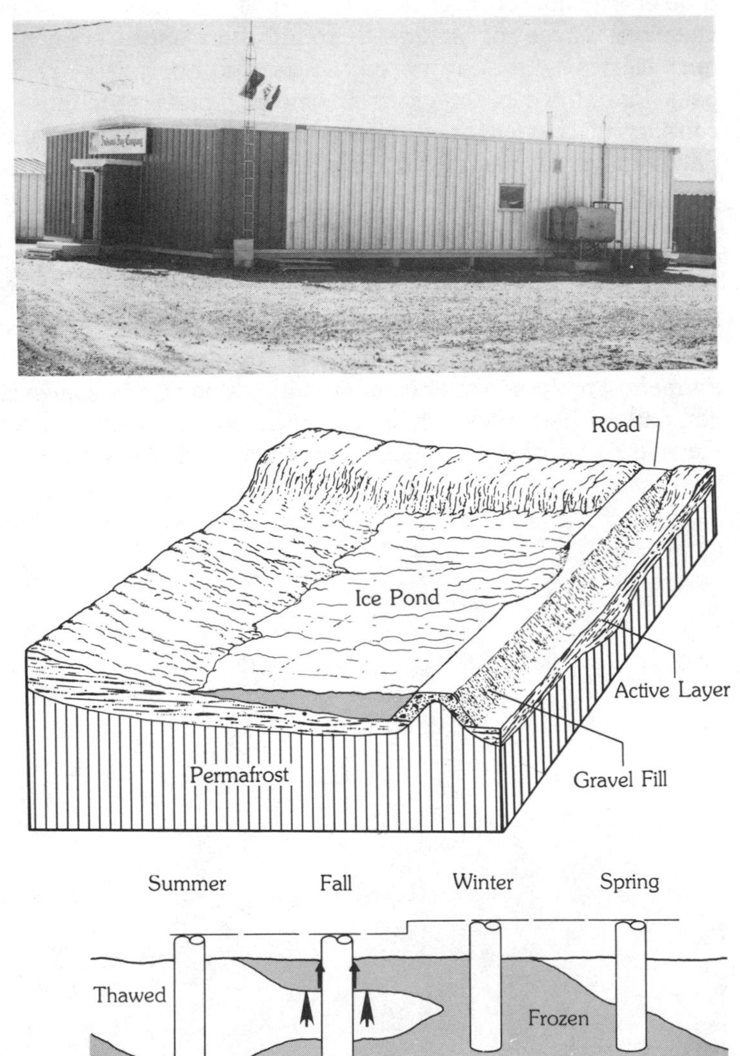

FIGURE 15–15 Building constructed on piles to protect the permafrost.

FIGURE 15–16 Ice dam developed in gravel fill under a road because of the rise in the permafrost table.

FIGURE 15–17 Frost jacking of a shallow pile in permafrost terrain. Uplift pressures develop as the active layer freezes in the fall, where the frozen surface layer grabs the pile and is forced upward during the freezing of the middle section.

CASE IN POINT

When Three Poles Replace One

The first telegraph and telephone lines strung in the Arctic followed the gold seekers. Problems developed as soon as the line reached the permafrost areas. Each spring found many poles down, and extensive maintenance was required. In order to have a more reliable system, the single pole was replaced with three poles that form a tripod, so that frost jacking could not remove the wire support system.

Written by L. D. Dyke, Engineering Geologist, Geological Survey of Canada, Ottawa, Canada.

subsidence of the fill may occur, as the active layer compacts, freezes, and thaws. Pile construction must consider the uplift forces that develop as the active layer freezes (Figure 15–17). Frost heave can easily remove shallow piles, point bearing on the permafrost table, and cause deep piles frozen into the permafrost to fail in tension. Piles should be insulated from the active layer materials to prevent frost heave. Any heated space in permafrost areas must be constructed with floor insulation to prevent melting the permafrost and to conserve energy. Variations on these two concepts include such designs as horizontal pipes set through a gravel pad and thermal piles that actively freeze the permafrost as were used along parts of the Alaska Pipeline.

Subsurface Openings

sixteen

Natural or undetected man-made subsurface openings can have a significant impact on any engineering project, including foundation instability, groundwater flow, or collapse problems. Limestone ($CaCO_3$) is soluble and is susceptible to the formation of **solution cavities** or **caves** (Figure 16–1). Volcanic terrain can contain **lava tubes** where basalt flows out from below solidified lava (Figure 16–2). Mining activities have in the past left behind unmapped and essentially unknown subsurface openings in many types of rock material (Figure 16–3).

The engineering geologist also plays an important role in the exploration, design, and construction of subsurface openings. The basic philosophy of design varies with the ultimate objective of the opening. In civil engineering projects, the opening is intended to serve some purpose: tunnel, storage, or shelter, and the rock or soil removed is a waste product. Mining engineering projects, however, leave the opening as a waste product and utilize the excavated material as a product: coal, copper, silver, or whatever. Roof support and the design life of the opening reflect these basic differences.

NATURAL OPENINGS

LIMESTONE SOLUTION CAVITIES

Limestone solution cavities should be expected in any limestone terrain until proven otherwise. Since limestone is an organic sedimentary rock composed of calcium carbonate, dissolution can occur where groundwater moves along fractures. Variations in the chemistry and environment of deposition of the limestone and in fracture density can have a profound influence on the solubility of the rock. Naturally occurring acidic rainwater,

FIGURE 16–1 Cave in limestone.

the most common dissolving agent, reacts with the limestone in a simple reaction:

$$H_2O + CO_2 \rightleftharpoons H_2CO_3 \rightleftharpoons H^+ + HCO_3^-$$
$$2H^+ + CaCO_3 \rightleftharpoons Ca^+ + 2HCO_3^-$$

This reaction continues as long as fresh, acidic rainwater moves through the system. As fractures and soluble areas enlarge, groundwater flow velocities increase, and clastic sediments can be carried into the openings and deposited. The combination of differential dissolution of the limestone and sediment transported into openings produces a unique subsurface geology (Figure 16–4). In some cases the opening is protected from sediment accumulation by fine fractures or other "filtering mechanisms" and is preserved as a cave.

Frequently, a unique surface drainage pattern will form in limestone terrain. The high permeability of fractures and solution cavities allows large quantities of water to flow as groundwater. Whenever one of these fractures intersects the surface, surface water may disappear underground or groundwater may appear at a spring. Limestone regions are frequently **internally drained,** and surface streams are small or nonexistent (Figure 16–5). The hydrogeology of limestone aquifers is controlled by the fractures in the limestone and not by the pore spaces in the rock. Very high flow velocities and discharge rates can exist in the limited area of a fracture with impermeable conditions between fractures.

FIGURE 16–2 Lava tubes.

If the regional water table is lowered by either natural or artificial means, active dissolution of the limestone ceases but the increased gradient of water moving downward or through sediment-filled voids can erode the fill. Loss of the supporting fill or excessive erosion along fractures can result in the collapse of the limestone. If this collapse reaches the surface, **sinkholes** form. Excessive surface collapse produces a land surface resembling a "bombing range," called **karst topography.**

Engineering geology in limestone terrain must always anticipate the existence of caves and other solution features. In areas where the limestone is near the surface, the irregular soil-rock contact produced by differential

FIGURE 16–3 Ground surface destruction caused by subsidence into an abandoned underground mine.

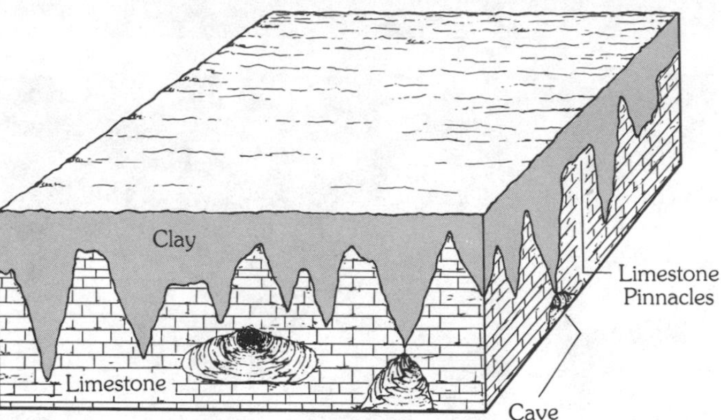

FIGURE 16–4 Subsurface conditions that can be formed in limestone terrain.

CASE IN POINT

The Disappearing Parking Lot

A large manufacturing corporation desired to establish a small plant in the East. Studies were conducted to determine the optimum location for this plant, so that it could serve to distribute expendable portions of their product to a large market. The usual variables of labor, taxes, utilities, and transportation were evaluated. An area was selected, options were taken on properties, a final analysis made, and a site purchased. The property was between a major highway and a railroad. It was just outside a small but adequate community. Labor was available, and the people were industrious. The community agreed to annex the site and make water and sewer connections available. An architect was engaged. Exploratory drilling disclosed an irregular limestone rock surface and some bedrock solution cavities.

Problems developed early when grading to level the building area disclosed a fluted, grooved, pinnacled surface that made work difficult and costly. Furthermore, portions of the area were found to be underlain by a zone of huge, loose, semidetached boulders in such a state that the placement of fill could not be satisfactorily accomplished. It was essential to have a uniform, well-compacted soil base, as the building would have a slab foundation on soil. It was decided to remove as much of the soft soil from around the rocks as possible, and to pour concrete in the pockets and voids to stabilize the mass and make a more uniform surface on which to construct. Nearly 3000 yd^3 of concrete were poured, and the project cost was increased by an estimated $100,000 from the originally planned $2,500,000. The severity of the limestone problem resulted in increased cost and unexpected delays in completing the facility.

Some months after the plant became operational, following a few days of heavy rain, sinkholes began appearing in several locations. Some were in the lawn where only the aesthetics were disturbed while others occurred in the parking lot and the access road to the loading dock. Finally, one developed beneath the fire protection water main that encircled the building. In all, more than twenty sinkholes appeared in the twenty-acre plot. Most were only a few feet in diameter at the surface, but on probing and excavation, many were found to be much larger farther down. All were simply voids in the soil overburden caused by migration of material into the cavernous bedrock. Nearly all the sinks were adjacent to clusters of dry wells constructed to dispose of the storm water runoff from the building roof and the parking area.

A comprehensive study of the problem showed that the plant occupied twenty acres of the lowest part of a forty-seven-acre depression that had no surface outlet for runoff. Prior to construction, rainwater had accumulated at low points and been absorbed through the soil into the bedrock cavities. The new facilities had placed nearly half the site under roof or pavement, thus eliminating this portion of the tract from the natural process. The preliminary study indicated that concentrated storm-water runoff, introduced by dry wells into small underground areas, was carrying overburden into bedrock cavities and causing sinkhole development at an accelerated rate. The possibility of structural damage to the plant or its appurtenances appeared possible.

Several schemes to correct the problem was considered, including repairing each new sink as it appeared. Because property damage seemed imminent, however, proposals to reduce or eliminate further development were favored. Most promising were those schemes that would collect all possible runoff from the site and conduct it to the nearest stream, thus reducing concentrated flow downward into the rock cavities. Cost estimates for the design, purchase of rights-of-way, and the construction of the least expensive of these plans were nearly $1,000,000. This bitter pill was compounded by the possible preclusion of future expansion of the facility.

Reprinted from F. J. Knight, Geologic problems of urban growth in limestone terrains of Pennsylvania, *Bulletin of the Association of Engineering Geologists* 8 (1971): 94–95. Used by permission.

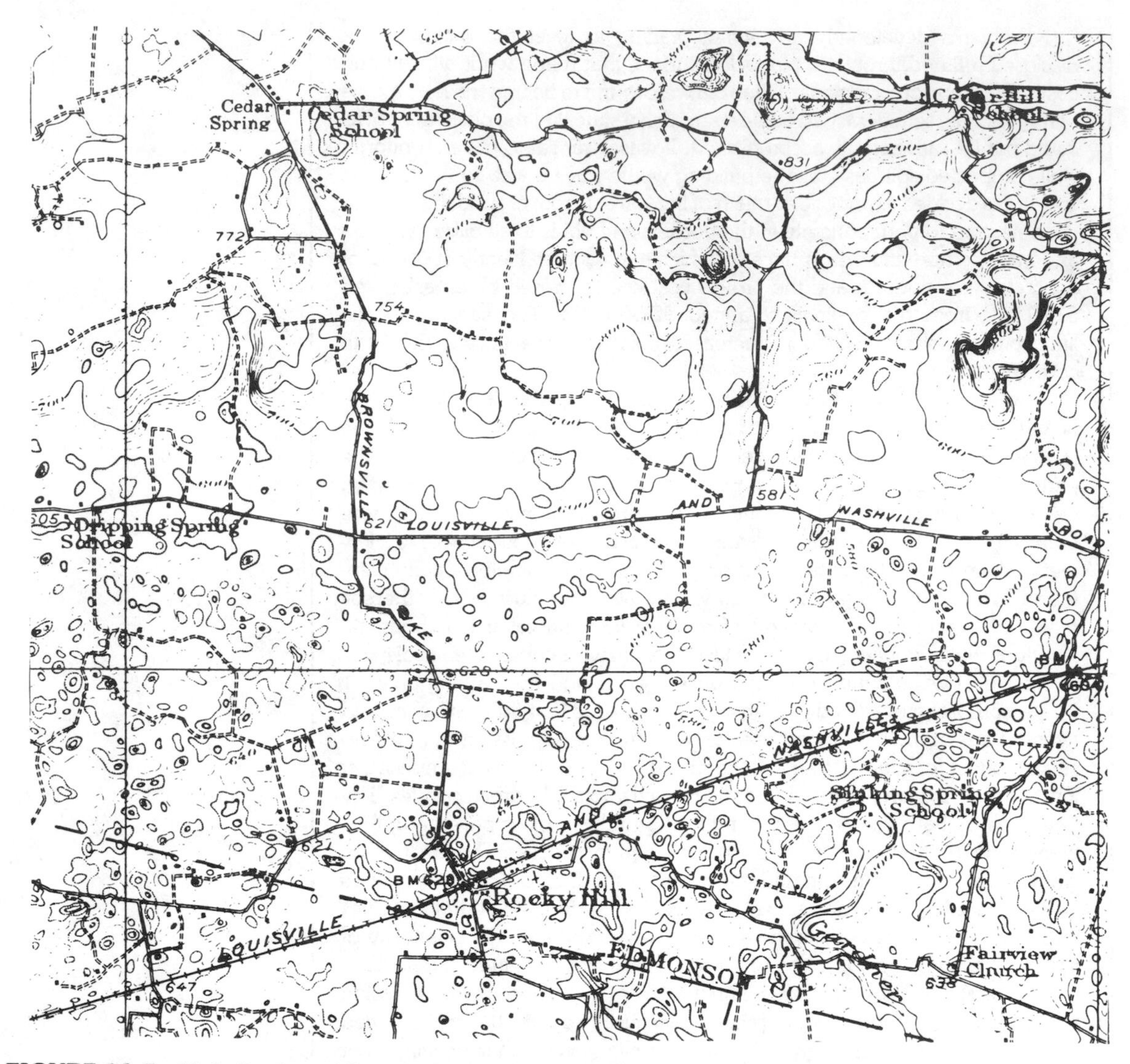

FIGURE 16–5 U. S. Geological Survey topographic map of limestone terrain showing the lack of surface drainage because of internal drainage in the limestone along the lower portion of the map.

dissolution produces difficult excavation, foundation settlement, and bearing capacity problems. Buried, sediment-filled openings may give the impression of "sound bedrock" that "mysteriously" fails when subjected to a load or exposed in a roadcut. Erosion of sediment fill from voids through groundwater dewatering operations can also lead to collapse. Because groundwater conditions in limestone terrain are controlled by the fracture and void geometry, proper care must be exercised in any subsurface operation to recognize and prevent flooding by, or the pollution of, groundwater. Surface drainage of polluted water from construction or other operations can rapidly enter the groundwater system through sinks. The shallow topographic expression of many sinks makes them difficult to identify without surveying. As a result, the uninformed builder or contractor can suffer serious flood losses either by direct runoff flooding or through subsurface channelization of runoff through fractures.

LAVA TUBES

Lava tubes are long, thin, sinuous to irregular-shaped tubes in a flow basalt. They form when lava escapes from below a solidified crust and are similar to solution openings in limestone in that they are potential collapse sites, irregular in their location in a flow, and potential groundwater channels.

ARTIFICIAL OPENINGS

TUNNELS

People's use of underground space dates back to prehistoric times when the first caves were inhabited. Since then, people have excavated underground space when necessary, beginning about 3000 B.C. for religious purposes. The first true tunnel, excavated in Babylon around 2180 B.C. to connect the palace with the temple, was about 3.5 × 4.5 m in size, stretched about 914 m, and passed under the Euphrates River. Since that time, tunnels have been constructed for religious, military, mining, water-supply, canal, railroad, and highway uses (Figure 16–6). Other underground projects include the excavation of temples and burials, storage areas, machinery spaces, military defenses, and mines.

FIGURE 16–6 Highway tunnels driven through igneous rocks.

Tunneling and other underground excavation operations are all basically similar, the difference being the shape and size of the opening and the corresponding stresses in the surrounding rock. Underground support, or **roof support,** is a primary concern in any underground excavation. Two basic conditions exist in the subsurface: **soft ground**, where continuous support is required; and **hard ground**, where the rock can support the load until supplemental support can be placed. Soft-ground conditions include raveling, running, flowing, squeezing, and swelling ground. Hard-ground conditions include massive, intact, moderately fractured, blocky, and seamy conditions. In subsurface operations, the rock strength and rock-mass properties are equally important to the success of the project. Table 16–1 shows classification of tunnel conditions for purposes of engineering geology.

TABLE 16–1 Tunnel condition classification

Classification	Description	Geologic Characteristics
1	Hard and intact	Massive igneous, metamorphic rocks, very well cemented sandstone, massive limestone
2	Hard stratified or schistose	Massive schist, gneiss, slate, hard sandstone, shale
3	Massive, moderately fractured or very firm ground in soft rock	Widely spaced fractures in hard rocks, sands, and gravels in soft rock
4	Moderately blocky and seamy or firm ground in soft rock	Moderately spaced fractures, seams filled with clays
5	Very blocky and seamy or potentially raveling ground in soft rock	Closely spaced fractures and seams
6	Crushed rock, chemically intact, potentially running or flowing ground	Shattered rock that is not altered
7	Squeezing rock (moderate depth)	Shattered rock with clay alteration, silts and clay shales
8	Squeezing rock (great depth)	Shattered rock with alteration
9	Swelling rock	Shattered rock extremely altered, high plastic clays

Source: After R. J. Proctor, Mapping geologic conditions in tunnels, *Bulletin of the Association of Engineering Geologists* 8 (1971):1–43. Used by permission.

FIGURE 16–7 Conventional tunnel driving method.

Tunnel construction methods are controlled by the geology and hydrogeology of the site and the method of excavation. In many cases, the geology along the route of the tunnel varies, and more than one method will be required. Tunneling methods fall into two basic groups: **conventional** methods (Figure 16–7), where drilling and blasting operations are used to break up the rock; and **boring** or **continuous excavation** methods, where the rock material is excavated by a machine without blasting. Tunnel boring methods include two types: **shield** methods, where a shield is forced ahead of the working face, thereby providing continuous roof support, and actual **boring** methods, where the boring machine cuts into the working face. Selection of the tunneling method is controlled by the rock strength and rock-mass properties (Table 16–2).

Conventional tunneling methods include **full face, heading and bench, top heading and bench, top heading,** and **multiple heading** methods (Figure 16–8). Tunnel support is related to the excavation method utilized and the geology of the site. Boring and shield methods frequently carry tunnel support along with the machine and therefore save the costs of

TABLE 16–2 Tunneling methods and geologic conditions

Shield—soft rock, must be able to push the shield ahead of the working face

Boring—medium strength rock units

Conventional—high and very high strength rock, complex structures and fracture density to massive.

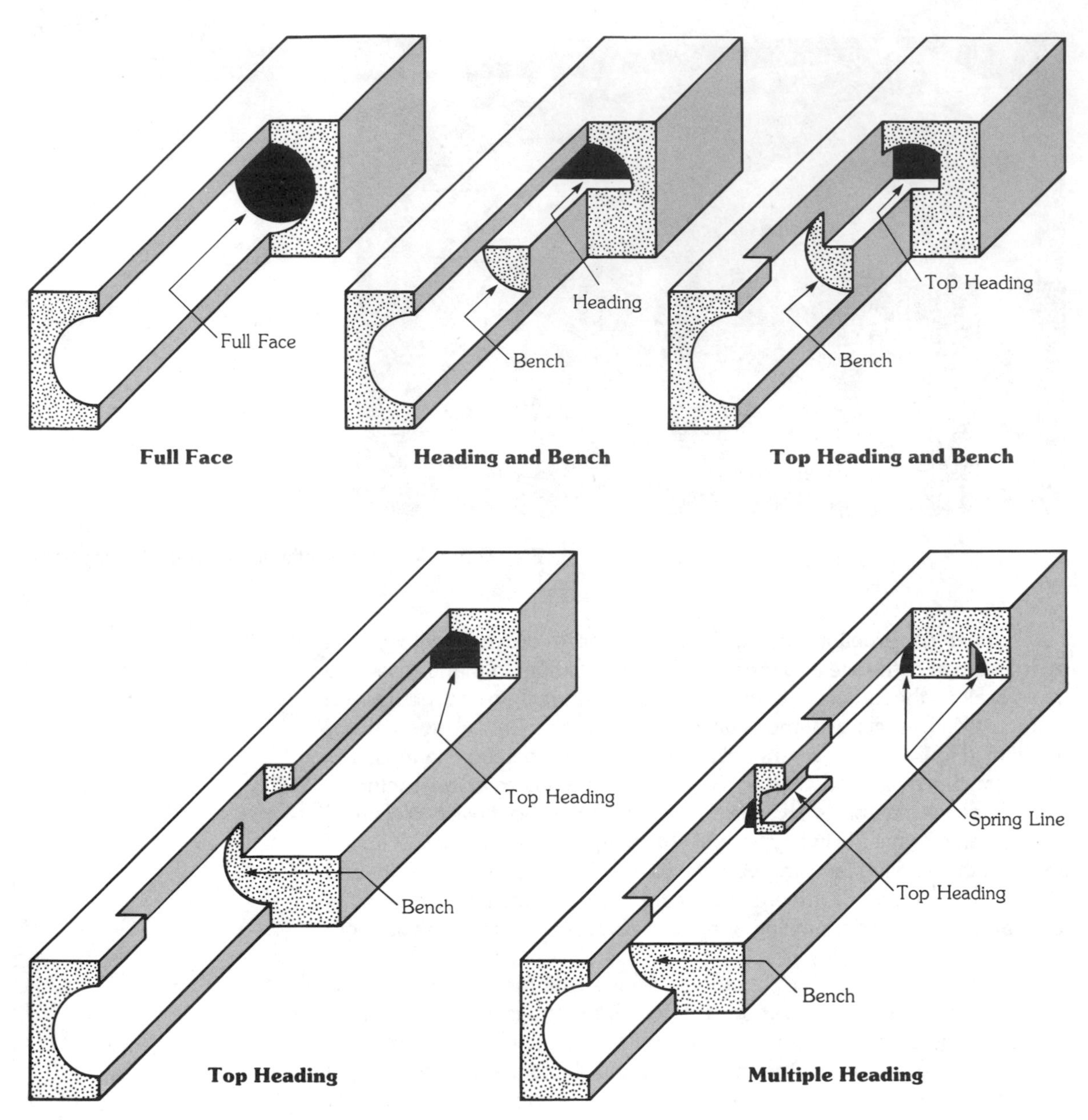

FIGURE 16–8 Common conventional tunnel driving methods.

temporary supports. Tunnel support in conventional methods falls into two basic types: **active support,** where the supporting member is strained to the rock, and **passive support,** where the rock must strain to the supporting member before any support is provided (Figure 16–9). Both types of support are frequently employed in the same tunnel, particularly when a finish lining is planned.

FIGURE 16–9 Underground support: active support (rock bolt) and passive support (timber or steel arches).

UNDERGROUND ROOMS

Underground room construction must consider the strength characteristics of the pillars and floor as well as the roof. As the size of the room is increased, the load carried by the walls or pillars and the load transferred to the floor rocks are also increased. Pillar performance is controlled by the rock strength and mass properties of the pillar and is related to their compressive

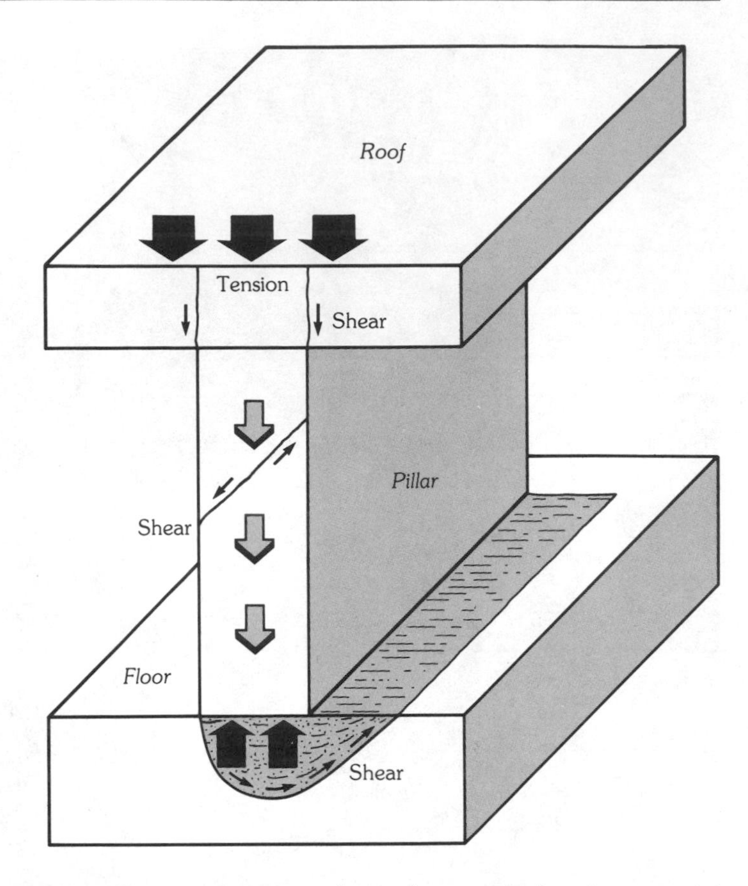

FIGURE 16–10 Rock stresses developed around a pillar in an underground room. The roof rock must resist bending and shear, the pillar must resist shear, and the floor must bear the pillar and roof load.

strength properties. Floor performance depends upon the bearing capacity (shear strength) properties of the floor rocks (Figure 16–10).

UNDERGROUND EXCAVATIONS

Since tunnels and underground rooms are engineering projects designed to meet a specified use in the most economical manner, the excavation operations must be carried out so that only the intended amount of rock is excavated and surface damage is prevented. This single project requirement is probably the biggest problem in subsurface excavations. When using the shield method in soft or squeezing conditions, the excavation contractor must be sure that only the material within the shield is removed, and that material is not squeezing into the opening (Figure 16–11). In competent ground, where a boring machine is used, the shape of the opening is controlled by the machine, and few problems are encountered.

Conventional methods are probably the most susceptible to problems of **overbreak** (where the opening is too large) and **underbreak** or **tight** (where the opening is too small). Rock-mass properties control the shape of the opening that naturally forms after a blast before support is installed

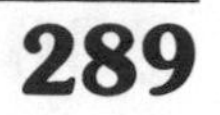

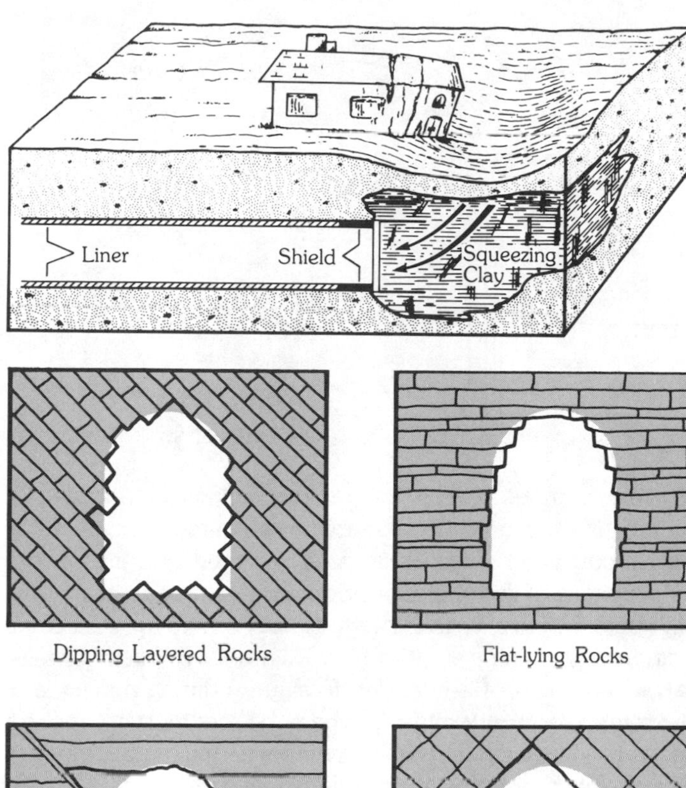

FIGURE 16–11 Influence of squeezing clay on surface subsidence during tunnel driving in soft ground.

FIGURE 16–12 Influence of geologic structure on the shape of a tunnel.

(Figure 16–12). These problems must be recognized by both the project owner and contractor before excavation starts if legal problems are to be avoided later. Once accepting the design and cross-sectional shape of the opening, the contractor is bound by the contract to excavate within specific limits (Figure 16–13). Any overbreak must be filled and any tights removed at the contractor's expense.

MINES

Engineering geology for underground mining differs from underground civil engineering projects because the objective of the opening is different. The shaft and main haulageways are civil engineering projects in that these openings must remain open and passable throughout the life of the mine.

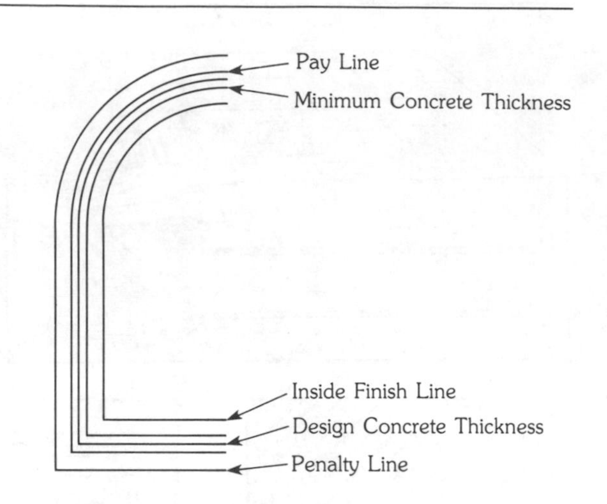

FIGURE 16–13 Design and construction lines used in tunnel construction. The tunnel contractor is paid to excavate to the pay line, must remove any material that extends beyond the minimum concrete thickness line, and must pay to fill anything beyond the penalty line.

Mine **drifts, rooms, stopes, raises,** and other openings (Figure 16–14), however, must remain open only until the ore has been removed. In many mines, passive support is removed as an area is mined out, in order to maximize the economics of the operation. Some mined-out areas may be back-filled with waste rock to save the energy of hauling it to the surface for disposal. Underground mining methods fall into two basic classes: **conventional,** where the roof is intended to stay up during mining, and **long wall, short wall,** or **block caving,** where the roof must collapse for the system to work. Roof support in conventional mining methods is short-term support, needed only during mining, and requires the same engineering geologic information as a civil engineering project. Caving methods, however, must pay particular attention to the successful collapse of the roof rock.

Land surface subsidence over abandoned conventional underground mines or modern caving mines is a significant problem. Collapse of the roof

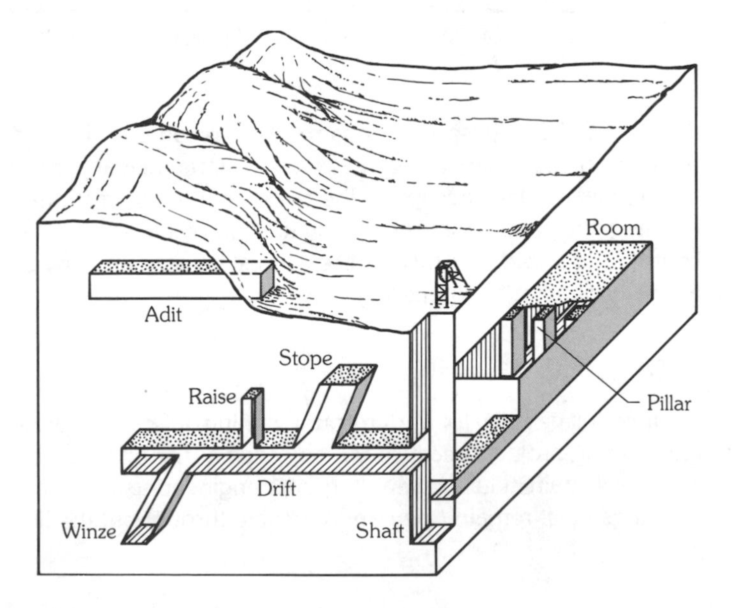

FIGURE 16–14 Common underground mining terms.

FIGURE 16–15 Subsidence over a block caving mine.

depends upon the shear-strength properties of the roof rock and overburden, and the aerial extent of subsided land can be significantly larger than the size of the original opening. The size, shape, and amount of surface subsidence are controlled by the geometry of the subsurface opening, overburden rock strengths and mass properties, and tectonic features. A fault, for example, can control the limits of subsidence by either reducing or enlarging the subsiding area (Figure 16–15). Analyses of abandoned-mine subsidence are further complicated because the geometry of the openings may be unknown. This unknown alone can have a serious impact on land use options and land values.

part III

Engineering Geology in Practice

Engineering Geology in Practice

seventeen

ENGINEERING GEOLOGY IN CIVIL ENGINEERING

Responsibilities of the engineering geologist in civil engineering projects that involve the earth or earth materials are (1) the identification and evaluation of the physical environment of the site, and (2) the analysis of the impact of the geologic processes on the proposed project. The engineering geologist must delineate the physical constraints and opportunities of each site that are imposed on the engineering design. The previous chapters have discussed the relationship between single geologic processes and specific engineering projects. Unfortunately, most engineering projects affect more than one geologic process. This chapter introduces civil engineering projects and their engineering requirements.

CONSTRUCTION

Civil construction projects can be divided into two basic groups: light and heavy. **Light construction** involves such projects as one- to three-story homes and apartments, local utility systems, and small shopping or commercial buildings. **Heavy construction** projects include major multistory structures, industrial plants, harbors, major transportation systems, and dams. Regardless of the size of the project, the performance of the earth materials must be such that the project can be safely and economically constructed and the completed project will perform as intended.

Excavations, whether one-meter utility trenches or deep basements, must consider the stability of the walls, groundwater drainage or flooding, and excavation methodology. Trench cave-ins are one of the most serious excavation risks and cause many construction injuries and deaths. The characteristics of the material that will be encountered along the entire route of the trench must be known to determine if some type of shoring will be required (Figure 17–1). Excavation projects also consider the effect of the

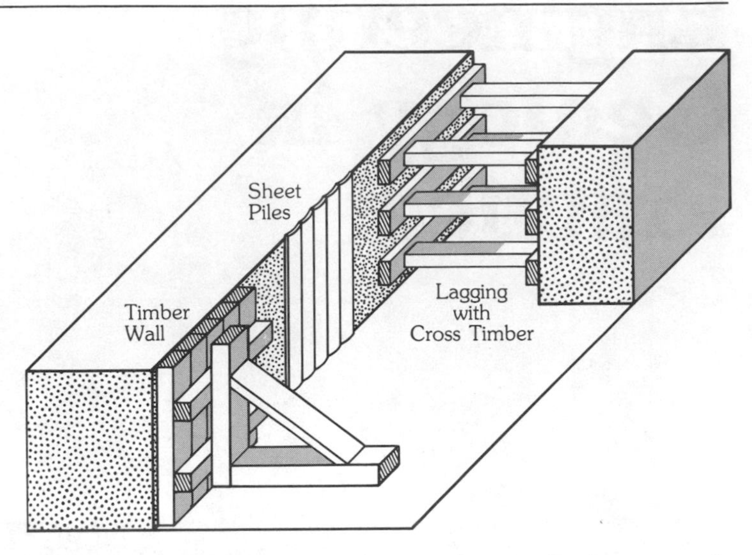

FIGURE 17–1 Common types of shoring used in excavations.

operation on the stability of surrounding structures. In low-strength, fractured material, **retaining walls** are frequently required to protect nearby structures (Figure 17–2).

Cut-and-fill operations, necessary to establish the grade (percent slope of the roadbed) of a highway, runway, or railbed, are similar to trenching projects in that the engineer needs to understand the characteristics of the earth materials in these project areas. The construction problems, however, are different. Roadcuts are designed to stand open for the life of the roadway; thus, naturally stable slopes should be produced or

FIGURE 17–2 Retaining wall constructed to protect surrounding buildings and to keep the excavation safe.

FIGURE 17–3 Naturally stable versus stabilized slopes.

permanent slope-retention structures or support should be constructed and installed (Figure 17–3). Long-term impacts on surface and groundwater drainage are incorporated into the design. Other problems such as rock weathering, erosion, and ice and frost heave also require evaluation. Fill materials are evaluated to determine their suitability. Frequently, material removed in a cut will be used for fill. If there are no cuts nearby, fill material will be obtained from nearby **borrow areas.** The cost of transporting fill precludes selective borrow, and consequently, borrow materials may vary greatly along a route (Figure 17–4).

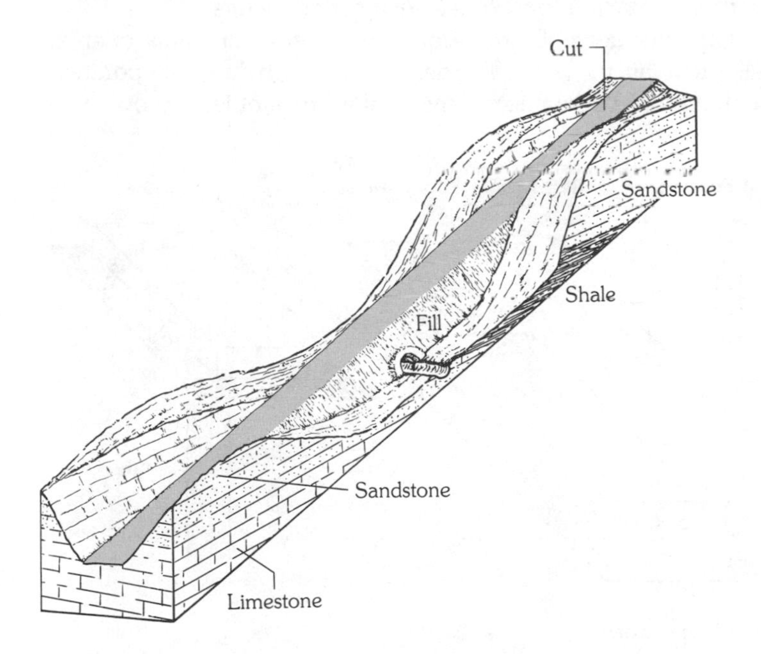

FIGURE 17–4 Cut and fill section of a transportation route. Notice the variation in the geology along the route.

Foundations are intended to support the structure above them and to transfer the structural loads to the underlying earth materials, so that the earth materials can support the loads without excessive **settlement** and without exceeding the **bearing capacity.** Allowable settlement and, more importantly, **differential settlement** values are controlled by the type of project. Large amounts of total settlement of a structure can be compensated for during construction. Differential settlement, however, is a serious problem with **rigid structures,** structures where differentials of less than one centimeter can cause serious structural damage. In **flexible structures,** such as earth fills or tanks, differential settlement may not pose any threat. The type of foundation selected is based on project economics and the physical characteristics of the site (Figure 17–5).

Dams have special foundation and abutment requirements because the earth materials support the weight of the dam and also retard the flow of water around and below the structure. The type of dam constructed is controlled to a large degree by the geology of the site. **Earth dams** (Figure 17–6) are constructed at sites where the foundation materials are compressible, because the structure is flexible and can deform as the foundation deforms without failing, or where large volumes of suitable earth materials are available. Masonry or concrete **gravity dams** (Figure 17–7) are large structures that resist overturning or sliding forces with their weight. These structures, however, are rigid and must therefore be founded on a stable material. **Thin-arch dams** (Figure 17–8) are "thin-shell" structures that are designed to transfer the force to the floor and walls of the valley. These thin concrete structures must be constructed in competent rocks, because differential movement of the shell can lead to catastrophic failure. The final selection of the dam design is based on the geology of the site, economics of construction, and other engineering factors.

Earth dams are generally trapezoidal earth embankments oriented across a valley to retain the water. The design of an earth dam incorporates a central, low-permeable core material, an earth or rock outer section, and a

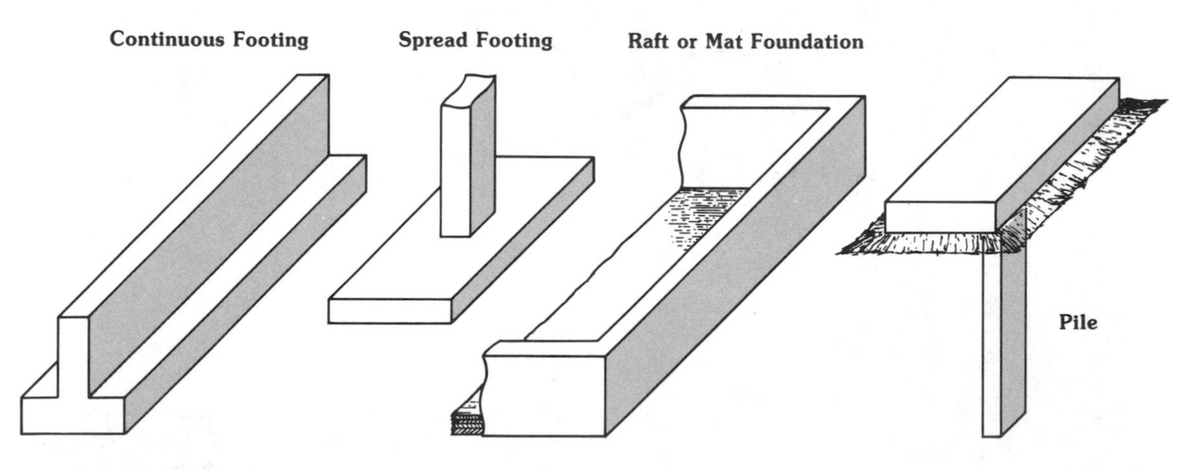

FIGURE 17–5 Common types of building foundations.

FIGURE 17–6 Earth fill dam.

FIGURE 17–7 Masonry gravity dams.

FIGURE 17–8 Concrete arch dam.

rip-rap surface to protect against wave erosion (Figure 17–9). Geologic factors that influence the design of an earth dam are:

1. The strength and hydraulic conductivity of the contact between the dam and foundation.
2. The strength, compressibility, and hydraulic conductivity of the foundation materials.
3. The physical characteristics of the abutment materials.
4. The availability, suitability, and transportation costs of the earth material for the core, outer sections, and surface protection.

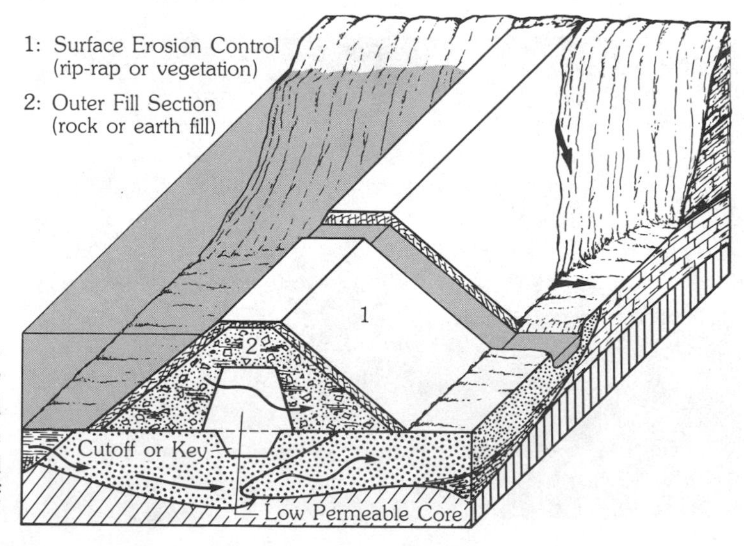

FIGURE 17–9 Earth or rock fill dam. Geological conditions for this type of dam are that seepage through the dam, below the dam, between the dam and the wall rocks, or through the wall rocks must be minimal and that suitable erosion control materials must be available for the face of the dam.

CASE IN POINT

Stream-Bank Erosion Below Dams, Missouri River

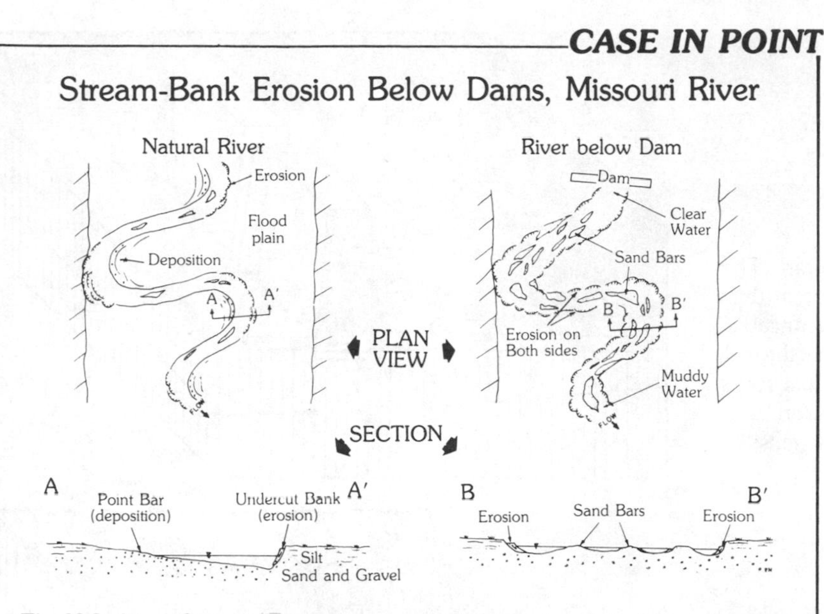

The U.S. Army Corps of Engineers constructed a series of six dams across the main stem of the Missouri River in North and South Dakota and Montana during 1937 to 1963. These dams provide flood mitigation, electric power generation, and water supplies for river navigation. The operating procedure used on the system is to release the water above the conservation pool soon after the flood peak has passed in order to maintain as constant a water level as possible. River flow conditions below the dams are therefore highly irregular with high bank full flood releases.

The Missouri River, the "Big Muddy," used to carry a large natural silt load prior to the construction of the dams. Water released from the dams carries a very low silt load; as a result excessive bank erosion is occurring. The river that once was a meandering stream is now becoming a braided channel as it erodes both river banks.

Modified from P. H. Rahn, Erosion below main stem dams on the Missouri River, *Bulletin of the Association of Engineering Geologists* 14 (1977). Used by permission.

Dams constructed of concrete or masonry blocks include gravity dams (Figure 17–10), buttress dams, and arch dams (Figure 17–11). Since these structures depend on the earth materials to carry most, if not all, stresses developed by the water behind the dam, the engineering geologic factors are critical. Geologic factors that affect selection include the following:

1. The foundation and abutment rock should be sound and able to resist the expected stresses including earthquake loading if applicable.
2. The foundation materials should resist sliding and differential settlement when loaded.

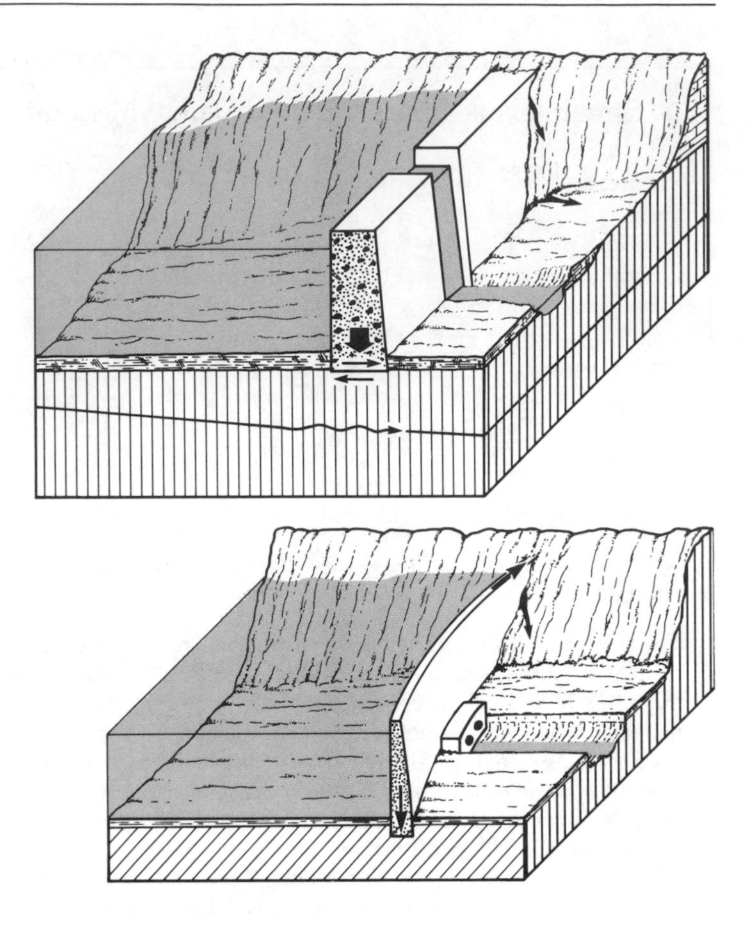

FIGURE 17–10 Gravity dam. The geologic conditions for this type of dam are that wall rocks must be impermeable; the junction between the dam and the wall rocks must be watertight; the floor rocks must resist sliding and compression loading by the dam; and floor rocks must be impermeable.

FIGURE 17–11 Concrete arch dam. Geological conditions required for this type of dam are that wall and floor rocks must carry the entire load of the dam and water behind it and that the contact between the dam and the wall and floor rocks must be watertight.

3. The geology of the site should be as uniform as possible to prevent differential strains.
4. The foundation and abutment rocks should resist weathering, frost heave, solution, and erosion.
5. The rocks at the site should be watertight.
6. The structure of the foundation and abutment materials should be favorable for the stability of an arch dam.
7. The engineering geologic characteristics of the surrounding rocks should be favorable for construction of support facilities, spillway, diversion tunnels, powerhouse, and others.

CONSTRUCTION MATERIALS

The engineering geologist plays a significant role in the search for and evaluation of construction materials like sand, gravel, crushed and dimension stone, fire and clay brick, and fill. These materials all have high place value and must be located close to the site because they are used in large quantities. Over 80 percent of the tonnage of new mineral materials mined in any year are the construction materials. Their bulk volume and low unit cost make transporation costs the most significant economic consideration.

Sand and gravel, or **aggregate**, are used in concrete and asphaltic concrete to provide structural strength and to reduce the volume of cementing agent. A clean, well-graded aggregate that will form a strong bond with the cementing agent is most desirable. Some aggregates react chemically with the cement, leading to the failure of the concrete and possibly of the structure. Aggregates are also produced by crushing rock **(crushed stone).** Such aggregates must be brittle materials that will fracture when crushed without degrading the strength of the individual particles, and must contain no reactive minerals that will degrade the cement-aggregate bond.

Aggregates and crushed stone are also used without a cementing agent as highway **base courses** and **wearing surfaces,** railroad **ballast,** and foundation **pad** materials (Figure 17–12). For these uses, the materials must also be clean and well graded so that they can be compacted to a high density **(unit weight)** and carry the intended loads. In addition, the material must form a flexible base that will deform as the **subgrade** deforms. Flexing must not cause the particles to degrade. Clay, silt, and fine sand-sized particles will cause frost heave and ice lens problems.

The geology of the aggregates and the environment of deposition of the aggregate source are primary factors that affect the suitability of the material. Geologic factors such as the mineralogy, existence of a coating, abrasion resistance, and soundness of the individual particles are significant. The mineralogy of the aggregate affects its suitability in concrete significantly, because curing (hardening) generates heat and alkali fluids, which can have deleterious effects on the strength of the concrete. Minerals susceptible to an alkali-aggregate reaction are hydrated silicates, opaline shale, chert, siliceous limestones, rhyolite, and dacite. Nonreactive minerals include quartz, feldspar, calcite, and the ferromagnesium minerals. The existence of secondary, postdepositional coatings, such as iron oxides, clay, and calcite, can have an undesirable effect, because the bond between the coatings and the aggregate particle may be weak, producing a weak concrete.

Resistance to abrasion is important in aggregates used on roadways, because a degree of surface roughness is necessary to produce a

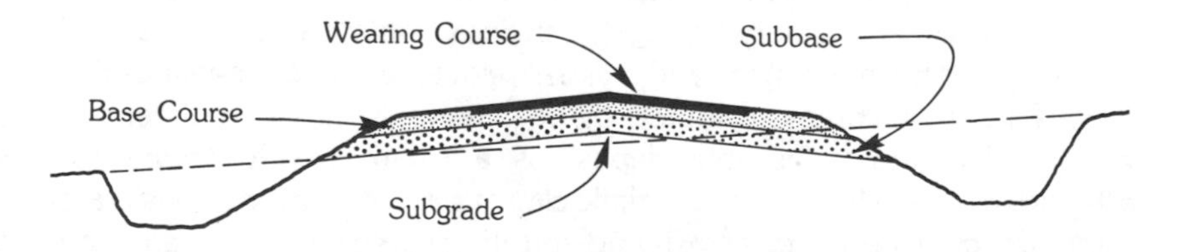

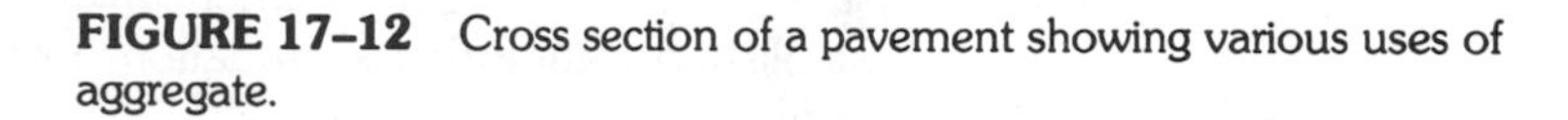
FIGURE 17–12 Cross section of a pavement showing various uses of aggregate.

CASE IN POINT

The Dam That's Not a Dam

An earth-fill dam was constructed at the base of a southwestern basin-range mountain to provide the center of attraction for a housing project. The site for the dam had been investigated using exploration drilling and geophysical surveys, but the details of the local geology were overlooked. Based on the investigation, the dam was built. After the dam began to fill, however, muddy water started to boil up into the river channel downstream of the dam. Investigation of the cause of the boils determined that the dam crossed a major fault that had been reported in studies of the local geology. The owners of the dam have been ordered to remove the structure because it represents a serious threat to the safety of residents downstream from the site, should the dam fail. This lesson was expensive!

skid-resistant pavement surface. Abrasion-resistant materials are quartz and crushed quartzite, granite, and dense, clay-free limestone and dolomite. Soundness is a measure of the fractures and other discontinuities and weathering of the particles. Their lowered strength makes fractured or weathered aggregates poor materials.

The environment of deposition, or formation, controls the grading or sorting of clastic sedimentary aggregates, the clay content in limestones and dolomites, and the mineralogy and crystal size of igneous and metamorphic rocks. The most common source of aggregates is clastic sedimentary deposits. The most desirable deposit is a clean, poorly sorted (or well-graded) gravel (GW) with very little clay or silt. Since these deposits are rare, many gravel operators wash and sort the aggregate in a preparation plant (Figure 17–13). The American Society for Testing and Materials has established standards for gradation, mineralogy, and other properties of aggregates.

FIGURE 17–13 Gravel preparation plant.

Artificial aggregates are made from clay or other materials that can be fired to a hard, brittle substance that in turn can be crushed and used to replace the aggregate in a concrete mix. These materials are manufactured in areas where suitable, naturally occurring aggregates are not available. Waste materials like slag, incinerated urban trash, broken glass, power plant ash, or cinders can also be used as an aggregate for some purposes.

Mortar sand plays the same role as the aggregate in a concrete, but on a finer scale. Mortar sands range from clean, well-graded, fine sand in finish mortar, to clean, well-graded sand in general mortar. These sands must also form a strong bond with the cement. Surface coatings, reactive minerals, and clay or silt contamination will diminish the quality and performance of the mortar.

Dimension stone is used for an exterior or interior finish, for flooring, and for monuments. Granite, marble, and indurated sedimentary rocks are common dimension stones because of their natural beauty or unique qualities. The architectural value of these materials compensates for their higher cost compared to other materials. Excellent wearing properties of some rocks make them valuable in heavy traffic areas like steps to public buildings (Figure 17–14). Usually the architect or client selects a dimension stone for a particular project. The engineering geologist, however, often has the responsibility of predicting the performance of the stone with respect to weathering and decay and the general performance for other wearing or use-related processes.

Clay is used in the manufacture of bricks and artificial aggregate. Materials for bricks are selected on the basis of their strength, color, and texture after firing. Low-strength, porous, and textured brick, frequently used for veneers and other architectural uses where it carries very low loads,

CASE IN POINT

Material for Beach Renourishment

Ediz Hook is a spit projecting eastward and subparallel along the shore of the Strait of Juan de Fuca in Washington. The hook, which is about 5.6 km long and from 27.5 to 275 m wide, provides a natural breakwater for Port Angeles Harbor. Construction of a dam across the Elwha River in 1911 and an artificial shoreline west of the spit in 1930 sharply reduced the sediment supply to the spit, and shoreline erosion problems developed. The U.S. Army Corps of Engineers was authorized to study the problem in 1970 and proposed a project to reconstruct the existing shoreline-protection system and to institute a sediment nourishment program. The selection of suitable sediment for the nourishment project required something that would match those on the hook as closely as possible. The selected source was a glacial upland plain, containing 170,000 m^3 of select gravel and cobbles, with 48 to 70 percent of the material ranging from 0.5 to 30 cm in size. The project specifications call for 50 percent of the select material to be larger than 7.5 cm.

are made from silty, sandy, and calcareous clays. High-strength brick, used in load-bearing situations, are made from low-plasticity clays. Fire brick or furnace bricks are quartz-rich bricks used as an insulation material in furnaces or fireplaces. The selection of a brick for the exterior of a building is based on its weather resistance, porosity, strength, and density. The American Society for Testing and Materials has mapped weathering regions for the United States, an important factor in the selection of exterior materials (Figure 17–15).

Fill material ranges from large rock to silts and clays depending upon the availability of material and the project needs. The material is used to fill in

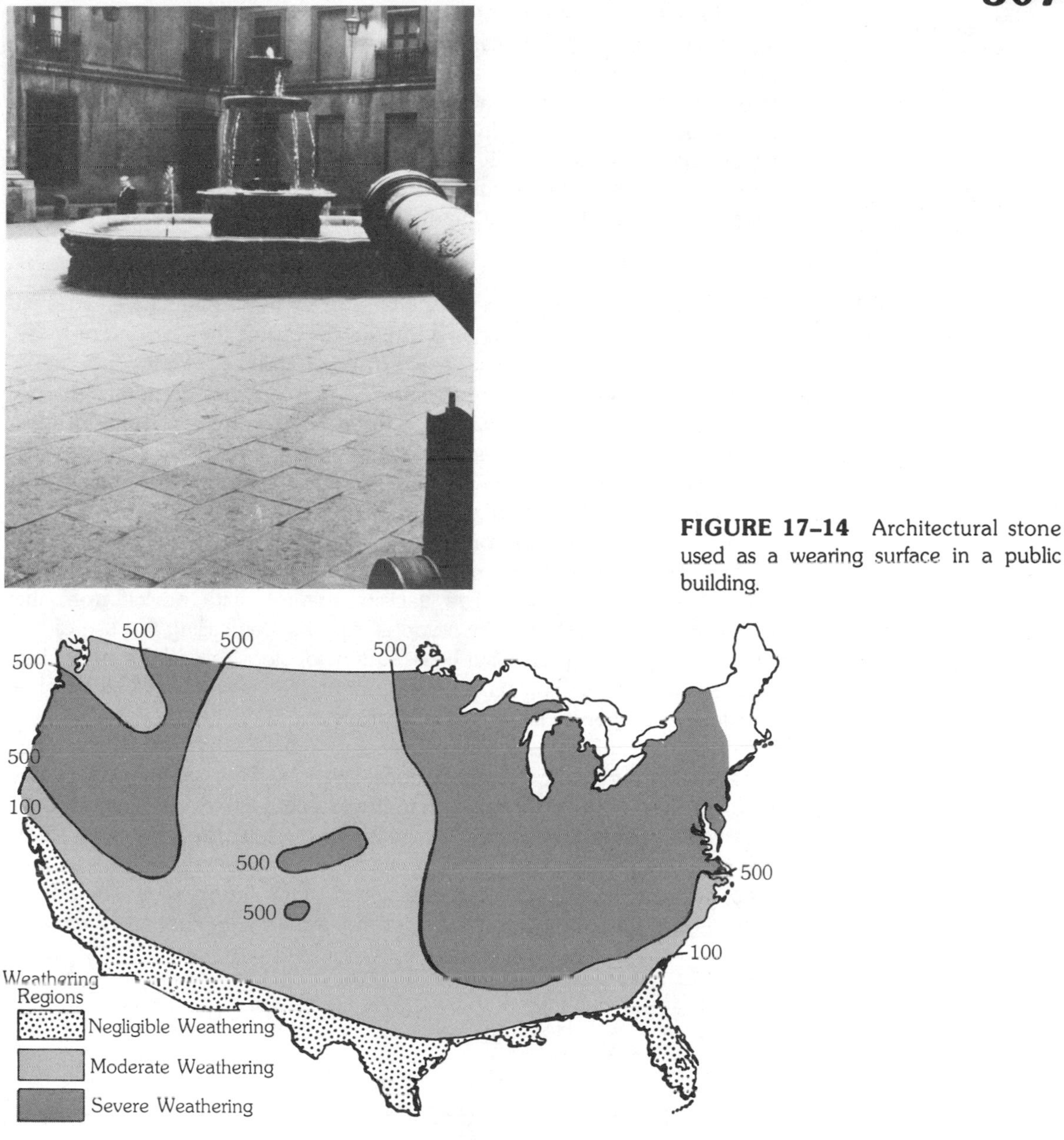

FIGURE 17–14 Architectural stone used as a wearing surface in a public building.

FIGURE 17–15 Weathering regions in the United States. (Adapted with permission. Copyright, American Society for Testing and Materials, 1916 Race Street, Philadelphia, PA 19103.)

low spots and raise the ground surface to the desired elevation and in the construction of earth dams. Fill may vary widely along a transportation route as the geology along the route changes. The high cost of transportation often limits the project engineer's selection of fill material to the most suitable material nearest the site. In many cases, material removed in a cut becomes the fill. The construction of earth dams requires the selection of fill materials that meet the design of each section of the dam. The engineering geologist must consider the mechanical behavior of the material, settlement, compaction, hydraulic conductivity and strength, and the weathering and frost susceptibility in the evaluation of any fill material. Limited processes to improve the material properties can sometimes be justified if other, more suitable fills are more costly than the improved fill.

ENGINEERING GEOLOGY IN URBAN PLANNING

Engineering geology is playing an increasing role in the planning and design of urban areas. As the value of land increases, sites that for one or more geological or physical reasons were previously bypassed are being developed. In addition to the increase in land values, the losses due to geologic processes, flooding, shoreline erosion, storms, slope processes, and subsidence have increased dramatically. As a result, the engineering geologist is asked to provide information to the decision maker about the suitability of sites for urban purposes.

Geology does not control the use of the site; it influences the economics of the project by defining constraints or opportunities of the physical characteristics of the site. The decision maker, with advice from the engineering geologist, has the responsibility for controlling the use of the site. Urban planning follows a basic system of steps, called the **planning process** (Figure 17–16), in the development of an urban plan. Geology is a constraint or opportunity within the process.

Urban geology is the application of the principles of engineering geology to an urban area, without consideration of the type of structure or type of construction. The approach to the problem differs from that used on a particular civil engineering construction job, because the scale of the study has changed. The planner needs geologic information about an area that may not be developed for many years. This information should be presented in the form of *land use opportunities* and *land use constraints*. A

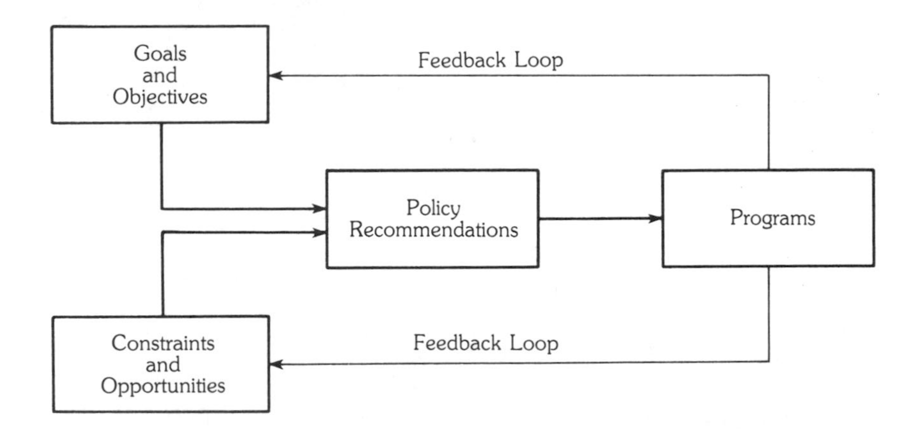

FIGURE 17–16 The planning process.

FIGURE 17–17 Flood-compatible construction. The floodplain is used for a golf course, and homes are constructed on the flood-safe terraces.

categorical statement that a tract of land is unsuitable because it is a floodplain assumes that the intended land use is not compatible with flooding. However, many safe and flood-compatible structures or land uses, such as parks, parking lots, and elevated structures, can be built within the constraint that the land is susceptible to flooding (Figure 17–17). The engineering geologist must develop techniques and methods to present geologic information in a form that people untrained in engineering geology can use in the planning process.

ENGINEERING GEOLOGY IN ENVIRONMENTAL CONSIDERATIONS

The National Environmental Policy Act of 1969 reflects the environmental policy of the United States. This policy is that development must consider the adverse impacts on the environment and quality of life. Many other federal and state environmental laws have been enacted since 1969—all of them requiring an assessment of the impact of a project on the environment. The engineering geologist is becoming an important professional in environmental assessment studies and impact statements. Questions regarding the impacts on groundwater, erosion and sedimentation, stability of slopes, and other geologic hazards and processes are asked.

In recent years, developers and engineers have turned to engineering geologists for help in evaluating the impact of the environment on a proposed project. Some projects, such as nuclear power stations and coal mines, must address this question before an operating or mining permit will be granted. Once again, the engineering geologist's ability to interpret geologic processes and predict process responses makes the profession valuable in the analysis of environmental conditions.

CASE IN POINT

The Impact of Active Sedimentary Processes on Land Use Planning Along a South Texas Barrier Island

The town of South Padre Island, Texas, located at the southern portion of a 177-km barrier island, has experienced rapid growth during the last fifteen years. A study team of urban planners and geologists found that an awareness of geological processes is required to prevent coastal disasters. Historical air photography identified the island's susceptibility to the destructive force of hurricanes as a major hazard.

In 1967, Hurricane Beulah filled back-island channels and canals with sediment and debris, and reopened numerous washover channels, isolating portions of the island from each other. Severe erosion and scour occur within the beach, foredunes, and washover channels; therefore development should avoid these areas. The foredune ridge should be maintained, as it protects back-island areas from storm-surge scour.

Land use planners have disregarded active coastal processes, creating a dangerous situation for residents and tourists: portions of the foredune ridge have been removed for hotel construction, and housing has been built in washover channels. These areas will experience flooding, scour, and erosion during future hurricanes; thus it is recommended that high-cost commercial buildings be placed behind the foredune ridge and residential areas be restricted to back-island areas not exposed to washover channels. Land use planners should consider the geologic problems arising from human alteration of natural systems. Geologic input can be incorporated into any land use plan at low cost and with little effort.

Written by W. F. Cole, Engineering Geologist, McClelland Engineers, Ventura, Ca.
Photo by Texas Department of Highways and Public Transportation.

SANITARY LANDFILLS

Sanitary landfill or **solid waste management** is one of the most common environmental problems facing any community. At present the economics of waste recycling on a large scale are not favorable, and most solid waste is buried in a landfill.

Since the beginning of time, people have disposed of solid wastes by burning or depositing them in open dumps. When land was plentiful and waste was natural materials, location of the dump posed little problem or danger (Figure 17–18). But in today's highly urbanized civilization, such procedures are no longer acceptable (Figure 17–19). Today's sanitary landfill is a controlled dump—controlled in that environmental factors that would tend to return the waste to free circulation within the environment have been determined and countered. A sanitary landfill is intended to isolate the wastes from the environment.

FIGURE 17–18 Ancient solid waste dump composed of oyster shells.

FIGURE 17–19 Modern solid waste, consisting of almost everything.

One of the first concerns in designing and locating a sanitary landfill site must be the nature of the waste materials. Modern solid wastes do not consist simply of rotten food, useless and broken clay bowls, wooden utensils, and stone knives, as they did in ancient times. Today's great surge of throw-away conveniences generates more than 250 million tons of urban solid waste per year. Landfills serve only this fraction, which pales beside the total 4 billion tons of municipal, industrial, and agricultural leftovers now produced annually in the United States (Turk, 1970).

Waste materials can be subdivided into six general classes: urban or municipal, nonhazardous and hazardous industrial, agricultural, and low-level and high-level radioactive wastes. Table 17–1 gives their general characteristics.

The problem arising from the disposal of solid wastes is not the material itself, but rather the types of pollution that can be released into the environment when air and water come into contact with it. The four main types of pollution are water, gas and air, soil, and aesthetic pollution—all problems that a sanitary landfill must control if it is to be successful (Table 17–2). The degree of pollution depends upon a variety of factors, including the type of waste material deposited and its physical stability, the geologic and topographic characteristics of the site, the depth to the water table, and the climate.

Solid waste has been compared to a natural sediment, and like sediment, a landfill compacts or settles, is permeable, and can erode and change chemically. Compaction takes place, because much of domestic

TABLE 17–1 Characteristics of wastes

Type	Sources
Municipal	
Garbage	Food
Combustible rubbish	Paper, wood, cloth, plastic
Noncombustible rubbish	Metals, glass, bricks, stone
Bulk wastes	Crates, appliances, tires
Animal	Dead animals
Industrial	
Construction/Demolition	Lumber, bricks, roofing, pipe
Nonhazardous	Nontoxic or nonflammable materials from manufacturing
Hazardous	Toxic or flammable or chemical wastes from manufacturing
Agricultural	
Animal	Manure, slaughterhouse
Plant	Food-processing operations
Radioactive	
Low-level	Hospitals, laboratories
High-level	Nuclear power plants

TABLE 17–2 Pollution potential from a landfill

Water Pollution

1. Depletion of dissolved oxygen through aerobic decay.
2. Bacteriological and virus contaminations.
3. Pollution of surface streams with runoff from the site.
4. Chemical alteration of groundwater and surface waters.

Gas and Air Pollution

1. Noxious odors.
2. Releases of methane from anaerobic decomposition, ammonia, hydrogen, hydrogen-sulfates, and other gases.
3. Dust released during operations and smoke caused by fires.

Soil Pollution

1. Release of ions that can react with the soils.
2. Release of chemical compounds or ions that are taken up and concentrated in plants.

Aesthetic Pollution

1. Unsightly or odious aspects of an active landfill.
2. Increased traffic and truck spillage.

waste is hollow (plastic bottles) or of low density (packing materials) and collapses under a load. Surface water passes through and over it, exerting the most significant influence, depending on the energy of the water. Since it is the universal solvent, water needs no energy to stimulate chemical decomposition. Decay takes oxygen from the water, in an amount known as the biological oxygen demand (BOD). As a result, the water leaving a site is not the same as that entering it, and by virtue of its newly acquired chemicals, is called **leachate.**

The control of leachate movement is of primary importance in the design of a landfill. Leachate generation and movement are controlled by the hydrologic characteristics of the area, and the hydrology is controlled by the geology of the site. Specifically the underlying rock or soil unit and its permeability, structure, attitude, and lithology as well as the cover material available are the factors that decide the suitability of a site as a landfill.

HYDRAULIC CONDUCTIVITY

The interconnection of voids or pores in a rock and soil governs the movement of fluids through them. Rock types with typically high hydraulic conductivities include sandstone, vesicular volcanics, and poorly cemented conglomerates. Rock types with typically low hydraulic conductivities include shale, slate, siltstone, mudstone, dolomite, nonvesicular volcanics, and granite and other igneous intrusives. The lower the hydraulic conductivity, the more desirable is the material.

STRUCTURE

A rock, which normally has a low hydraulic conductivity, may have certain structures which make it highly conductive and therefore undesirable. These structures include fracturing, faulting, bedding planes, and solution cavities.

ATTITUDE

The angle at which a rock, or soil, might dip is an important factor. If a rock has been tilted toward the vertical, its bedding planes may act as a ready avenue for the transport of leachates. For example, a shale may have a much greater permeability parallel to the bedding planes than perpendicular. This also applies to the laminations and foliations found in metamorphic rocks, such as gneiss and schist.

LITHOLOGY

The effect of these three factors differs for different rock types. For example, a limestone is readily soluble in a weak solution of carbonic acid, whereas a granite is not. Soils and alluvial fill show characteristics similar to rocks. Unjointed clays have very low hydraulic conductivities, whereas sands and gravels commonly are highly conductive. Structures can be induced in soils by desiccation, plant roots, boring animals, sliding, and subsidence.

TOPOGRAPHY

The topography is an important parameter. Depressions, where water may accumulate, and floodplains, where water plus leachate may enter a stream or the near-surface water table during flooding, are examples of areas that should be avoided. Hilltops and flood-free areas are more suitable sites.

HYDROLOGY

The dispersion rate of pollutants is increased if leachate enters the groundwater system or is incorporated into surface water and carried to streams. Hydrologic criteria for a landfill site places the site above the groundwater table, near the crest of hills, and in material having very low hydraulic conductivity.

COVER MATERIAL

An important factor in the design of a landfill is the type of available cover material. Many states require that a layer of soil, at least fifteen centimeters thick, be placed over the exposed waste at the end of each operating day. The necessity of obtaining adequate cover material can often remove a prospective site from consideration as a landfill. The best cover material is a clay or silty soil because it allows a minimum of moisture to reach the wastes.

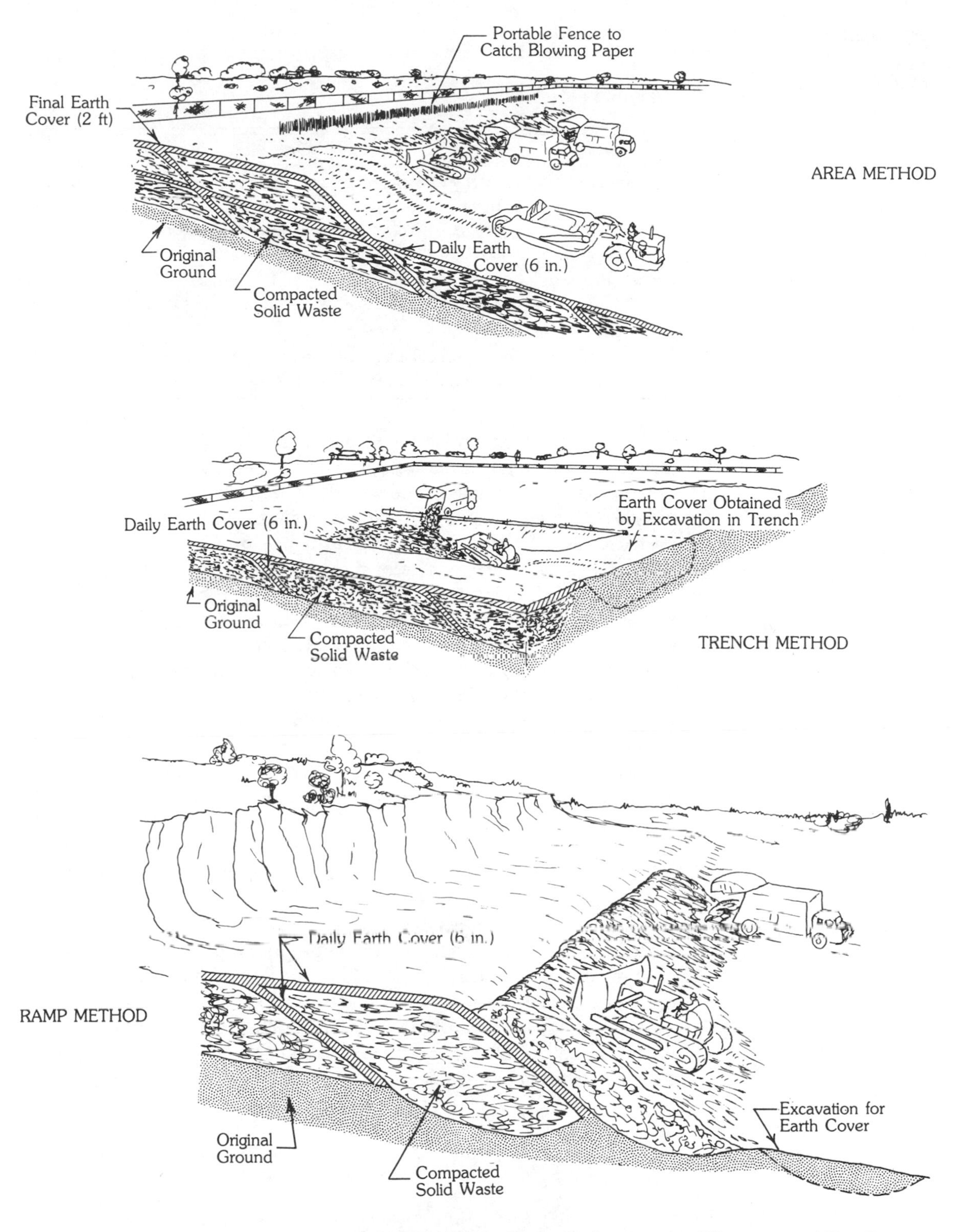

FIGURE 17–20 Methods of sanitary landfill operation. (After Environmental Protection Agency, 1972.)

TABLE 17–3 Landfill designs

Design	Operation
Area method	Spread and compact refuse; cover with compacted fill. Suitable for flat or gently sloping areas and in old quarries, ravines, valleys, or other suitable land depressions. Cover material must be hauled in.
Trench method	Cut trench; fill, spread, and compact refuse; cover with compacted fill. Suitable for flat land above water table. Cover material is excavated from the trench.
Ramp or slope method	Spread on existing slope and compact refuse. Cover material is excavated just ahead of the toe of the slope.

The purpose of a landfill is to dispose of solid waste materials safely at a minimum cost. Thus, it is necessary to plan its operation once a geologically suitable site has been defined. Landfill design and operating procedures will usually be carried out by a professional engineer. Three basic designs have been established: **area, trench,** and **ramp** or **slope** (Figure 17–20). Table 17–3 shows the characteristics of these three designs.

The landfill, after completion, may become a problem unless adequate planning and maintenance practices are followed. Compaction can cause surface collapse, and long-term chemical reactions within the waste may produce dangerous gases, such as methane. Precautions taken during the active life of the landfill can minimize these dangers. Some artificial compactions can be accomplished in the home using electric trash compacters and during dumping simply by running heavy machinery over the waste. Gas generation will be enhanced by entry of water into the body of the fill. To prevent a buildup of gas pressure, a venting system may be necessary, the simplest being gravel partitions between cells of waste. The erosional capability of running water poses the threat of exposing buried refuse—another reason for establishment of fills away from watercourses and for proper site maintenance.

Disposal of hazardous and radioactive wastes by burial is a common practice. The significant difference between municipal solid waste disposal in a sanitary landfill and industrial solid waste disposal is the geologic criteria used to evaluate the site. Toxic and hazardous wastes must be further separated from the environment by increasing the distance between the waste and water. Radioactive and toxic materials can be buried deep in salt, shale, or other impermeable rocks.

The Practice of Engineering Geology

eighteen

The engineering geologist is involved in engineering projects from their original conception through acceptance of the structure or project by the owner. Selection of the site, evaluation of construction materials, investigation of the geology of the site, field evaluation of construction practices and design, and contract monitoring all fall within the scope of work for an engineering geologist. The professional may work for the project owner or contractor or play an independent role; in each case, the highest professional ethics must be followed because these services carry great responsibility and involve liability for each action.

Professional practice can be subdivided into two general areas: **site characterization** and **construction monitoring**. In the first, the professional applies basic geologic principles to the goal of predicting the physical characteristics of the site in order to provide information that will lead to the most economic design and construction of the project. In the second area, the engineering geologist is actively involved in the construction of the project, starting with the preparation of construction bids and ending when the project is completed, with the objective of identifying geologic conditions that were not expected or that changed, and describing the revised geologic characteristics of the site in order to minimize adverse economic impacts.

These two roles are frequently filled by two different persons: the **owner's geologist**, who is responsible for the preconstruction site characteristics, and the **contractor's geologist**, who is responsible for successful completion of the project within the economic and performance standards of the contract. The engineering geologist, regardless of the employer, must guard against becoming involved in the project politics at the expense of professional ethics.

SITE INVESTIGATION

A site investigation is a geologic study that provides a description of the site's physical characteristics in the most economical manner. The investigation can only establish the *expected* conditions based upon a sound evaluation of the *available* data. A site investigation may range from a generalized overview to a complete and extensive series of field tests depending on the needs of the project owner. A **preliminary investigation** is designed to evaluate the site for obvious adverse geologic conditions and frequently is carried out as part of the initial selection of the specific site. The **design investigation** is conducted at the selected project site to gather detailed geologic data for evaluation and inclusion in the engineering design of the project. The objective of the site investigation is to define the site characteristics and *not* to control project site's suitability. The engineering geologist must remember that economics, *not* the geology of the site, controls the project, and that the geologist is a staff person along with the architect, design engineer, financiers, and others.

Site investigations can be subdivided into three basic components: **initial studies, field investigations**, and **laboratory testing**. Each is equally important, and any poor-quality work in one can lead to problems. Initial studies include a literature survey of the regional and project site geology, logistics, and planning for the field investigation and laboratory testing phases. The engineering geologist plays an important role in the initial and field phases of an investigation and an advisory role in the laboratory testing phase. The responsibility of the field investigator is to assure that all samples delivered to the laboratory for testing are representative samples, because the most extensive laboratory testing program and detailed design evaluation are only as good as the sample tested. Failure to identify unusual geologic conditions or failure to sample critical materials can only lead to excess costs, construction problems, or structural failures in the future.

INITIAL STUDIES PHASE

The purpose of initial studies in any site investigation is to maximize the economics and efficiency of the following phases. All too often the critical jobs of literature search, review, and evaluation are cut short because the project owner or the engineering geologist sees no value to them. The literature study provides the data base and starting point for the field investigation. Methods, techniques, and equipment selection for the field investigation must depend upon some geologic knowledge of the expected site characteristics.

The logistics and planning component are also economically based. Field investigations are expensive! Thorough planning must, therefore, be carried out prior to the field investigation in order to maximize the cost/benefit ratio of the study. This component considers everything from defining the field operations to selecting personnel and even providing their life support if necessary. The literature survey defines the present state of geologic knowledge; the engineering geologist now designs the investigation to fit this information.

FIELD INVESTIGATION (SURFACE)

The objective of the field investigation is to provide a three-dimensional characterization of the geology of the site. Based on this site characterization, critical and representative samples are collected for the laboratory testing phase. Unfortunately, the three-dimensional character is not exposed for all to see, and neither are geologic data in three dimensions available to the geologist. Thus, every site investigation is an interpretation and prediction of the expected three-dimensional character of the site based upon the data collected.

Geologic data can be collected as **point**, **one-dimensional**, and **two-dimensional** information. *Outcrops* are single points, *cores* are one-dimensional lines of data, and *river cuts, aerial photographs,* and *geophysical surveys* are two-dimensional data (Figure 18–1). Evaluation of these data produces the three-dimensional characterization of the site. Any interpretation is based on a framework of geologic knowledge. Consider a core sample, for example, that shows a fining upward sequence of quartz-rich clastic sediments ranging from gravel to fine sand, and containing ripple and dune sedimentary structures. The interpretation of this section of core is a *fluvial* or river-channel sequence. The interpretation of the site geology is a fluvial environment of deposition, and direct correlation between any two cores would not necessarily be expected. The sediments within the area of the site can be expected to range from overbank, floodplain clays to channel gravels.

MAPPING A map is a two-dimensional presentation of a surface. **Topographic** maps present the surface elevation (Figure 18–2). **Geologic** maps present an *interpretation* of the "bedrock" or geologic units at the site (Figure 18–3). The engineering geologist must use geologic maps with caution, because most such maps present generalized information about rock types, not specific information about their engineering geologic

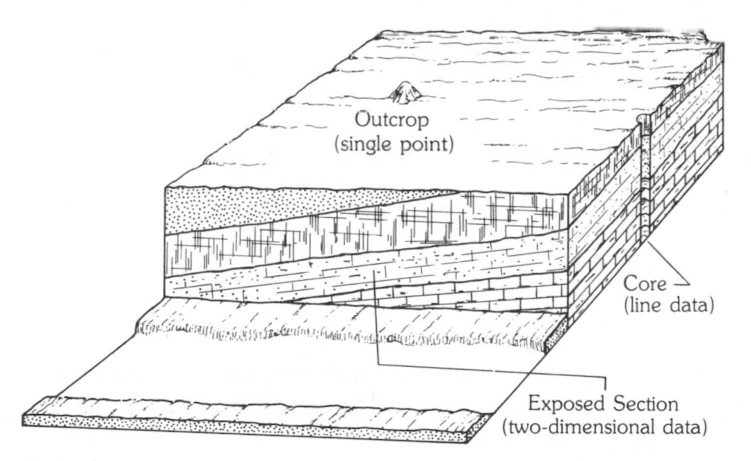

FIGURE 18–1 Characteristics of types of geologic data as they appear in the field.

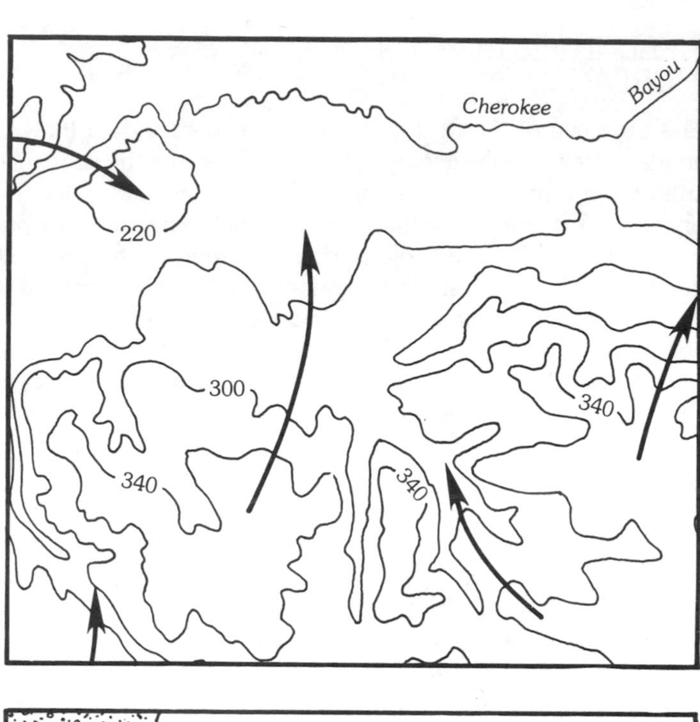

FIGURE 18–2 Topographic map of a project site. The arrows show the direction of surface runoff.

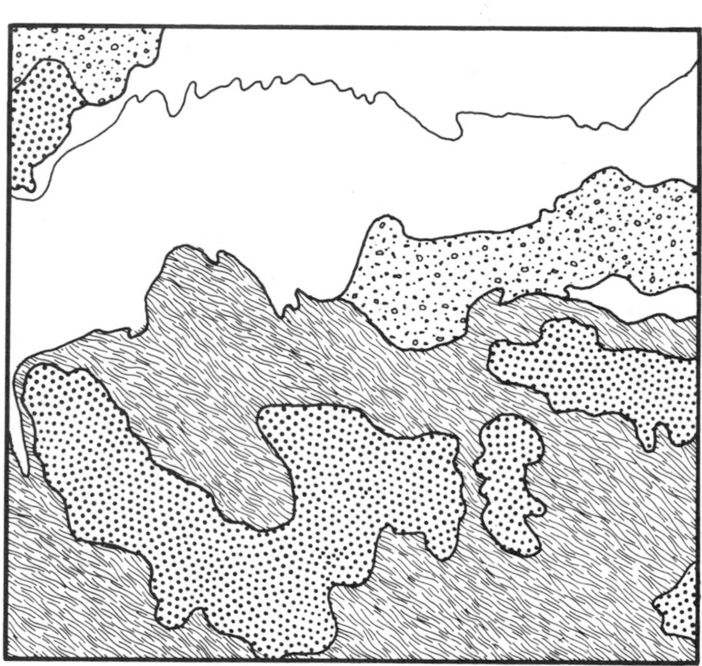

FIGURE 18–3 Surface geologic map.

properties. An **isopach** map is a subsurface map showing the thickness of a particular unit or rock type (Figure 18–4), while a **structure** map shows the elevation of some subsurface marker bed or unit (Figure 18–5). **Special-purpose** maps present specific geologic information (Figure 18–6).

FIGURE 18–4 Isopach map showing the thickness of aquifer sands in the material above the base of an excavation.

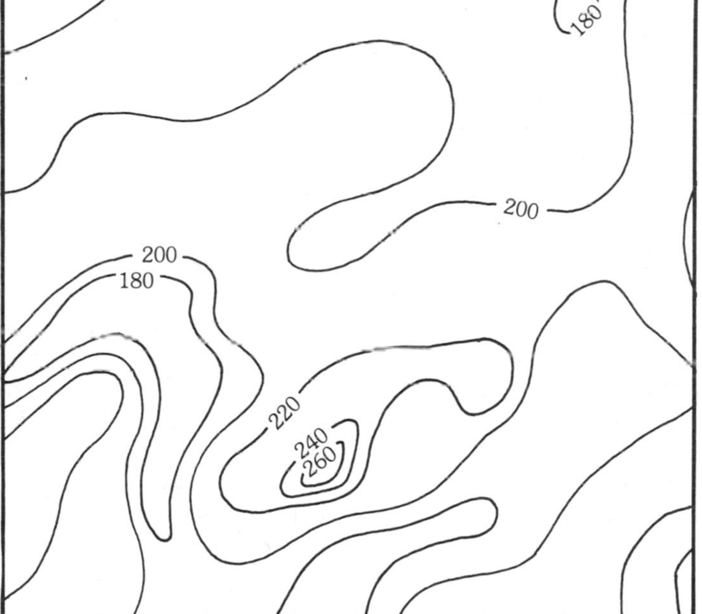

FIGURE 18–5 Structure map showing the elevation relative to sea level of a marker bed.

Most topographic maps are prepared by an engineer or surveyor for the engineering geologist. This type of map is a useful and valuable tool in the preparation of a geologic map because the topography frequently reflects the underlying geology as modified by the active geologic processes. River, or **drainage**, patterns and topographic trends are strongly controlled

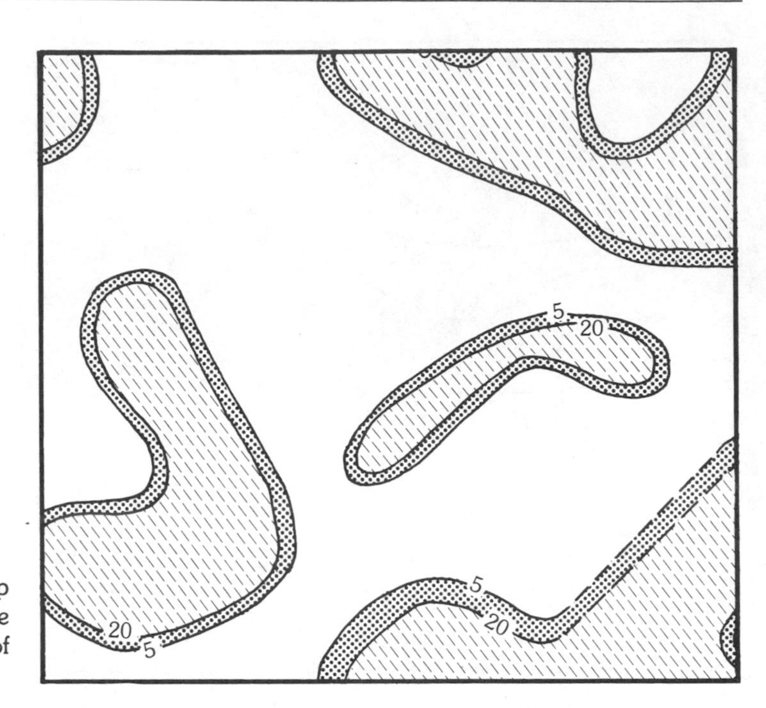

FIGURE 18–6 Special-purpose map showing the thickness of clay below the excavation to identify potential areas of floor heave.

by the near-surface geology (Figure 11–22). In some regions, younger geologic deposits may completely mask the subsurface geology, and the topography will reflect recent surface processes.

Geologic mapping depends upon the combination of outcrop and topographic or aerial photographic data. Landforms are defined by their topographic and surface expression. Rock types are then determined by outcrop studies. A long, tree-covered ridge, for example, is mapped as a landform. The rock or rock units that support the ridge are then determined at stream cuts through the ridge. The landform is reinterpreted to reflect the rocks observed at outcrops on the ridge and mapped accordingly. Boundaries between rock units are usually controlled by the topography because different rock types weather and erode differently. Therefore, when a geologic map is constructed, changes in topography are used as a guide. (see Appendix 3).

Aerial photographs are one of the most efficient geologic mapping tools because when viewed in stereo, they provide a three-dimensional view of the land surface. Stereo viewing is possible when two overlapping photographs are taken of the same site from two different camera locations (Figure 18–7). The appearance of relief seen in the set of photographs is related to the **parallax** of any two points. These relationships can be used to calculate the difference in elevation between any two points. Thus, stereo aerial photography can be used to make topographic maps for a project site (Concept 18–1).

CONCEPT 18–1

Stereo Viewing—Parallax

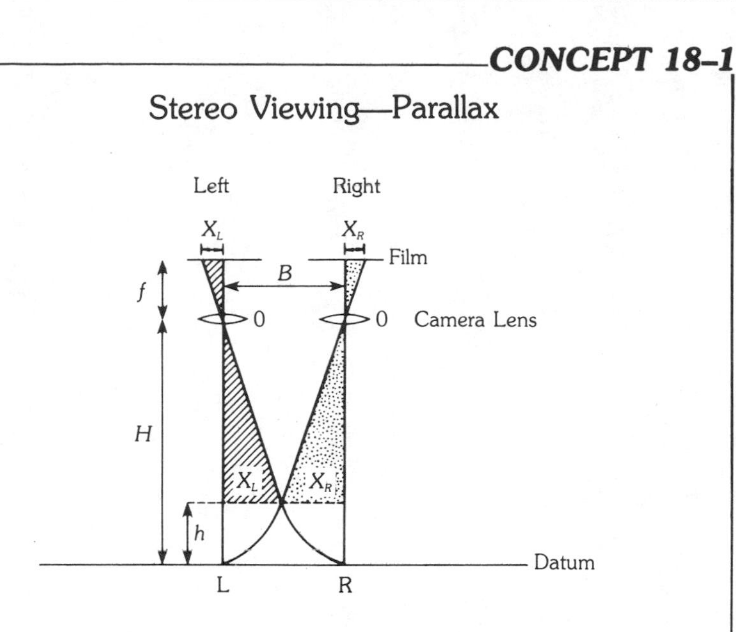

Overlapping aerial photographs, taken from two different locations (left and right) that are B units apart, can be used to measure the height of the ground by measuring the parallax of the image on each photograph. Parallax is the total distance from the center of each photograph to the point ($X_L + X_R$). The height h is determined from the relationships between similar triangles as follows:

1. The cross-hatched and dotted triangles are similar.
2. The ratio of parallax ($X_L + X_R$) to base (B) equals the ratio of focal length (f) to flying height (H) minus height of feature (h):

$$\frac{(X_L + X_R)}{B} = \frac{f}{(H - h)}$$

3. Since everything except h is known, then

$$h = H - \frac{fB}{(X_L + X_R)}$$

Photogeology, or geologic mapping on aerial photographs, has many advantages over mapping on topographic maps. Since the photograph is original data and a topographic map is interpreted data, the photograph contains all the topographic information, and subtle or unique topographic features can be identified. In addition to providing data for a landform analysis, aerial photographs are also useful for pattern analysis, land use analysis, and other special studies. Photogeologic mapping draws upon all the terrain data and therefore provides for a more accurate geologic interpretation. In addition to providing a more complete data bank for

FIGURE 18–7 Overlapping aerial photographs that can be viewed in stereo. (Photos courtesy of U. S. Department of Agriculture, Soil Conservation Service.)

analysis, aerial photography has a significant economic advantage. Photographs can be taken and printed at any scale needed for the study, thereby providing the engineering geologist with the necessary accuracy.

Landform analysis is the study of terrain features that are formed by natural processes and always have a definite composition and range of physical and visual characteristics wherever they occur. The analysis technique assumes that each landform is the result of a set of unique conditions that will *always* produce the same landform if the same set of conditions is established. From a geologic viewpoint, each landform can be related to specific geologic conditions. Landform analysis keys are an important aid to interpretation. With experience, each engineering geologist will develop his or her own set of keys based upon photogeologic and field studies.

Pattern analysis is a companion component to landform analysis because many geologic features produce unique patterns just as they produce landforms. Topographic patterns can easily be seen on aerial photographs. These patterns are related to surface features such as vegetation, soil type, rock type, and human activities or land use patterns. The photogeologist utilizes these patterns in interpreting the site by relating the observed pattern to the underlying geologic conditions. Tonal patterns are produced by variations in moisture and surface textural conditions. Wet areas will appear darker than dry areas of the same material, while smooth surfaces will appear brighter than rough surfaces of the same material.

Aerial photography is one component of the **remote sensing system** using the electromagnetic spectrum (Table 18–1). Recent advances in space and electronic technology have made other components of the electromagnetic spectrum available to the engineering geologist. **Near infrared** photographic techniques (Figure 18–8) and **far infrared** thermal scanning techniques have added methods to sense thermal radiation transmitted from the earth. **Microwave** systems (**radar**) are active sensing systems that transmit and receive radio frequency radiation and do not depend upon solar or earth radiation as a source. Active sensing systems are particularly valuable for topographic studies in areas under extensive cloud

TABLE 18–1 Electromagnetic spectrum

Wavelength	Wave Name
Below 0.3Å	Gamma rays
0.3–30Å	X rays
30Å–.3μ	Ultraviolet
.3μ–.78μ	Visible light
.78μ–200μ	Infrared
200μ–30 cm	Microwave (radar)
30 cm–5 m	Ultrahigh frequency
5 m–300 m	Radio
300 m–30 km	Low frequency
30 km and above	Audio

FIGURE 18–8 Normal (top) and near-infrared (bottom) photographs of an inlet. Notice that the shallow water features are enhanced in the infrared photograph.

cover because clouds do not affect them. Imagery using the **Landsat** satellite system provides visible light and infrared data (green, band 4; red, band 5; near infrared, band 6; and near infrared, band 7). These very high-altitude images of the earth's surface are useful for regional studies but cannot be used for site-specific analyses because of their very small scale (Figure 18–9). A remote sensing study for a project like a large dam and reservoir would start with satellite data and end with low-level aerial photographs, thereby providing information on regional features down to site-specific features.

Geological maps, which are two-dimensional representatives of the surface geology, can be interpreted to provide a three-dimensional presentation. Geological **cross sections** are constructed by projecting the structure of the geologic units, as determined from surface outcrops, to the subsurface. Structure depends upon determing the type and orientation of the rock units and their relative age and then creating a realistic interpretation of the subsurface geology (see Appendix 3). The key word here is *realistic*, because there may be more than one interpretation. Since the objective of the site investigation is to provide the most accurate interpretation of the geology of the site, all subsurface geological interpretations should be field checked and confirmed.

FIGURE 18–9 LANDSAT image of part of the Appalachian Mountains. The image is scanned in the near-infrared, band 7. (Photo courtesy of EROS Data Center.)

FIELD INVESTIGATION (SUBSURFACE)

Subsurface investigations include two basic methods, **direct observation** and **remote sensing** or **geophysical analysis**. Direct observation involves the collection of a core or the excavation of trenches, pilot tunnels, and shafts (Figure 18–10). Each of these observations is one-dimensional data giving point or line information for the geological map or cross section. Coring and other direct-observation techniques provide samples for laboratory studies and ground-truth information for geophysical or remote sensing analysis. These methods are, however, expensive to carry out, and therefore, the number of such observations is limited for any project study.

Geophysical techniques are based upon the physical properties of the rock units, including velocity of sound (**seismic velocity**), which is related to rock density and fracturing; **electrical resistivity** or **conductivity**, which is related to the availability of free electrons; **magnetic properties**, which are related to the amount and orientation of iron-bearing minerals; **gravity**, which is related to rock density and depth of burial (distance from the **gravimeter**), and natural radioactivity (**gamma radiation**), which is related to the availability of radioactive elements in the rock system. Each of these physical properties must be correlated with the rock type at the site, because more than one rock type can have the same physical property.

Seismic velocity, used in engineering geology to interpret subsurface geologic conditions, generally includes **seismic refraction** on land and **seismic reflection** in marine problems (Figure 18–11). Seismic methods require that a **velocity contrast** exist in the area to be studied. Seismic refraction surveys determine the seismic velocity of the rock units on a

FIGURE 18–10 Exploration tunnel (pilot tunnel) used to investigate conditions for a larger tunnel, which appears in the lower portion of the photograph.

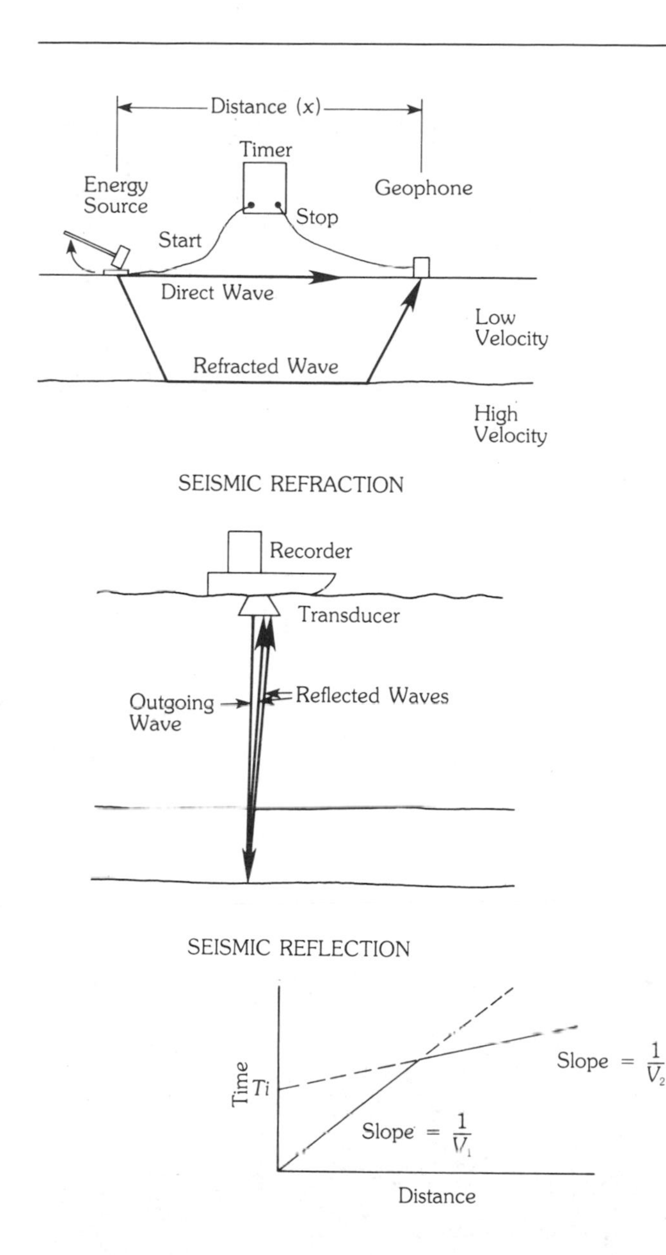

FIGURE 18–11 Comparison of seismic refraction and seismic reflection methods.

FIGURE 18–12 Seismic refraction time-distance plot.

time-distance plot (Figure 18–12). The relationship between the orientation, thickness, and seismic velocity of each unit is determined from the time-distance plots (Figure 18–13). The resolution of a seismic refraction survey depends upon the spacing between the observation points. The engineering geologist must have a general idea of the subsurface geology prior to running a seismic refraction survey in order to plan the survey and to interpret the data, because structural features can produce complicated data that can lead to incorrect interpretations.

Seismic reflection profiles, or soundings, provide **travel time** data that must be corrected for velocity before depth interpretations can be made

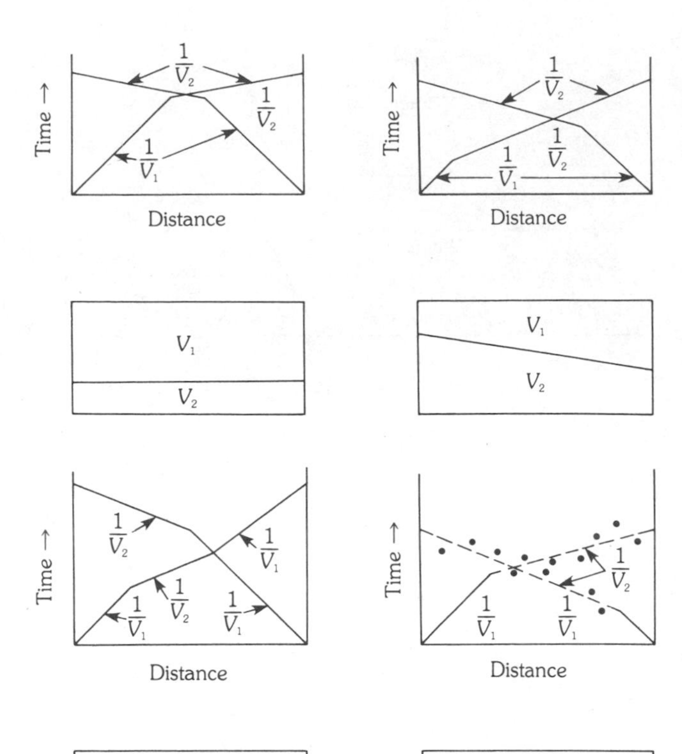

FIGURE 18–13 Seismic refraction data and the corresponding interpretation.

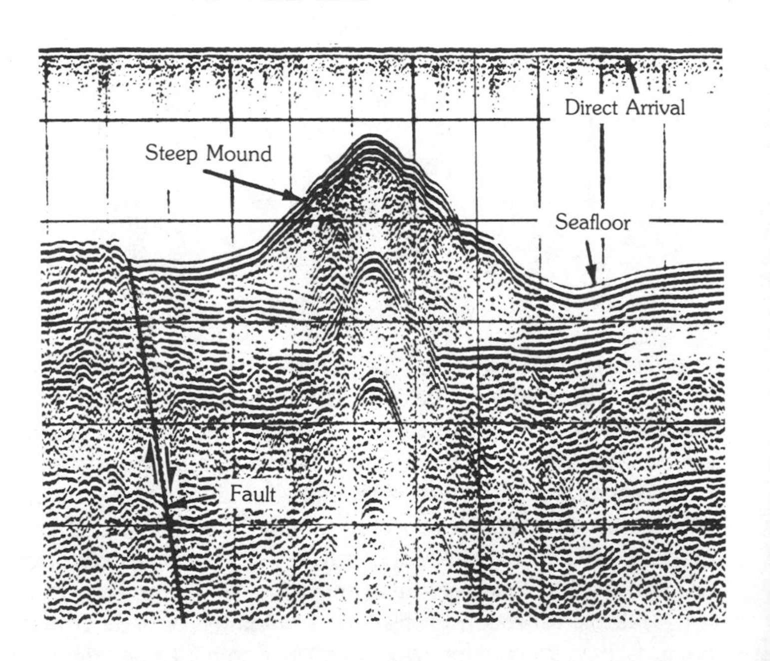

FIGURE 18–14 Copy of a seismic reflection record. (Courtesy of McClelland Engineers, Ventura, California.)

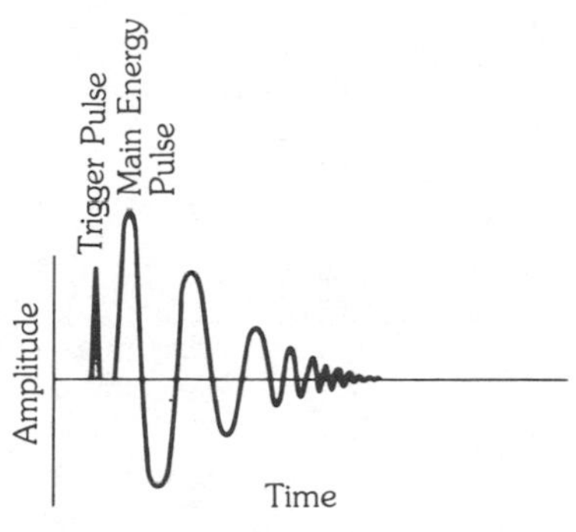

FIGURE 18–15 Wave train generated by some types of seismic reflection energy sources.

(Figure 18–14). A seismic refraction survey is sometimes run to measure seismic velocity values to calibrate seismic reflection surveys. The **resolution** of a seismic reflection survey depends upon the wavelength of the energy source/receiver system used, with resolution increasing as wavelength decreases. **Penetration**, or depth of the survey, also depends upon the wavelength of the energy source, with penetration increasing as wavelength increases. As a result of these relationships, high resolution seismic reflection surveys are usually limited to shallow depths of penetration. Interpretation of seismic reflection surveys requires an understanding of both the equipment used and the geology of the site. Some of the high-power seismic energy sources produce their energy in the form of a wave train rather than as a single energy pulse. The wave train produces a series of reflections from each reflector (Figure 18–15). The characteristics of the reflecting surface (velocity contrast) and the material also affect the reflected signal and interpretation of the data. Computer processing, filtering, and signal enhancement techniques can be applied to digitized seismic reflection data.

Electrical properties can be utilized to interpret subsurface conditions (Figure 18–16). The common methods used in engineering geologic studies

FIGURE 18–16 An electrical resistivity survey.

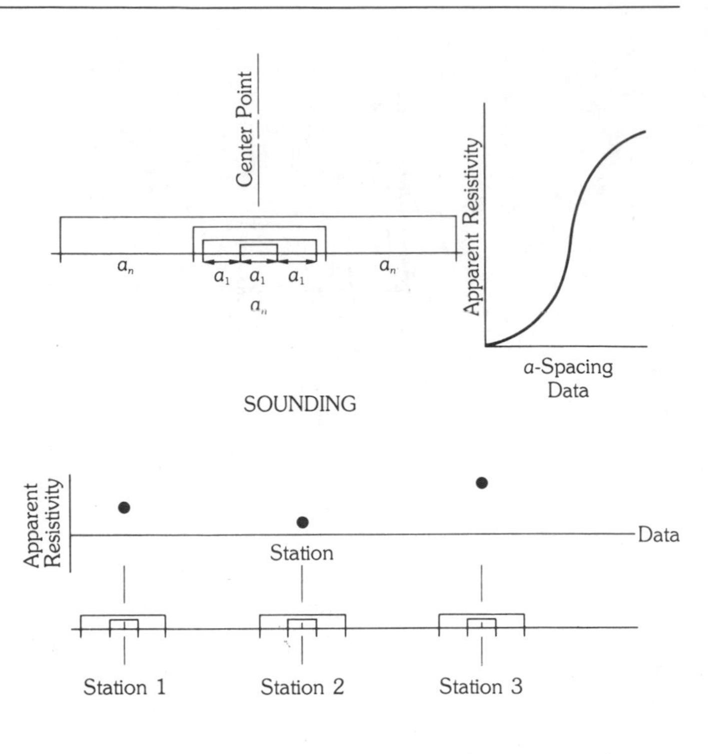

FIGURE 18–17 Comparison of an electric resistivity sounding and a survey.

are either a **sounding** or a **profile** or **survey** (Figure 18–17). The interpretation of electrical resistivity data is more complicated than interpretation of seismic data, because it is based on matching curves. The field data are plotted against theoretically generated data from estimated geologic conditions. Electrical resistivity surveys in areas having units with distinctly contrasting electrical properties (gravel or clean sand versus clays) can be run and plotted in the field. When combined with selected direct observations from cores or pits, the subsurface geology can be mapped because the resistivity survey is interpreted by direct observation.

Magnetic and gravity surveys are generally used for regional studies because the anomaly—magnetic or gravity (mass) contrasts—is often masked by larger and deeper features that are not involved in the engineering project. This masking effect exists because the magnetometer or gravimeter is measuring the total magnetic force or gravitational force of the earth. In order to evaluate smaller-scale features, such as intrusive bodies or crustal displacements, the regional average or total earth force must be removed from the recorded data. The limitation on the resolution of these geophysical techniques lies in the accuracy of the determination of regional values.

Techniques that measure changes in the magnetic or gravity **gradient** have recently been developed and applied to engineering studies. The basis for these techniques is the fact that both magnetic and gravity forces are a function of the inverse of the distance (R):

$$\text{Magnetics:} \quad F = \frac{1}{\mu}\left(\frac{P_0 P}{R^2}\right)$$

$$\text{Gravity:} \quad F = G\left(\frac{mM}{R^2}\right)$$

where

μ = magnetic permeability ($\mu = 1$ in air)
P_0, P = magnetic strength
G = universal gravitational constant
M, m = mass of body
R = distance between the two bodies

A small change in distance (R) relative to a deep-seated feature is negligible but may be significant to a nearby shallow feature.

Natural radioactivity is a technique used in **borehole logging**. This technique, along with electrical resistivity or electrical conductivity, should be used in all subsurface coring programs to maximize the information collected during the investigation (Figure 18–18). In soil and soft-rock areas the low strength of the rock units may complicate coring and consequently increase the cost. **Wash borings** or **rotary holes** can be rapidly and

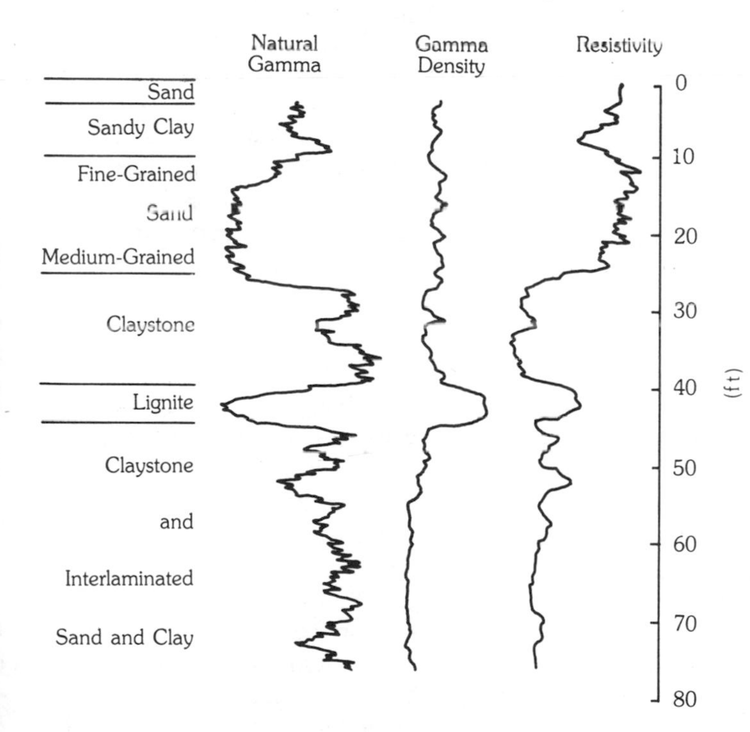

FIGURE 18–18 Geophysical (electric) logs with the corresponding geologic description of the core.

inexpensively drilled and the "electric" logs interpreted. Then these logs can be used to select significant sites where the more expensive cores should be collected for laboratory testing. Electric logging techniques should be used in all core holes in order to confirm that the core recovered is representative of the subsurface conditions. In many cases, thin, loose sand beds will be missed during coring, only to be discovered as "surprises" during construction. In hard rocks, the expense of drilling a rotary hole destroys its advantage. As a result, hard rocks are usually cored; however, "electric" logs should be obtained as a check against the core.

Surveying is usually provided on a major job; however, on many small projects, a surveyor may not be available to determine the location of observations. The engineering geologist should have a basic knowledge of how to find the vertical and horizontal position of each data point. If aerial photographs of the site are available at the desired scale, the position can be determined by **photogrammetry**, where the field location is plotted on the photograph and then scaled off at a later time. If the position of the observation cannot be determined on the photograph, or if the position is on a body, as for a seismic reflection survey, the engineering geologist must resort to some surveying.

Numerous techniques and modifications of the principles of surveying can be applied to **marine position fixing**. Marine position fixing differs from navigation, because the marine survey has to maintain a continuous record of the position of the survey vessel in order to relate the survey data, while the purpose of navigation is just to get from one place to another. For example, a survey of a dredged channel ultimately will relate the survey data to a geographic location. Therefore, position data may be recorded every five minutes. The navigator, on the other hand, needs to keep the vessel in the marked channel and is not concerned about the exact position throughout the trip. Marine position fixing uses either visual or electronic techniques. Both determine the position of the survey vessel relative to known points on land. Visual techniques rely on some form of triangulation while electronic techniques use distance measuring.

SAMPLING

The engineering geologist has the responsibility for characterizing the physical properties of the site and identifying the geologic units that may adversely affect the engineering design and construction of the project. A significant aspect of any site investigation is the collection of representative samples of the materials at the site for laboratory testing. Laboratory testing is expensive and time-consuming, and therefore, the engineering geologist must assure that units with physical properties that make them liable to affect the project are adequately sampled, described in the field, and preserved for laboratory work. The geologic information about the site gained from field study must be applied to sample selection and collection.

A large portion of the design of a foundation, excavation, or other engineering activity that deals directly with the earth will be based on the samples sent to the laboratory. The design engineer may visit the site only during an inspection tour, because he or she is counting on the professionals in the field to be professional. "Sitting a drill rig" may not appear to be an exciting job, but it is critical. Field notes, outcrop and core descriptions, and sample selection and handling reflect the quality of the engineering geologist and the products produced.

Core and sample handling are critical to the success of laboratory testing. The properties of soils and some rocks change when the moisture content changes; therefore, every effort must be made to assure that the moisture conditions are preserved. A perfectly preserved sample, without adequate sample identification, has absolutely no value in the laboratory. Sample identification should include:

1. Sample number.
2. Project name or number.
3. Location or core number.
4. Top and bottom on all cores.

Oriented outcrop samples should be marked before they are removed.

CONSTRUCTION MONITORING

Regardless of how extensive a site investigation may have been, the geologic characteristics of the site are still an interpretation of the available data. The engineering geologist must keep in mind that the "contractor will dig up your mistakes" during construction, because construction excavations expose the entire subsurface. The engineering geologist involved in construction monitoring has the responsibility of reviewing the site investigation report and determining whether the geology exposed during construction is what was expected. If the geologic conditions vary from the expected, the engineering geologist must revise these expected conditions and provide this information to the contractor in time to avoid construction accidents or delays.

Accurate records, maps, descriptions, and occasionally a photographic record of the geology as it is exposed should be kept. Detailed geologic maps may be required as part of the record for major projects, such as nuclear power plants and large underground chambers. In some cases, these records are required by law. Geologic techniques such as **planetable** mapping (Figure 18–19) and frequent low-level aerial photographs and photogeologic maps may be used. The engineering geologist's work schedule and conditions are drastically different from those during the site investigation. Heavy equipment, blasting, and other construction activities are going on at the same time. Time spent mapping is often time lost to construction, and therefore the engineering geologist is a member of the construction staff and works with the engineers on construction scheduling. Seismic surveys may have to be scheduled prior to start-up, during lunch, or after shutdown each day in order to obtain "noise-free" data. The granting of an operating permit or the rapid settlement of claims for changed

FIGURE 18–19 Planetable survey of a construction excavation to record foundation conditions.

conditions may depend upon the accuracy and thoroughness of construction-monitoring geology.

In addition to recording the geology of the site as it is exposed, the engineering geologist frequently monitors the impact of construction on the surrounding area. Deformation and creep of materials into an excavation, settlement of the ground surface above an underground opening, groundwater drawdown and movements, and blast-related ground vibration are all part of this responsibility. By monitoring these phenomena, the engineering geologist tries to protect the contractor from unwarranted damage claims from nearby property owners and to protect nearby property owners from damages caused by poor construction practices. The geology of the site, as predicted during the site investigation and modified during construction monitoring, forms the basis for any monitoring effort. The use of any protective structure, such as a retaining wall, rock bolt pattern, or underground support system, is related to the geology of the site. Numerous types of support are available to the contractor, but their performance and costs vary widely. Deformation monitoring systems must be selected to match the type of support used, the geology of the site, and the rate and amount of expected deformation.

Groundwater monitoring programs must be preplanned based on the site geology in order to preserve the resource if it is used in the area, to prevent construction flooding, and to protect the resource from construction-related pollution. Groundwater **dewatering** programs may be necessary when excavations extend below the water table. These programs will cause a cone of depression around the construction site, which could temporarily deplete a local water source. The replacement of this water source will probably be the contractor's responsiblity.

CASE IN POINT

Explosion in the San Fernando Tunnel

The California Aqueduct distribution system called for 5.5-meter finished diameter water-pressure tunnel across the northern San Fernando Valley. The 9-kilometer-long tunnel was excavated from east to west using a 6.7-meter tunnel boring machine in soft sandstone. The site investigation for the project warned of potential natural gas areas near the western end of the tunnel route. The geologic evidence for the warning was based on the existence of a producing field only 2.75 kilometers away, known oil and tar seeps in the area, the fact that two other tunnels in the area encountered gas, knowledge that the tunnel route crossed the Santa Susana fault which is a structural trap in the nearby Cascade oil field, and the presence of the Pico Formation sandstone, which is a known oil and gas reservoir rock, along the tunnel route.

In spite of this geologic evidence warning of potential gas and warning signs stating to "expect explosive gas ahead," an explosion occurred at 12:30 A.M. on June 24, 1971, that killed seventeen of the eighteen tunnel heading crew. The investigation following the explosion determined that for two days prior to the fatal explosion, work had been halted numerous times because of the presence of gas, that a core smelling of oil had been recovered from the face one day before the explosion, and on the same day a minor explosion had injured four workers.

The significance of this disaster for engineering geologists is that the contractor, who was found guilty of violating California Labor Code provisions, did not continue to maintain a program of construction monitoring of geologic conditions. The contractor did not halt the project when faced with the gas problem and call in geologic or engineering consultants. A site investigation that warns of a potential hazard as part of the contractor's bid package information and then is not heeded is as good as no investigation at all.

Bibliography

City of Los Angeles, *Criminal Court Trial*, Transcript of Record, 1971.

"Disaster and Aftermath," *Western Construction*, August 1973, pp. 58–60.

Richard J. Proctor, "San Fernando Tunnel Explosion," *Underground Space* (1979).

Written by Richard J. Proctor, Consulting Engineering Geologist, Arcadia, California.

Whenever hard rock that must be blasted is encountered, the engineering geologist is involved in monitoring the performance of the blasting round and assuring that nearby property is protected. Understanding the regional geology and the orientation and distribution of the rock to be blasted must be combined with a knowledge of the foundations of surrounding buildings. In some cases, blast-related accelerations will not, or did not, cause damage. Once again, an accurate geologic interpretation and complete geologic records are necessary to protect both the contractor and the public.

Professional Practices

nineteen

RECORD KEEPING AND REPORT WRITING

Accurate record keeping and clear, concise, and complete reporting are vital professional skills for the engineering geologist. In many cases, the report will form the basis for significant decisions. A poorly documented field study can lead to an incomplete or erroneous report, which in turn can produce a liability suit against the engineering geologist or the client.

FIELD NOTES

Field notes must include time, date, location, and names of people involved in sufficient detail for a third party to read them and understand the material or discussion. Field notes should be kept in an organized and orderly manner and should be recorded on the spot, not from memory sometime later, using a standardized notebook system. Some engineering geologists use bound notebooks and keep everything in them, while others use loose-leaf notebooks and keep the past days' notes in a file in the office. The bound book user runs a risk of losing everything if the book is lost in the field, unless copies are maintained in the office, while the loose-leaf notebook user does not have notes from previous days available as a field reference.

Every geologic description should include at least the following:

1. Rock or soil type and definitive adjectives.
2. Color and range of color, moist and dry.
3. Grain size and sorting.
4. Prominent minerals, type of cementation.
5. Groundwater conditions.

6. Structure, fractures and faults in rock, structure in soil.
7. Degree of weathering.
8. Orientation of structures and rock mass.
9. Rock or soil conditions; soundness in rock, hard to soft; dry, moist, wet consistency in soil.
10. Other pertinent observations, such as slope stability.

This information should be recorded in both outcrop and core descriptions. In many cases, the engineering geologist should include a color photograph with the description, being sure that every photograph includes at least a scale, even if it is a knife, lens cap, rock hammer, or scale bar, and an identification mark so that the photograph can be identified later. A useful tool is a color card with the photo number and scale on it. The color card is used later to assure that the colors in the photograph are true to the colors in the field. All special data sheets, forms, or records must be identified when they are filled out or removed from a recorder. As a minimum, the name of the engineering geologist, time, date, and location, or instrument number, plus any unusual information, should be recorded. Data identification, ordering, and handling are the first components of any complete and accurate job.

WRITTEN REPORTS

Written reports take many forms, depending upon their intended use. Regardless of the form of the report, they all contain the same basic components:

1. *Who* asked for the report?
2. *Why* was the report written?
3. *What* was investigated or done?
4. *How* was the investigation or work carried out?
5. *What* are the results?
6. *What* do the results mean?
7. *What* are your conclusions and recommendations?

A **letter, memorandum**, or **short report** is just what it sounds like—a short summary of a larger study or a complete report on some particular project or problem. These reports are used for internal information, such as a report on an existing foundation or the performance of a construction technique. The short length of the report does not excuse the engineering geologist from accuracy or completeness. It simply saves time both for the writer and the reader. The **who** is answered by the address given at the top of the report. The **why** and **what** are answered in the first few sentences followed by the **how** in the first paragraph. The results may be an attached figure or graph or a verbal description in the second paragraph. Discussion of the results are given in the last paragraph. If the report requires more than four or five pages, it is generally not suited for a short report form.

The **formal report** is longer and is frequently submitted at the conclusion of a project or project component. For example, weekly field investigation reports may be submitted in short-report form, with the formal report being submitted at the end of the field investigation. This type report is the professional transmittal of yourself and is therefore the primary source or review of your work. The report should follow the form and organization outlined below (when possible):

I. Letter of Transmittal

II. Prefatory Material
   A. Title page
   B. Abstract
   C. Table of Contents

III. Body of Report
   A. Introduction
   B. Investigation
      1. Field
      2. Laboratory
      3. Analysis or problem solution.
   C. Results of the investigation
   D. Discussion of the results
   E. Conclusions and recommendations

IV. Appended Material
   A. List of references
   B. Field and/or laboratory data sheets
   C. Other material (such as photographs, exhibits, and so on)

WRITING THE REPORT

GENERAL The report should be written in correct, grammatical English, using technical phrasing where necessary and avoiding the use of slang and colloquial expressions. It should be written in the third person singular (impersonal) whenever possible, although the modern trend is toward accepting "I" or "we" in expressing opinions. No abbreviations should appear in text material, although they may be used in tabulations, graphs, charts, and the like. With the exception of the conclusions and recommendations, the entire report should be phrased in the past tense. Simple, straightforward expression is preferred over lengthy, involved, or overly complex sentence structure. Remember that you are trying to convey ideas and information to another person through your writing. Simple, clear, direct statements will accomplish this end best. The following paragraphs expand this brief outline, indicating content of each section and preferred style of presentation.

LETTER OF TRANSMITTAL The letter should be set up in standard business letter format. Its purpose is to transmit the report formally to the person or organization for whom the work was done. Accordingly, the content of the letter should be: (*a*) a statement that the report is being transmitted, (*b*) a statement referring to the authorization under which the work was done, and (*c*) an offer to clarify any portion of the report, if needed.

PREFATORY MATERIAL:

1. *Title page.* The title page should include the title of the report, name and address of the person or company receiving it, the name and address of the person submitting it, and the date it was submitted.
2. *Abstract.* The abstract is a short one- or two-paragraph statement of the project results, conclusions, and recommendations. Do not use phrases like *is reported* or *will be shown.* Be specific—the abstract is intended to give the reader an immediate summary of the important information. In many large organizations, the abstract may be the *only* part of the report that management reads, because their time is limited. As a result, the abstract is the most important indication of the quality of your work.
3. *Table of contents.* The table of contents should list all the headings and subheadings with their page numbers. Figures or tables in the text should be listed in the list of figures and list of tables.

BODY OF REPORT:

1. *Introduction.* This section sets forth the scope of the investigation and frequently also contains the background and reasons why the investigation was undertaken (need). The magnitude and scope of the investigation should be clearly specified; general statements should be avoided.
2. *Investigation.* This section should open with a statement outlining the program for the investigation. The aim of the opening statement is to give the reader an overall picture of the way in which the problem was attacked. Following the opening statement, details of the investigation (usually in chronological order) are given. Sampling operations and techniques and details of laboratory testing procedures should be fully described. The equipment used in both field and laboratory investigations should

be described. Standard field and laboratory test procedures can best be described by giving a complete reference to the standard. Where the equipment used must be described, it is sufficient to give the proper name of the item used and give the name of the manufacturer. Nonstandard tests or equipment must be described in sufficient detail that the reader will have a clear picture of the procedure or the piece of equipment. Frequently, a step-by-step outline of the procedure is adequate and probably preferable to an essay-type description.

3. *Results of the investigation.* This section of the report should present the final results of the investigation. Do not present laboratory or field data here. Such material belongs in the appendix. Do not discuss the results in this section; this is the function of the following section. It is proper, however, to note the significant points of the final results. A reference to the portion of the report where the test data may be found usually concludes this section.
4. *Discussion of the results.* In this section, discuss the validity and accuracy of the test results and their applicability to the problem. If applicable, include a discussion of any modification of the results that might be needed to apply them to the solution of the problem, including the reasons for such modification. The aim of this section is to establish in the reader's mind the precision, accuracy, validity, or applicability of the results to the problem undertaken in the investigation.

Occasionally sections 3 and 4 are combined into one section called "interpretation of results" in order to simplify the report. This combination is particularly useful if several different field studies and laboratory tests were carried out during the study and the final conclusion of the work is based on the combined interpretation of the results. A good rule to follow is: If the results and discussion sections are separated, will the reader have to shift back and forth to understand the significance of the results? If so, then the sections should be combined in an Interpretation of the Results section.

5. *Conclusions and recommendations.* Conclusions on whether the investigation was adequate, on whether a satisfactory solution to the problem has been found, and on whether or not additional work needs to be done are presented here. Recommendations on adoption of the results of the investigation should follow. (It should be noted that, should the investigation show that the solution is not satisfactory or feasible, a negative recommendation is appropriate.) A research-type project is usually concluded with a recommendation for further investigation, including a brief statement of the lines along which it should proceed, if further work would appear to be fruitful.

APPENDED MATERIAL

1. *List of references.* It is customary to include a list of pertinent references here. Note that this list includes *only* those works to which direct reference is made in the body of the report.
2. *Field and/or laboratory data sheets.* Include original data sheets or good quality copies.
3. *Other material.* In the formal report supporting evidence, such as extra photographs or maps, usually is not included in the text, but appear in the Appendix. If photographs, figures, and maps are required in the discussion of results, they then belong in that section.

The style, the appearance, and the organization also reflect on the professional. A sloppy report implies a poor job!

CONTRACTS

In many cases, the engineering geologist is engaged as a consultant by the project owner or by the contractor, and a formal, legal, often binding agreement is drawn up between the two parties. This agreement is in the form of a **contract** that sets out the conditions of the agreement. The engineering geologist will perform specific duties, surveys, tests, and submit reports for which the client will pay a specified amount of money. The details of the contract vary in the way the work and payment schedules are set up. Common contract types are given in Table 19–1.

In any contractual agreement, both parties must understand all conditions of the contract. A contract should always be a written document unless the work is easily carried out as a service function. Contracts may take the form of a *purchase order, letter of agreement,* or a *multipage document*, depending upon the scope and complexity of the project.

Since contracts are legal documents that are binding agreements between two parties, forming a contract must be based upon specific legal grounds as follows:

TABLE 19–1 Common contracts

Type	Conditions
Fixed fee	The work will be accomplished for a fixed and agreed upon price or fee.
Cost plus fixed fee	The client pays all costs plus a fixed fee.
Cost plus fee	The client pays all costs and a fee based upon a percentage of costs (for example, cost plus 10 percent).
Cost plus premium	Client pays all costs plus a fee based on cost savings.

1. The work must be legal.
2. The parties to the contract are competent (sane and legal adults).
3. Money or services must change hands.
4. The contract is by mutual agreement of both parties.
5. The contract must conform to legal requirements of the job.

Equally, the grounds for breaking or dissolving a contract are also specified as follows:

1. Acts of war, government, or God.
2. Death of a principal in the contract.
3. Bankruptcy of one of the parties to the contract.
4. Mutual agreement of both parties.
5. Statement of the termination of ability (for example, after starting the work the contractor finds that the work exceeds the ability or capability of the contractor).
6. Court order, if the court finds that discharge of the contract is impossible.
7. Major or significant change in conditions from what was expected.
8. Completion of the work in the contract.

Contracts are not intended to "put it over" on either party. They are agreements between two parties that outline *what is expected.* The engineering geologist is involved with contracts, either as a principal to the contract for services or as an expert witness for either party when geologic conditions are found to be different from what was expected. These disputes can often be expected in complex geologic environments, because the site investigation sees only a limited view of the geology while the construction phase uncovers all the geology. The engineering geologist should work with the client whenever contracts are being drawn up in order to keep them geologically realistic. A geologically unrealistic contract requirement can only lead to project delays, cost overruns, and court battles.

ETHICS AND REGISTRATION

In some states, geology and engineering geology are recognized as a profession that has a direct effect on public safety and welfare, and therefore these professionals are **registered** or **certified** by a state board. These state actions, plus the activities of some professional societies that **certify** professional geologists, have resulted in an increasing awareness of registration or certification.

Professional registration and/or certification in geology and engineering geology is relatively new compared to registration as a professional engineer. However, as this entire text has shown, engineering geology affects the life and property of the public—a public that has grown aware of the responsibility of any professional to provide the safest and most cost-efficient service or product. The engineering geologist must be a complete *geologist* who can analyze geologic data and interpret how the geologic system will impact on human activities and how human activities will impact on the geologic system. This professional responsibility cannot be carried out unless the engineering geologist can communicate these findings to engineers, architects, planners, other professions, and most of all, to the public.

The engineering geologist must also remember that *engineering* is only the adjective and not the noun. The professional is a member of the staff, providing geologic advice to the ultimate decision maker. The byline of this advice is that the *geology does not control the suitability of a site—the economic requirements, necessary to complete a safe project based on the geologic constraints of the site, control.* This byline is restated in the definition of an engineering geologist given in the Preface: the engineering geologist is a geoscientist denied the scientific privilege of being wrong, because an error can lead to the loss of life or property for which the courts may find the professional liable. The willingness to accept liability as a component of the profession is to admit that *geology affects people*!

The primary objectives behind registration/certification programs are (1) to assure that the professional geologist is adequately educated and trained, and (2) to assure that the professional believes in and follows a **code of ethics**. A code of ethics is a set of basic commonsense practices that are used as a guide to the way the professional conducts business. The following examples are from the code of ethics of the American Institute of Professional Geologists and the code of ethics that applies to a registered professional geologist in Arizona.

AMERICAN INSTITUTE OF PROFESSIONAL GEOLOGISTS CODE OF ETHICS

Article I. General Principles

1. Geology is a profession, and the privilege of professional practice requires professional morality and professional responsibility, as well as professional knowledge, on the part of the practitioner.
2. Each Member of the Institute shall be guided by the highest standards of business ethics, personal honor, and professional conduct.
3. Honesty, integrity, loyalty, fairness, impartiality, candor, fidelity to trust, inviolability of confidence, and honorable conduct are incumbent upon every Member, not for submissive observance, but as a set of dynamic principles to guide a way of life.

Article II. Relation of Members to the Public

1. A Member shall avoid and discourage sensational, exaggerated and unwarranted statements that might induce participation in unsound enterprises.
2. A Member shall not knowingly permit the publication of his reports, maps or other documents for any unsound or illegitimate undertaking.
3. A Member having or expecting to have beneficial interest in a property on which he reports must state in his report the fact of the existence of such interest or expected interest.
4. A Member shall not give a professional opinion or make a report without being as thoroughly informed as might reasonably be expected, considering the purpose for which the opinion or report is desired; and the degree of completeness of information upon which it is based should be made clear.
5. A Member may publish dignified business, professional, or announcement cards, but shall not advertise his work or accomplishments in a self-laudatory, exaggerated, or unduly conspicuous manner.
6. A Member shall not issue a false statement or false information even though directed to do so by employer or client.

Article III. Relation of Members to Employer and Client

1. A Member shall protect, to the fullest extent possible, the interest of his employer or client so far as is consistent with the public welfare and his professional obligations and ethics.
2. A Member who finds that his obligations to his employer or client conflict with his professional obligations or ethics should have such objectional conditions corrected or resign.
3. A Member shall offer to disclose to his prospective employer or client the existence of any mineral or other interest which he holds, either directly or indirectly, having a pertinent bearing on such employment.
4. A Member shall not use, directly or indirectly, any employer's or client's confidential information in any way which is competitive, adverse, or detrimental to the interest of employer or client.
5. A Member retained by one client shall not accept, without client's written consent, an engagement by another if the interests of the two are in any manner conflicting.
6. A Member who has made an investigation for any employer or client shall not seek to profit economically from the information gained, unless written permission to do so is granted, or until it is clear that there can no longer be conflict of interest with original employer or client.
7. A Member shall not divulge information given him in confidence.
8. A Member shall engage, or advise his employer or client to engage, and cooperate with, other experts and specialists whenever the employer's or client's interests would be best served by such service.

9. A Member shall not accept a concealed fee for referring a client or employer to a specialist or for recommending geological services other than his own.

Article IV. Relation of Members to Each Other

1. A Member shall not falsely or maliciously attempt to injure the reputation or business of another.
2. A Member shall freely give credit for work done by others to whom the credit is due and shall refrain from plagiarism in oral and written communications, and not knowingly accept credit rightfully due another geologist.
3. A Member shall not use the advantages of salaried employment to compete unfairly with another member of his profession.
4. A Member shall endeavor to cooperate with others in the profession and encourage the ethical dissemination of geological knowledge.
5. A Member having knowledge of unethical practices of another Member, shall avoid association with that Member in professional work.

Article V. Duty to the Institute

1. Every Member of the Institute shall aid in preventing the election to membership of those who lack moral character, who have not followed these standards of ethics, or who do not have the required education and experience.
2. It shall be the duty and professional responsibility of every Member not only to uphold these standards of ethics by precept and example but also, where necessary, to encourage, by counsel and advice to other Members, their adherence to such standards.
3. It shall be the obligation of any Member having positive knowledge of deviation from these standards on the part of another Member, to bring such deviation to the attention of the Institute so that the necessary steps can be taken for the correction or elimination of these unethical practices.
4. By applying for or continuing membership in the Institute, every Member agrees to uphold the ethical standards set out in this Code of Ethics.

STATE OF ARIZONA CODE OF ETHICS

I hereby subscribe to and agree to exemplify the following Code of Ethics:

It shall be considered unprofessional and inconsistent with honorable and dignified bearing for an Architect, Assayer, Professional Engineer, Geologist or Land Surveyor:

1. To act for his client, or employer, in professional matters otherwise than as a faithful agent or trustee, or to accept any remuneration other than his stated recompense for services rendered.
2. To attempt to injure falsely or maliciously, directly or indirectly, the professional reputation, prospects or business of anyone.
3. To attempt to supplant another Architect, Assayer, Engineer, Geologist or Land Surveyor after definite steps have been taken toward his employment.
4. To compete with another Architect, Assayer, Engineer, Geologist or Land Surveyor for employment by the use of unethical practices.
5. To review the work of another Architect, Assayer, Engineer, Geologist or Land Surveyor for the same client, except with the knowledge of such person or unless the connection of such Architect, Assayer, Engineer, Geologist or Surveyor with the work has terminated.
6. To attempt to obtain technical services or assistance without fair and just compensation commensurate with the services rendered.
7. To advertise in self-laudatory language, or in any other manner derogatory to the dignity of the Profession.

1: Properties of Common Rock-Forming Minerals

appendices

A mineral is a natural, usually inorganic substance that has distinctive physical properties and a characteristic chemical composition.

STRUCTURE

In the solid state, nearly all minerals possess crystalline structure, in other words, a special arrangement of the atoms that comprise the mineral. The atoms assume this orderly relation to each other during the formation of the solid.

If the crystallization of the matter takes place under conditions sufficiently free from outside influences, the internal arrangement of the atoms may be expressed in the external form of the mass. The result is a crystal, a solid body bounded by plane surfaces intersecting along straight edges, the directions of which bear an intimate relation to the internal structure. The study of crystals is the province of the science of crystallography, the importance of which is now recognized by chemists and physicists as well as by geologists and mineralogists.

COMMON CRYSTAL FORMS

Some of the common forms are

1. Cube: six square faces at right angles to one another.
2. Rhombohedron: an equal six-sided form in which the sides have parallel but not right-angled edges. Such a form is produced when a cube is distorted by pushing at one of the corners.
3. Prism: faces usually rectangular, with parallel edges.
4. Pyramid: a crystal form with triangular faces which meet at a point; a peaklike form.
5. Octahedron: an eight-sided figure with equal triangular faces.

6. Tetrahedron: a four-sided form with equal triangular faces.
7. Dodecahedron: a twelve-sided figure bounded by diamond-shaped faces.

Only under very favorable circumstances does a mineral form good, easily recognizable crystals. Those inexperienced in mineral determination will find this property of little use.

CRYSTAL SYSTEMS

Crystals are grouped into six systems and thirty-two classes, depending upon their degree of symmetry and the number and relation of their crystallographic axes, as follows:

1. Isometric system—Three crystal axes of equal length at right angles to each other.
2. Tetragonal system—Three crystal axes at right angles to each other. Two, taken as horizontal, are of equal length, and the third, the vertical axis, is of a different length.
3. Hexagonal system—Four crystal axes, three of which are horizontal, equal in length, and lie in the same plane, making angles of 60° and 120° with each other. The fourth is vertical, longer or shorter than the other three but at right angles to the other three.
4. Orthorhombic system—Three crystal axes of unequal length but all at right angles to each other.
5. Monoclinic system—Three crystal axes of unequal length. Two of the axes intersect at right angles, and the plane formed by these two axes is oblique to the third axis.
6. Triclinic system—Three crystal axes of unequal length, all oblique to each other.

This classification is not, however, of much assistance in an elementary study of geology, because most of the minerals that compose common rocks were crystallized under conditions that did not permit the formation of large, perfect crystals. Physical properties other than crystal form will, therefore, be used more frequently in the identification of minerals.

DESCRIPTIVE STRUCTURAL TERMS

When minerals form, the particles arrange themselves in various ways that are usually described by familiar terms. The following terms have been used:

1. Crystalline: a structure that results from the orderly arrangement of atoms throughout the whole mass. A property not easily recognized unless definite crystals are present. Most minerals are crystalline.
2. Amorphous: not crystalline, without crystal form or orderly internal atomic arrangement.
3. Acicular: slender, needle-like growth of crystals.
4. Radiating: crystals arranged about a common center like the spokes of a wheel about the hub.
5. Dendritic: crystals, often minute, arranged in a treelike branching structure.
6. Filiform: in forms of thin wires or hairs.
7. Granular: a crystal aggregate of fine or coarse individuals held like the grains in lump sugar.
8. Drusy: a crystalline incrustation in which a surface becomes coated by a layer of small crystals.
9. Botryoidal: semiglobular kind of structure giving a surface that resembles a bunch of grapes cut through the middle.
10. Reniform: a structure with a surface much like that of a kidney.
11. Foliated: a thin leaf- or booklike structure.
12. Micaceous: a platy structure in which the mineral can be split into exceedingly thin sheets.
13. Lamellar: consisting of plates of appreciable thickness.
14. Stalactitic: iciclelike aggregates.
15. Concentric: onionlike structure composed of layers with a common center.
16. Concretionary: consisting of rounded particles.
17. Oolitic: consisting of small rounded particles, like bird shot.
18. Pisolitic: consisting of rounded particles about the size of a pea.
19. Nodular: the particles are large, rounded, and irregular in shape.
20. Compact: a uniform aggregate of minute particles.
21. Tabular: broad, flat surfaces.
22. Massive: a mineral composed of compact material and not showing any definite form is said to have a massive structure.
23. Disseminated: the mineral is rather uniformly distributed through the rock in which it occurs.

PHYSICAL PROPERTIES OF MINERALS

A great many of the various minerals can be identified by observing their physical properties. The more important of these properties are streak, hardness, color, luster, cleavage, fracture, specific gravity, diaphaneity, and tenacity; the less important are odor, taste, feel, and magnetism. The nature of these properties must be ascertained by examining a fresh specimen of the mineral, since alteration of the mineral causes its physical properties to vary.

COLOR

The color of some minerals is constant, but that of many minerals may range through various shades. Consequently, different specimens of the same mineral may show different colors, and in such cases the color of the specimen may be misleading.

STREAK

The streak of a mineral is the color of its powder. This may be determined by rubbing the mineral on a streak plate (a small piece of unglazed porcelain), by scratching the specimen to produce a powder, or by pulverizing a small bit of the mineral and noting the color of the powder.

HARDNESS

Hardness is the resistance a mineral offers to scratching. Hardness can be determined by attempting to scratch the specimen with another mineral or with some common object of known hardness. Table 1 gives the minerals of the standard hardness scale and other minerals that may be used to test hardness. The scale is an arbitrary division of all minerals into ten groups on the basis of hardness and does not indicate the actual ratio of hardness of one mineral to another.

CLEAVAGE

Cleavage is the tendency of minerals to break or split along certain parallel directions, forming smooth planes. These planes are usually parallel to some crystal face or some possible crystal face. Minerals may have 1, 2, 3, 4, or 6 cleavage directions. If a mineral has three or more directions, the cleavage fragments may be bounded on all sides by fairly smooth, plane faces and may resemble a complete crystal. Cleavage is frequently used as a diagnostic feature in the identification of minerals.

FRACTURE

Fracture is the term used to define the character of the surface obtained when a mineral is broken in any direction other than one of its cleavage

TABLE 1 Scale of mineral hardness

Grade	Standard Mineral	Other Minerals	Useful Objects
1	Talc	Graphite	
2	Gypsum	Sulphur Halite	
2.5			Fingernail
3	Calcite	Galena Barite	
3.5		Sphalerite	Copper coin
4	Fluorite	Pyrrhotite Chalcopyrite	
5	Apatite		Iron nail
6	Orthoclase	Magnetite	Glass Knife blade
7	Quartz	Garnet	File
8	Topaz		
9	Corundum		
10	Diamond		

directions. There are several terms used to describe the fracture of minerals, such as:

1. Even: a smooth but not flat surface.
2. Uneven: irregular surface without any special character.
3. Hackly: many irregular sharp edges in the broken surface.
4. Conchoidal: a shell-like surface with curved and often concentrically ribbed markings.
5. Splintery: a woodlike fracture.
6. Earthy: broken surface like that of clay or chalk.

Notice that cleavage and fracture in minerals may be compared with the tendency of a piece of wood to split with fairly smooth faces in a direction parallel to the grain and to fracture with a splintery surface in any direction across the grain.

LUSTER

Luster is defined as the appearance of the surface of a mineral in reflected light. Minerals are divided into two large groups on the basis of their luster: first, metallic luster (reflecting light like a piece of polished metal), and second, nonmetallic luster. In the nonmetallic group there are several subdivisions: vitreous (reflecting light like glass), greasy (reflecting light like a piece of greasy glass), adamantine (reflecting light like a diamond), earthy, waxy, resinous, and so on.

DIAPHANEITY

Diaphaneity is the term applied to the degree of transparency, the amount of light that will pass through a mineral. A mineral is described as transparent if the outline of an object seen through it is distinct, as translucent if light is transmitted through the specimen but objects cannot be seen, and opaque if no light passes through the specimen even on very thin edges (most minerals with metallic luster are opaque).

SPECIFIC GRAVITY

Specific gravity is the weight of the mineral as compared to the weight of an equal volume of water. Accurate determinations of specific gravity require apparatus that is not readily available in field work.

TENACITY

The tenacity of a mineral depends upon the cohesion of the particles, or the power they have to hold together. Various terms are used to describe this property:

1. Brittle: easily shattered by a hammer blow.
2. Malleable: can be hammered into thin sheets.
3. Ductile: can be drawn out into a wire.
4. Friable: granular minerals that can be crushed or worn off in the hand.
5. Sectile: can be cut into thin shavings by a knife.
6. Flexible: thin sheets of the mineral can be bent back and forth without breaking. Bends under stress and stays bent when stress is removed.
7. Elastic: thin sheets of the mineral can be bent within the so-called elastic limit but will not remain bent after the distorting force is removed.
8. Tough: resists shattering by hammer blows. Opposite to brittle.

ODOR

Some minerals have a distinct odor after they have been rubbed, moistened, or struck by a blow.

TABLE 2 Properties of common minerals

Mineral	Common Color	Luster	Streak	Hardness	Cleavage	Fracture	Relative Specific Gravity	Remarks
Native sulfur	Yellow	Resinous	—	1.5–2.5	—	Conchoidal to uneven	Light	—
Graphite	Black	Metallic	Black	1–2	1	—	Light	Greasy feel
Galena	Gray	Metallic	Black	2	3, forms cube	—	Very heavy	—
Sphalerite	Brown to black	Resinous	Light	4	6 planes	—	Heavy	—
Chalcopyrite	Golden yellow	Metallic	Black	4	—	Uneven	Heavy	Tarnished on old surface
Pyrite	Pale yellow	Metallic	Black	6	—	Uneven	Heavy	Sometimes cube-shaped crystals
Marcasite	Pale yellow	Metallic	Grayish black	6	—	Irregular	Heavy	Cockscomb structure
Hematite (specular)	Black	Metallic	Reddish brown	6	—	Platy	Heavy	Micaceous
Hematite (oolitic)	Red to brown	Earthy	Reddish brown	1–3	—	Uneven	Heavy	—
Magnetite	Black	Metallic	Black	6	—	Uneven	Heavy	Magnetic
Limonite	Brown	Earthy	Yellow brown	1–6	—	Conchoidal	Average to heavy	—
Bauxite	Red to white	Earthy	—	2	—	Uneven	Light	—
Halite	White to colorless	Vitreous	—	3	3, cubical	—	Light	Salty taste
Fluorite	Varies	Vitreous	—	4	4	—	Above average	—
Calcite	White to colorless	Vitreous	—	3	3, rhombohedral	—	Average	—
Dolomite	Varies	Vitreous	—	3.5–4	3	—	Average	—
Siderite	Colorless to gray	Vitreous to pearly	White or yellow	3.5–4.5	3, perfect rhombohedral	Conchoidal	Heavy	Brown due to alteration

TABLE 2 Properties of common minerals (continued)

Mineral	Common Color	Luster	Streak	Hardness	Cleavage	Fracture	Relative Specific Gravity	Remarks
Apatite	Green or brown	Vitreous	—	5	1, poor	—	Average	Can just be scratched by knife
Gypsum	White to colorless	Vitreous	—	2	1	—	Light	—
Anhydrite	White, gray	Vitreous to pearly	—	3	3 at right angles	—	Average	Granular
Quartz	Varies	Vitreous	—	7	—	Conchoidal	Average	—
Chalcedony	White to gray	Dull	—	6	—	Conchoidal	Average	—
Opal	Varies	Waxy	—	6	—	Conchoidal	Light	—
Orthoclase	White, gray to pink	Vitreous	—	6	2, at right angles	Irregular	Average	—
Microcline	White, gray to pink, rarely green	Vitreous	—	6	2, almost 90°	Irregular	Average	—
Plagioclase	White to gray	Vitreous	—	6	2, at right angles	Irregular	Average	Thin parallel lines on cleavage surface
Talc	Light	Pearly	—	1	1	Uneven	Average	Foliated
Serpentine	Green to yellow	Greasy	—	4	—	Irregular	Light	May be in fibrous form
Muscovite	Colorless	Vitreous	—	2	1, perfect	—	Average	—
Biotite	Dark brown	Vitreous	—	2	1, perfect	—	Average	—
Chlorite	Green	Vitreous	Green	2	1	—	Average	Foliated
Amphibole	Bright black	Vitreous	—	5	2, at 60° and 120°	—	Average	—
Pyroxene	Greenish black	Vitreous	—	5	2, at 90°	—	Average	Cleavage sometimes poor

TABLE 2 Properties of common minerals (continued)

Mineral	Common Color	Luster	Streak	Hardness	Cleavage	Fracture	Relative Specific Gravity	Remarks
Olivine	Green	Vitreous	—	7	—	Uneven	Average	Sugary appearance
Garnet	Red brown	Vitreous	—	7	—	Conchoidal	Average	—
Glauconite	Green	—	—	2	—	—	Average	Colitic to amorphous granular
Clay minerals	White	Earthy	—	1	—	Uneven	Light	Greasy feel, claylike odor when breathed upon

TASTE

Minerals that are soluble in water can be tasted. Common salt (the mineral halite) is a good example.

FEEL

Certain minerals feel smooth, soapy, clayey, or rough when touched.

MAGNETISM

A few minerals are attracted by a magnet and are termed magnetic. One variety of magnetite (lodestone) acts as a magnet and will attract small tacks or iron filings.

2: Simple Field Identification Tests of Soils

Most simple tests and examinations are indicator tests. In many cases, no single test or examination will be completely diagnostic, but several will usually identify the basic soil type. There are always some "borderline" soils that will be difficult to identify; in these cases the classification that is the more coarse-grained will usually be right. The coarse-grained soils are rarely identified incorrectly because grain size for sand (0.05 mm) is about the limit of resolution for the average person. Below this size, individual grains cannot be distinguished; hence, means other than visual examinations must be used.

The following simple tests will be found most useful for determining the proper descriptive terms:

1. Grain size and grain shape determinations.
   a. Visual examination—coarse-grained soils only.
   b. Feel test or abrasion test—feel of sand grains on palm when subjected to shear motion.
   c. Taste test—feel of grittiness between teeth.
2. Molding test—examination of plasticity.
3. Crushing test—cohesion in the dry state.
4. Dusting test—cohesion in the dry state.
5. Shaking test—pore water mobility.
6. Settling test—rate of sedimentation in water.
7. Free swell test—estimate of expansive nature of clays.

VISUAL EXAMINATION

All coarse-grained soils in the wet or dry state can be identified by visual inspection of grain size, grain shape, and uniformity of grain size. A small

10× hand lens will greatly enhance this test and may provide additional information relative to mineralogy and grain shape. If accurate grain-size descriptions of coarse-grained soils are required, a pocket microscope will allow accurate measurements.

FEEL TEST OR ABRASION TEST

Fine sand or very small quantities of sand can often be detected by feel, since these grains will scratch or grind when pushed over a surface. A wet soil pat is prepared and placed in the palm of the hand. Then the thumb of the other hand is forcefully pushed through the soil, providing shearing motion. If sand grains are present, they will grind or abrade the palm.

TASTE TEST

Identification of the fine-grained soils of the silt and clay sizes depends upon secondary tests, since these sizes cannot be detected by the unaided eye. The taste test detects silt by its grittiness and clay by its smoothness between the teeth. A small sample is placed between the front teeth, which are then gently ground together. Silt-size particles will feel gritty, much like toothpaste. Since a very small sample is used, the test should be run on three samples in order to detect small quantities of silt that may be present in a clay soil.

MOLDING TEST

The molding test can be used to help distinguish between silts and clays by rolling the sample into threads and observing whether they have wet strength or not. If silt is present, the thread will break into pieces when it is picked up by one end; if clay, the thread will remain intact when it is picked up and, in addition, will show some tensile strength when pulled apart. The relative amount of tensile strength will determine whether the clay is lean or fat (low or high plasticity).

Additional information can be obtained by continuing the molding test after the tensile strength of a soil thread has been determined. The thread is reworked by folding and rerolling repeatedly, resulting in the gradual reduction of the water content of the soil. As the water content decreases, the soil will stiffen until it finally loses its plasticity and crumbles. When this happens, the soil should be worked as a lump until it also crumbles. A tough thread and stiff lump indicate high plasticity, while a weak thread and quick loss of cohesion of the lump indicate low plasticity. Highly organic soils have a very weak thread strength and feel spongy as a soil lump.

CRUSHING TEST

The crushing test gives an indication of the properties of the soil material in the dry state. After a sample has been collected, allow a small lump (preferably undisturbed) to air dry, then test the soil strength by crushing the sample between fingers. Dry strength increases as plasticity increases. Silts will crumble and powder under light to moderate pressure, whereas clays will fracture or break (but not crumble or powder) under considerable pressure. Feebly plastic (lean) clays will break under moderate to heavy pressure; highly plastic (fat) clays require heavy pressure and may often be strong enough to resist fracture entirely. Fine silty sands and silts have about the same slightly dry strength; however, they can be distinguished by feel when powdering the dried specimen. Fine sands feel gritty, whereas a typical silt has a smooth feel, similar to flour. These results may also help in the determination of soil type.

DUSTING TEST

The dusting test consists of making a thin smear of wet soil on the heel of the hand, allowing it to dry, and then attempting to brush off the dried material with the other hand. If the soil is mostly silt, it will dust off, leaving the hand reasonably clean; if mostly clay, it will resist dusting off, and will scale rather than dust as it comes off the hand.

SHAKING TEST

The shaking test will readily and immediately identify the coarser silts and rock flours. In the test, a small amount of the soil is thoroughly mixed with water to make a fairly stiff, saturated slurry, the mixture is shaken rapidly in the hand from side to side, and an observation is made as to whether or not pore water is brought to the surface (pore water mobility). Silts will become quite shiny, and clays will show little or no change in appearance. The wet soil is then pinched between the fingers, and both feel and water intake (dull appearance, dilatancy) are observed. If the soil shows a noticeable increase in strength and the shiny appearance disappears, it is a silt.

SETTLING TEST

In soils where silt and clay (also very fine sand) may be present in about equal amounts, the settling test is most useful for identification. A small amount of the dry powdered soil is shaken with distilled water (to which a dispersant has been added to prevent flocculation) for about a minute in a 10-cm test tube and set aside in a rack to let the sediment settle out. All sand-size particles will settle out in 30 seconds or less. At the end of about 20 to 25 minutes all silt-size particles will have settled out. By the end of two hours all but the finest colloidal clay will be settled out. The sediments at the bottom of the tube will have reasonably well-defined layers of sand, silt, and clay (in that order from the bottom up). The relative percentages of sand, silt, and clay can be estimated from the thickness of each layer. Note that because the silt will be unconsolidated and almost liquid, the height of the silt layer should be divided by two for this comparison.

FREE SWELL TEST

In soils where high plasticity is indicated and an estimate of potential expansive-soil problems is desired, the free swell test provides a simple comparison of the volume of a dry powder to the volume of the sedimented soil material. The larger the volume of the sedimented sample, the greater is the potential for expansive-soil problems. Free swell (*FS*) is defined as:

$$FS = \frac{V_s - V_d}{V_d} \times 100\%$$

where

V_s = sedimented volume
V_d = dry powder volume

Initially a 25-cc glass test tube is etched at volumes of 5 cc, 7.5 cc, 8.75 cc, and 10 cc, or a 25- to 50-cc graduated cylinder can be used. An oven- or sun-dried sample of soil material is powdered, and the test tube is filled to the 5-cc mark. Distilled water is added to increase the volume to about 20 cc, and the test tube is vigorously shaken to suspend and disaggregate the clay particles. The test tube is allowed to stand until the clay settles out. The free swell is read directly from the marking on the test tube as follows:

Low: FS is less than 50 percent or V_s is less than 7.5.

Medium: FS is between 50 and 75 percent or V_s is between 7.5 and 8.75.

High: FS is between 75 and 100 percent or V_s is between 8.75 and 10.0.

Extreme: FS is greater than 100 percent or V_s is greater than 10.0.

3: Geologic Maps and Sections

GEOLOGIC MAPS

Geologic maps show the distribution of rocks exposed at the earth's surface. Some geologic maps do not include surficial deposits such as alluvium, glacial, or aeolian deposits. Geologic maps do not show soils.

INTERPRETATION OF GEOLOGIC MAPS

The interpretation of a geologic map should begin with a study of the map legend, because geologic time and the time relationships between the rocks are shown by the relative listing of each rock unit in the legend or key: the youngest rock unit is listed first and age increases downward to the oldest at the bottom. Rock types are given by a rock-type symbol or by a written description of the characteristics of the mapped units. A study of the map key provides a summary of the geologic history of the mapped area and outlines the general sequence of rock deposition or formation.

The determination of time relationships and the interpretation of a geologic map are based on four geologic laws:

1. **The Law of Horizontality:** Most sedimentary rocks are deposited on horizontal or very gently sloping surfaces.
2. **The Law of Lateral Continuity:** Sedimentary rocks are laterally continuous within their depositional environment. The units are bounded by rocks deposited in other depositional environments.
3. **The Law of Superposition:** Sedimentary rocks are always deposited on older rocks. In a stratigraphic section that has not been overturned (normal stratigraphic section), the oldest rock is at the bottom and youngest rock is on top.
4. **The Law of Transection:** Any rock or geologic feature that cuts across another rock is younger than the rock that is cut.

The basic laws can be restated for various geologic conditions as follows:

1. **Flat-lying sedimentary rocks:** A cross section through flat-lying sedimentary rock areas looks just like a layer cake with the oldest rock unit on the bottom.
2. **Dipping sedimentary rocks:** A cross section through dipping sedimentary rock areas is drawn so that the older units are below the younger units, and the direction of dip is toward the younger rock units.
3. **Folded sedimentary rocks:** A cross section through folded sedimentary rocks is drawn so that the older rocks are below the younger rocks. In a syncline, the youngest rock is in the center, and the rocks become older away from the center. In an anticline the oldest rock is in the center with younger rocks farther away.
4. **Intrusive igneous rocks** are always younger than the rocks that were intruded by the igneous body.
5. **Metamorphic rocks** are always younger than the rocks that were metamorphosed.
6. **Faults** have the younger rocks structurally below older rocks. In a normal fault the younger rocks moved down relative to the older rocks, while in a reverse or thrust fault older rocks moved up in relation to the younger rocks. Faults or fractures are always younger than the rock units that are faulted or fractured.

Time-rock relationships are also important features in the interpretation of sedimentary rock deposits because the environments of deposition of the rocks will vary with location and different rocks will be deposited at the same time. Regional changes, such as the gradual subsidence or filling of a depositional basin, will cause the depositional sites to move, and a particular rock type may become progressively younger without any noticeable change in the physical characteristics of the rock units. Time lines are lines across a geologic section that reflect the surface geology at any given time. For example, the surface of a modern geologic map represents a time line showing the erosional (source rock) and depositional sites that exist now. Time lines, therefore, can cross rock units. Time line determinations are used to construct paleoenvironment or paleogeographic maps that show the surface geology at some time in the past.

GEOLOGIC SECTIONS

Geologic cross sections show the subsurface geology on a vertical plane. Sections may be drawn from the interpretation of a geologic map, from subsurface geologic data, such as cores or geophysical data, or from a combination of the two. Accurate construction of geologic sections is important in many engineering geologic projects because the section is a prediction of the subsurface geology. Therefore, *geologic sections drawn for engineering projects should always be drawn at true scale.* Any vertical exaggeration will distort the visual relationships between the thickness and dip of any rock unit or discontinuity within the section, which can lead to improper or unsafe interpretations of the subsurface conditions.

CONSTRUCTING SECTIONS FROM GEOLOGIC MAPS

Since a geologic map shows time-rock relationships in the key and the geographic and structural relationships on the map, constructing geologic sections from geologic maps is based on a careful interpretation of the mapped data. Both the interpretation of the map and the resulting section must be in conformity with the basic laws of geology. Boundaries between igneous and metamorphic rocks and the surrounding sedimentary rocks are inferred from the age relationship between the rocks. If the sedimentary rocks were intruded by an igneous rock body, metamorphic rocks lie between the sediments and the igneous body, and the boundaries are inferred from an interpretation of the mechanism of intrusion. If the sedimentary rocks overlie the igneous or metamorphic rocks, they were deposited on an erosional surface; thus the boundary between the sedimentary rocks and the older, underlying rocks is inferred from the type of sediments deposited and the laws of horizontality and continuity.

Geologic sections drawn through sedimentary rock units are controlled by the laws of horizontality, continuity, and superposition. Constructing sedimentary rock sections is based on the following steps:

1. Determine the relative age between the rock units.
2. Determine the dip angle and direction.
3. Calculate or measure the thickness of the units.

The relative ages of the rock units is determined from the map key. The direction of dip is determined from strike-dip symbols on the map, the law of superposition, or the shape of the outcrop pattern following the "rule of V's." The most valuable tool in the interpretation of dipping sedimentary rocks is the character of the outcrop pattern, since the dip angle or direction may vary through the mapped area.

The shape of the outcrop of regularly dipping beds on absolutely plane topography would be straight parallel belts. As soon, however, as these belts are carved by erosion, each valley produces a reentrant or crenulation in the outcrop, commonly termed a V, and these V's constitute such an important item in map interpretation that they demand a very thorough study. They may be summed up as follows:

1. In horizontal rocks, the trace of the contact plane with the topography follows the contours; and since these bend upstream, in crossing a drainage line, the V will point upstream, its length depending on the magnitude of the valley.
2. If the beds dip upstream, the V will point upstream and in the direction of dip of the rocks.

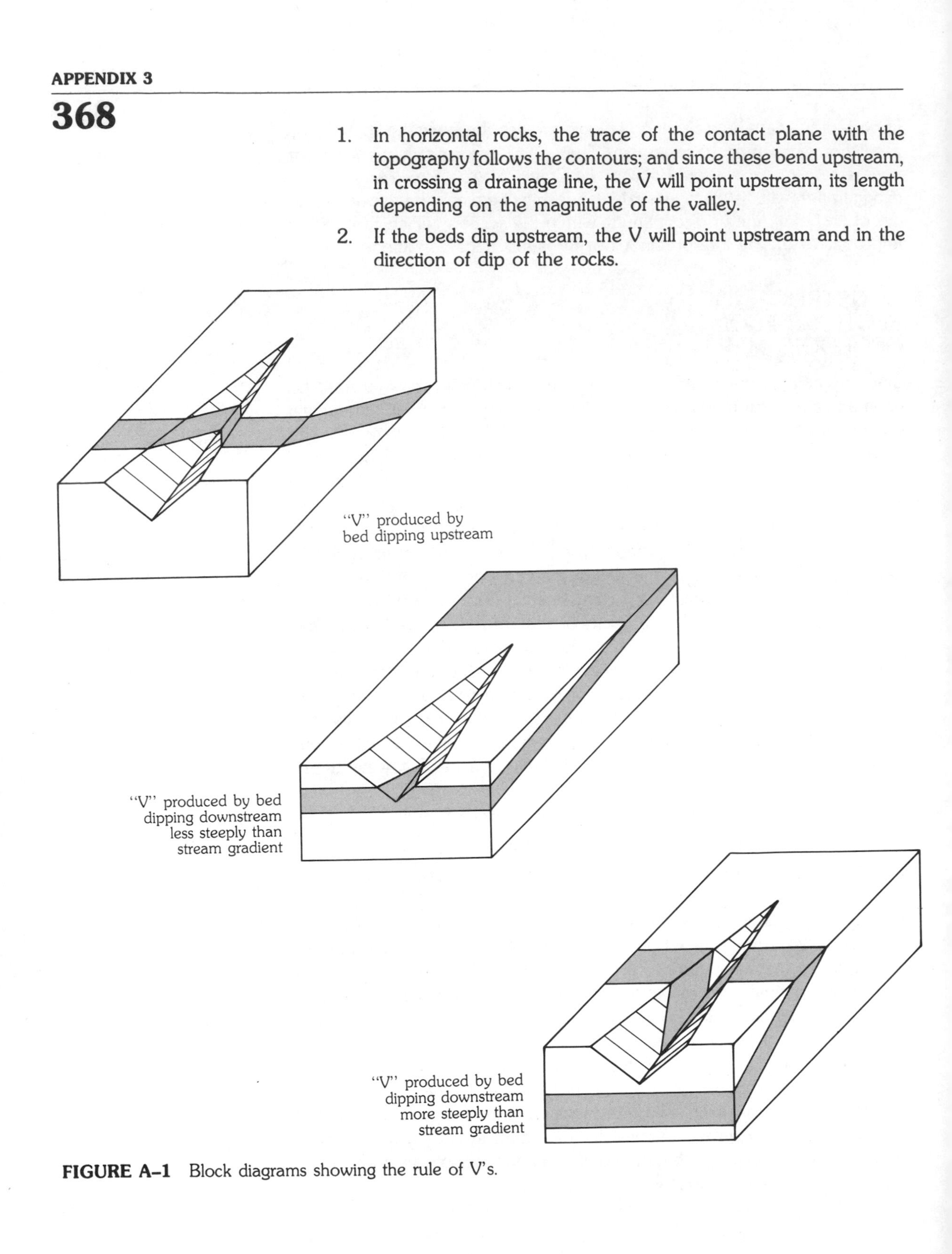

FIGURE A–1 Block diagrams showing the rule of V's.

3. If the beds dip downstream more steeply than the stream gradient, the V will point downstream, and in the direction of rock dip.
4. If the beds dip downstream less steeply than the stream gradient, the V will point upstream.
5. The V's will be infinity when rock dip and stream gradient have the same inclination; and as the dip increases beyond this point, the length of the V's decreases, becoming zero with vertical beds.

Perhaps the most outstanding feature of this summary is the fact that in only one case do the V's point downstream, where the beds are dipping downstream more steeply than the stream gradient. Therefore, all V's that point downstream give absolute evidence of the direction of rock dip. This statement is a dependable, invariable rule.

It will also be noted that there is only one case in which the V does *not* point in the direction of dip: when the beds dip downstream less steeply than the stream gradient. To appreciate the significance of this exception, it must be remembered that few streams of any considerable size have gradients exceeding 10 m/km, and most rivers do not exceed 3 to 5 m/km. Beds, therefore, that dip downstream more gently than the stream gradient of any large creek or river are essentially horizontal.

Since beds do not usually show appreciable parallel or linear outcrop, unless the dip is several degrees (a degree is about 17 m/km), one may safely conclude that on a map on which the outcrops are belted all V's on the larger streams point in the direction of dip, and only on the steeper headwater tributaries V's point in the opposite direction.

If beds are vertical, there are, of course, no V's where the streams cross the contacts between formations. However, the absence of V's where a stream crosses such a contact does not prove vertical beds. There are several reasons why such an inference would be unsafe: (1) If the stream has no appreciable valley, no crenulation will be formed. Obviously it is the valley, not the stream, that is responsible for the reentrant. Even though no permanent stream occupies the valley, the rules for V's still hold perfectly. (2) On many of the older maps careless field work or a failure to appreciate the importance of the V's has resulted in their omission, unless they were perfectly obvious in the field. Consequently, on certain of the older maps the absence of V's has little structural significance. On the contrary, in the best modern mapping, many V's are shown where outcrops are sparse or entirely lacking, the V's being drawn by inference from the known structure. (3) The V, though actually present in the field, may be so short that it cannot be shown on the scale of the map. To appreciate the great importance of this point, let us assume a case of beds dipping 45° and a valley 50 m deep will produce a V 50 m long. On a 1:50,000 scale, 50 m will show as 50/50,000 or 0.1 mm on the map. On a map with a scale of 1:100,000, the same V becomes even less conspicuous, so that the absence of V's, even on the modern, well-made maps, does not necessarily indicate vertical beds, but only very steep dips.

Once the geometry of the sedimentary rocks has been determined from the interpretation of the geologic map, the geometry of the sedimentary rocks in the section must be determined. In flat-lying sedimentary rocks the orientation of the section will not affect the thickness or dip of the beds, since the section is drawn perpendicular to the rock units. In dipping rocks, however, the orientation of the section with respect to the dip direction will produce different sections, in that the dip angle and thickness of the unit will vary with the orientation of the section.

The dip angle on a section will range from zero degrees when the section is perpendicular to the direction of dip to the true dip when the section is drawn parallel to the direction of the dip. Any dip angle that is not true dip is called the **apparent dip**. The thickness of a sedimentary rock unit, measured in a vertical direction, will always be greater than the true thickness if the rock unit is dipping. As the dip increases the **apparent thickness** increases.

Apparent dip is calculated from the following equation:

$$\tan \theta_A = \tan \theta \cos \phi$$

where

θ_A = apparent dip
θ = true dip
ϕ = orientation angle from the direction of dip

The apparent thickness is calculated from:

$$t_a = t/\cos \theta$$

where

t_a = apparent thickness
t = true thickness
θ = true dip angle

The construction of a geologic section begins with the selection of the location of the section on the geologic map and proceeds as follows:

1. Transfer the surface geology and topography to the section.
2. Measure the section orientation angle and calculate apparent thickness and dip.
3. Project each sedimentary rock unit into the subsurface using apparent dip and thickness and complying with the laws of geology.
4. Calculate the apparent dip of any fault or other discontinuity and project the discontinuity into the subsurface.

5. Draw the section, using freehand lines, with offsets or other displacements at the discontinuity, if one exists.
6. Check the accuracy of the section for compliance with the laws of geology and for compatibility with the original interpretation of the geologic map.

GEOLOGIC SECTIONS FROM SUBSURFACE DATA

In many engineering projects the engineering geologist or geotechnical engineer must construct geologic sections from exploration data because detailed geologic maps are not available. In these cases, the geologic map is interpreted to determine the regional geology of the site and to provide a basic framework for the detailed studies.

The interpretation of the exploration data is based on a analysis of the driller's logs, rock or soil samples collected, and if available, geophysical data. Prior to constructing a geologic section, the engineering geologist must determine the time and rock-type relationships at the site. An analysis of the stratigraphy, environments of deposition, igneous or metamorphic rock type, and the nature of the contacts between rock types is required to establish the time and rock relationships. This analysis replaces the study of the map key and must be made before the section can be constructed. For example, the determination that the sands were deposited in stream channels will lead to a different interpretation of the exploration data than would be reached if the sands were deposited in a beach environment. The lack of a metamorphic zone between a sedimentary rock section and an intrusive igneous rock suggests a different interpretation than one that included a metamorphic zone.

Once the time and rock relationships have been determined for each core, construction of a geologic section can be started. In structurally complex sedimentary rock regions, it may be easier to construct a stratigraphic section first in which some characteristic sedimentary rock unit is used as the datum. The unit selected as the datum should, if possible, be a continuous unit that was deposited in a large, nearly uniform environment of deposition, such as a shallow marine limestone or a coal seam. Once the stratigraphic section has been constructed, it should be checked against the core interpretations to assure that the section does not violate any geologic law.

The stratigraphic section is used to identify the sedimentary units and to provide a stratigraphic correlation between exploration sites. The section is now redrawn using sea level as the datum. Stratigraphic offsets or other discontinuities are interpreted as being related to deformation of the stratigraphic section by faulting or folding.

For example, Figure A-2 shows a stratigraphic section and the corresponding structural section constructed along the centerline of a

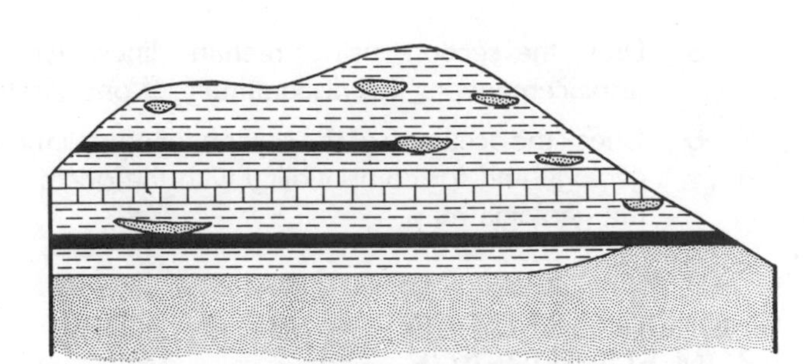

Stratigraphic Section

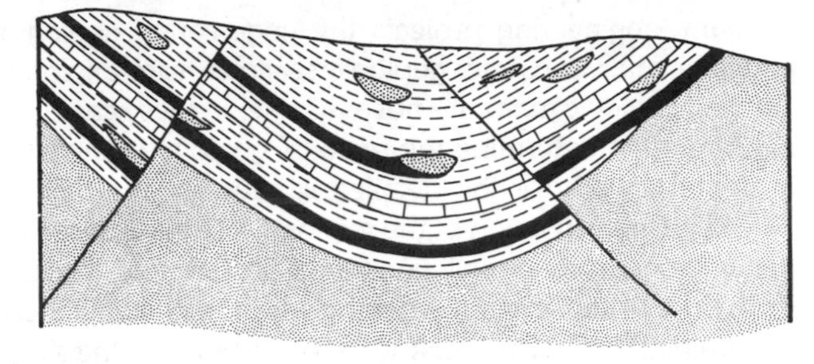

Structural Section

FIGURE A–2 Stratigraphic and structural sections along a proposed storm sewage tunnel. The marker bed used for stratigraphic section is a lower coal seam. (Vertical dimension has been exaggerated 10 times in order to fit the sections on a page.)

proposed storm sewer tunnel. The figure has been intentionally distorted in the vertical dimension so that it can be shown on a text page. It would be 65 inches long for the 10:1 vertical exaggeration to be removed. This figure also demonstrates why all engineering geologic sections should be drawn to true scale. After the structural section has been constructed, it should be checked again to verify that the section complies with the laws of geology.

A very important philosophy in the interpretation of cores and in the construction of geologic sections is to maintain multiple working hypotheses. All logical and realistic geologic interpretations of the initial exploration data should be defined in the beginning of a project in order to assure that the correct interpretation is finally made. New data obtained are used to disprove interpretations until a very few remain under consideration. The number of potential realistic interpretations that still exist at the end of any investigation are controlled by the magnitude of the engineering project. For example, an investigation for a highway roadcut may be economically able to accept a geologic report that states that the "cut to be made is in sedimentary rocks" while a nuclear power plant site may require a statement that a "fault has not moved during the past one million years."

Surficial

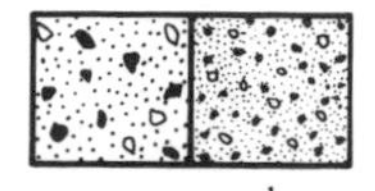

a b
Glacial till and moraines

a b
Glacial till and moraines

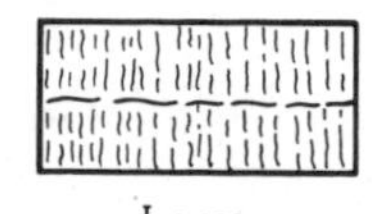

Loess

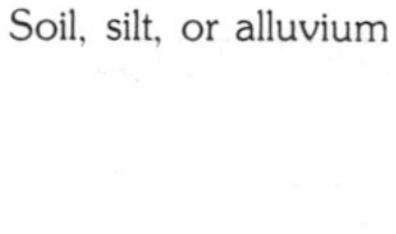

Soil, silt, or alluvium

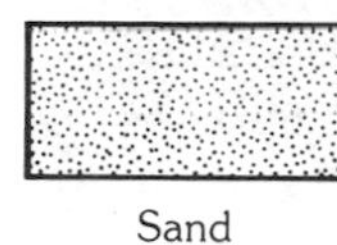

Sand

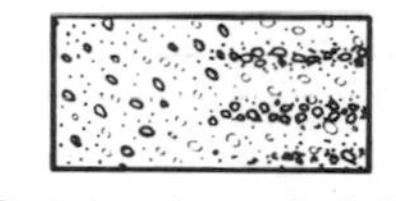

Gravel and stratified drift

Sedimentary

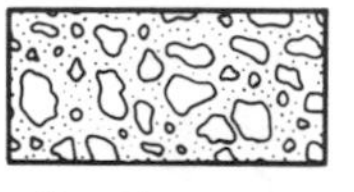

Conglomerate

Breccia

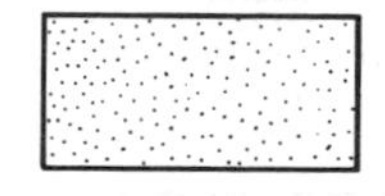

Massive sandstone

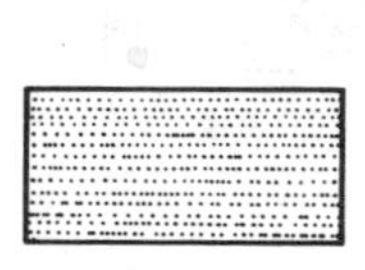

Bedded sandstone

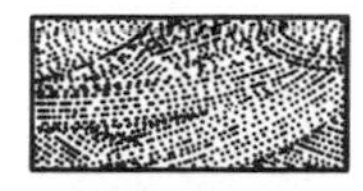

Cross-bedded sandstone

Thin-bedded or shaly sandstone

Calcareous sandstone

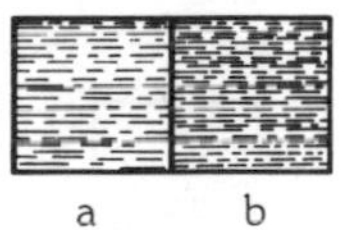

a b
Clay

Sandy clay

Fire clay or flint clay

Calcareous shale or shaly limestone

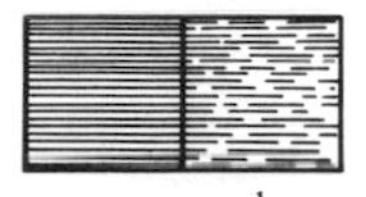

a b
Shale

Sandy shale

Clayey or argillaceous limestone

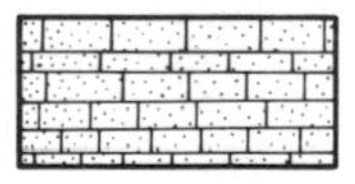

Sandy limestone

Massively bedded limestone

Thin-bedded limestone

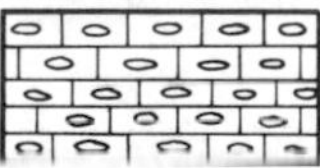

Limestone containing nodules of chert or flint

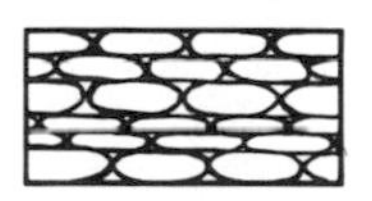

Bedded chert

Dolomite

Crystalline limestone

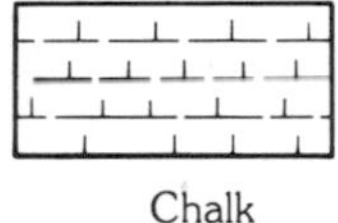

Chalk

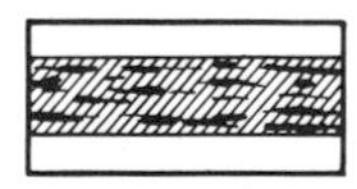

Peat

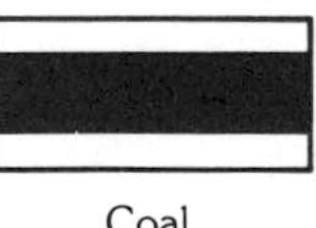

Coal

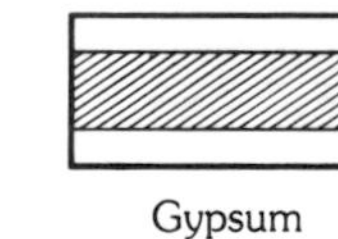

Gypsum

Salt

FIGURE A–3 Selected symbols for sedimentary rocks.

Metamorphic

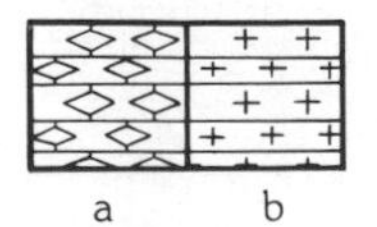
a b
Marble

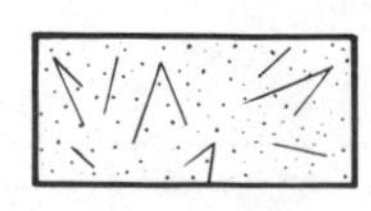
Quartzite

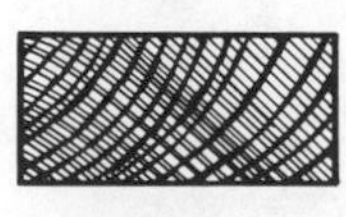
Slate

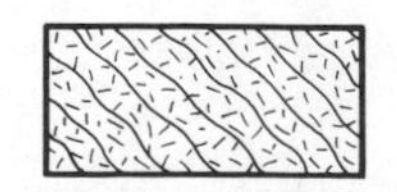
Schistose or gneissoid granite

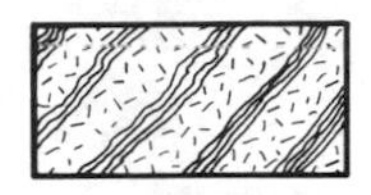
Gneiss

Contorted gneiss

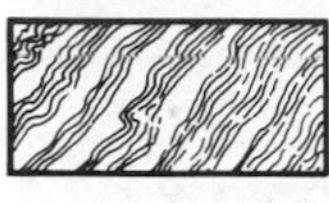
Gneiss and schist

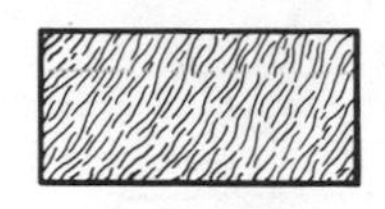
Schist

Contorted schist

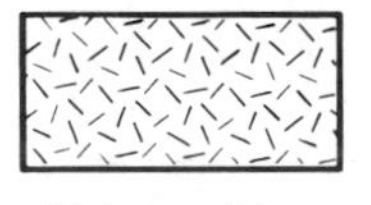
Metamorphism
(combined with sedimentary
and igneous patterns)

Igneous

Volcanic breccia and tuff

Basaltic flows

Bedded lava and tuff

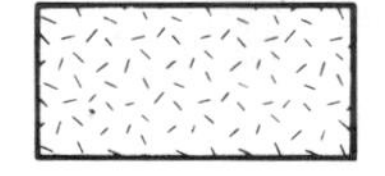
Granite

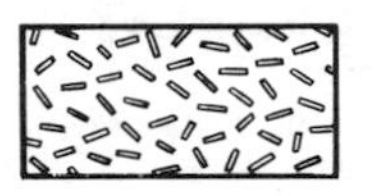
Porphyritic rock

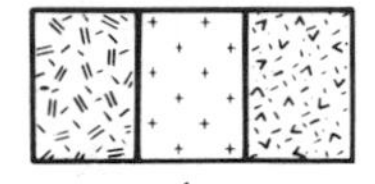
a b c
Massive igneous rock

a b
Brecciated rock
(sedimentary and igneous)

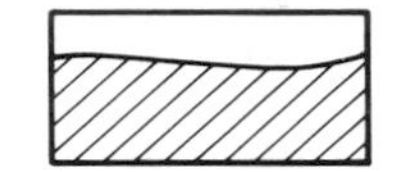
Bedrock
(kind not indicated)

FIGURE A–4 Selected symbols for metamorphic and igneous rocks.

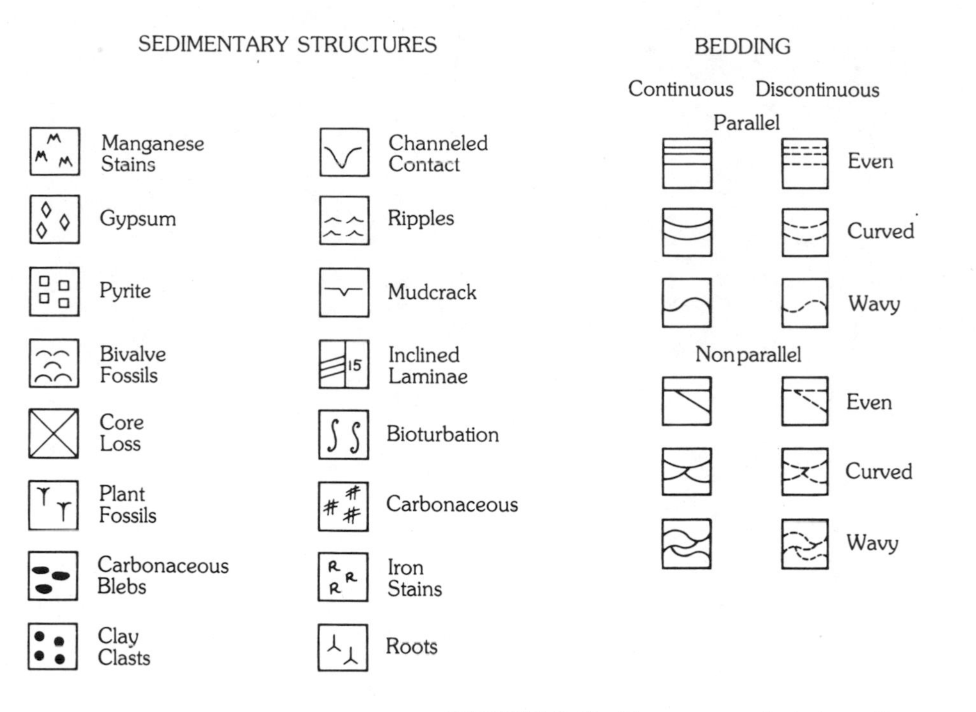

FIGURE A–5 Common symbols for sedimentary structures and bedding in sedimentary rock core descriptions.

4: References Cited and Selected Bibliography

REFERENCES CITED

Adams, F. S. 1973. Hurricane Agnes: Flooding versus dams in Pennsylvania. *Bulletin of the Atomic Scientists* 29:30–34.

Altmeyer, W. T. 1956. Discussion of engineering properties of expansive clays. *Transactions of the American Society of Civil Engineers* 121:669.

Barber, E. W. 1956. Discussion of engineering properties of expansive clays. *Transactions of the American Society of Civil Engineers* 121:669.

Boyd, M. B.; Saucier, R. T.; Kedey, J. W.; Montgomery, R. I.; Brown, R. D.; Mathis, P. B.; and Buice, C. S. 1972. *Disposal of dredge spoil-problem identification and assessment and research program development.* U.S. Army Engineer Waterways Experiment Station Technical Report H–72–8. Waterways Experiment Station, Vicksburg, Mississippi.

Bozozuk, M., and Burn, K. N. 1960. Vertical ground movements near elm trees. *Geotechnique* 10:19–32.

Bruun, P., and Gerritsen, R. 1960. *Stability of coastal inlets.* Amsterdam: North-Holland Publishing.

Building Research Advisory Board (BRAB). 1968. *Criteria for selection and design of residential slabs-on-ground.* Report no. 33. Washington, D. C.: National Academy of Sciences.

Burton, I., and Kates, R. W. 1964. The perception of natural hazards in resource management. *Natural Resources Journal* 3:412–41.

Casagrande, A. 1932. Research on the Atterberg limits of soils. *Public Roads* 13:121–30.

______. 1948. Classification and identification of soils. *Transactions of the American Society of Civil Engineers* 113:901.

Deere, D. H., and Miller, R. P. 1966. *Engineering classification and index properties for intact rock.* Technical Report no. AFWL–TR–65–116. Air Force Weapons Laboratories, Kirkland Air Force Base, New Mexico.

Deere, D. H. 1964. Technical description of rock cores for engineering purposes. *Rock Mechanics and Engineering Geology* 1:16–22.

Derbyshire, E. 1975. The formation and distribution of glacial materials. In *The engineering behaviour of glacial materials,* ed. Midland Soil Mechanics and Foundation Engineering Society, Department of Civil Engineering, pp. 6–17. Proceedings of symposium at the University of Birmingham, April 1975. Norwich, England: Geo Abstracts Ltd.

Donaldson, G. W. 1969. The occurrence of problems of heave and the factors affecting its nature. In *Proceedings of the Second International Research and Engineering Conference on Expansive Soils,* pp. 25–36. College Station, Texas: Texas A & M University Press.

Felt, E. J. 1953. Influence of vegetation on soil moisture contents and resulting soil volume change. In *Proceedings of the Third Internaional Conference on Soil Mechanics and Foundation Engineering,* vol. 1, pp. 24–27. Zurich: The Organizing Committee.

Friedkin, J. F. 1945. *A laboratory study of the meandering of alluvial rivers.* U.S. Army Engineers Waterways Experiment Station, Mississippi River Commission Report no. 12. Vicksburg, Mississippi.

Gillott, J. E. 1968. *Clay in engineering geology.* New York: American Elsevier Publishers.

Hammer, M. J., and Thompson, O. B. 1966. Foundation clay shrinkage caused by large trees. *Proceedings of the American Society of Civil Engineers* 92:1–17.

Hjulstrom, F. 1939. Transportation of detritus by running water. In *Recent marine sediments,* ed. P. D. Trask, pp. 5–29. New York: Dover Publications.

Holtz, W. G., and Gibbs, H. J. 1956. Engineering properties of expansive clays. *Transactions of the American Society of Civil Engineers* 121:641–63.

Jones, E. C., and Holtz, W. G. 1973. Expansive soils—The hidden disaster. *Civil Engineering* 8:49–51.

Keller, W. D. 1957. *Principles of chemical weathering.* Columbia, Mo.: Lucas Brothers.

Kennedy, J. L., III. 1980. Hydrogeology of reclaimed Gulf Coast lignite mines. Ph.D. dissertation, Texas A & M University, College Station, Texas.

Kenney, T. C.; Moum, J.; and Berre, T. 1967. An experimental study of bonds in a natural clay. In *Proceedings of the Geotechnical Conference on Shear Strength of Natural Soils and Rocks,* vol. 1, pp. 65–69. Oslo: Norwegian Geotechnical Institute.

Lane, E. W. 1937. Stable channels in erodible material. *Transactions of the American Society of Civil Engineers* 63:123–42.

Langbein, W. B. 1962. *Hydraulics of river channels as related to navigability.* U.S. Geological Survey Water Supply Paper no. 1539–W. Washington, D.C.

LeBlanc, R. J. 1972. Geometry of sandstone reservoir bodies. In *Underground waste management and environmental implications,* ed. T. D. Cook, pp. 133–89. American Association of Petroleum Geologists Memoir 18. Tulsa, Oklahoma.

Leopold, L. B., and Maddock, T., Jr. 1953. *The hydraulic geometry of stream channels and some physiographic implications.* U.S. Geological Survey Professional Paper 252. Washington, D.C.

Leopold, L. B., and Wolman, M. G. 1957a. *River channel patterns—Braided, meandering and straight.* U.S. Geological Survey Professional Paper 282–B. Washington, D.C.

______. 1957b. *River flood plains: Some observations on their formation.* U.S. Geological Survey Professional Paper 282–C. Washington, D.C.

______. 1960. River meanders. *Geological Society of America Bulletin* 71:769–93.

Mathewson, C. C., and McHam, R. M. 1977. Physical factors affecting the siting of dredged material islands. In *Oceans '77 Conference Record,* vol. 2, pp. G-1–G-6. New York: Institute of Electrical and Electronics Engineers, Inc., and Washington, D.C.: The Marine Technology Society.

McGowen, J. H., and Garner, L. E. 1970. Physiographic features and stratification types of coarse-grained point bars: Modern and ancient examples. *Sedimentology* 14:77–111.

Moon, C. F. 1975. The failure mechanism of quick-clay soils—A model approach. In *The engineering behaviour of glacial materials,* ed. Midland Soil Mechanics and Foundation Engineering Society, Department of Civil Engineering, pp. 83–88. Proceedings of symposium at the University of Birmingham, April 1975. Norwich, England: Geo Abstracts Ltd.

Pierce, J. W. 1970. Tidal inlets and washover fans. *Journal of Geology* 78:230–34.

Prescott, J. A. 1949. A climatic index for the leaching factor in soils. *Journal of Soil Science* 1:9–19.

Proctor, R. J. 1971. Mapping geological conditions in tunnels. *Bulletin of the Association of Engineering Geologists* 8:1–43.

Pusch, R., and Arnold, M. 1969. Recent quick-clay studies: The sensitivity of artificially sedimented organic-free illitic clay. *Engineering Geology* 3:135–48.

Rosenqvist, I. T. 1953. Considerations on the sensitivity of Norwegian quick-clays. *Geotechnique* 3:195–200.

Russam, K. 1965. The prediction of subgrade moisture conditions for design purposes. In *Moisture equilibria and moisture changes in soils beneath covered areas—A symposium in print,* ed. G. D. Aitchison, pp. 233–36. Sydney, Australia: Butterworth Scientific Publications.

Schneider, W. J. 1977. Analysis of the densification of reclaimed surface mined land. Master's thesis, Texas A & M University, College Station, Texas.

Seed, H. B. 1973. Landslides caused by soil liquefaction: The great Alaska earthquake of 1964. In *National Academy of Sciences Engineering Volume,* pp. 73–119. Washington, D.C.: National Academy of Sciences.

Smalley, I. J. 1971. Nature of quick-clays. *Nature* 231:310.

Smith, D. D. 1975. Dredging and spoil disposal—Major geologic processes in San Diego Bay, California, sedimentary processes. Paper presented at the Third International Estuarine Research Foundation Conference: Recent Advances in Estuarine Research, 7–9 October 1975, at Galveston, Texas.

Smith, D. D., and Wischmeier, W. H. 1962. Rainfall erosion. *Advances in Agronomy* 14:109–48.

Soil Survey Staff. 1960. *Soil classification—A comprehensive system—7th approximation.* Washington, D.C.: U.S. Department of Agriculture.

______. 1951. *Soil survey manual.* Washington, D.C.: U.S. Department of Agriculture.

Sutherland, A. J. 1967. Proposed mechanism for sediment entrainment by turbulent flows. *Journal of Geophysical Research* 72:6183–94.

Swanson, W. M. 1970. *Fluid mechanics.* New York: Holt, Rinehart & Winston.

Thornthwaite, C. W. 1948. An approach toward a rational classification of climate. *Geographical Review* 38:55–94.

Tourtelot, H. A. 1974. Geologic origin and distribution of swelling clays. *Bulletin of the Association of Engineering Geologists* 11:259–75.

Turk, L. J. 1970. Disposal of solid wastes—Acceptable practice or geological nightmare? American Geological Institute Short Course Lecture Notes on Environmental Geology, 9–10 November 1968, at Milwaukee, Wisconsin. Washington, D.C.: American Geological Institute.

Varnes, D. J. 1978. Slope movement types and processes. In *Landslides: Analysis and control,* eds. R. L. Schuster and R. J. Krizek, pp. 11–33. Transportation Research Board Special Report 176, National Research Council. Washington, D.C.: National Academy of Sciences.

Watts, G. M. 1962. Mechanical by-passing of littoral drift at inlets. *Journal of the Waterways and Harbors Division* 88:83–89. (American Society of Civil Engineers Paper 3058)

SELECTED PROFESSIONAL REFERENCE TEXTS

ENGINEERING GEOLOGY

Attewell, P. B., and Farmer, I. W. 1976. *Principles of engineering geology.* New York: Methuen, Inc.

Krynine, D. P., and Judd, W. R. 1957. *Principles of engineering geology and geotechnics.* New York: McGraw-Hill.

MECHANICS

Johnson, A. M. 1970. *Physical processes in geology.* San Francisco: W. H. Freeman.

Lambe, T. W., and Whitman, R. V.; with the assistance of H. G. Poulos. 1979. *Soil mechanics, SI version.* New York: John Wiley & Sons.

Means, W. D. 1976. *Stress and strain: Basic concepts of continuum mechanics for geologists.* New York: Springer-Verlag.

Mitchell, J. K. 1976. *Fundamentals of soil behavior.* New York: John Wiley & Sons.

Rouse, H. 1978. *Elementary mechanics of fluids.* New York: Dover Publications.

Terzaghi, K., and Peck, R. B. 1967. *Soil mechanics in engineering practice.* 2d ed. New York: John Wiley & Sons.

Wu, T-H. 1976. *Soil mechanics.* 2d ed. Boston: Allyn and Bacon.

ROCK MATERIALS

Jaeger, C. 1979. *Rock mechanics and engineering.* 2d ed. New York: Cambridge University Press.

Jaeger, J. C., and Cook, N. G. 1976. *Fundamentals of rock mechanics.* 2d ed. London: Chapman and Hall.

Stagg, K. G., and Zienkiewicz, O. C., eds. 1968. *Rock mechanics in engineering practice.* New York: John Wiley & Sons.

Winkler, E. M. 1975. *Stone: Properties, durability in man's environment.* 2d rev. ed. New York: Springer-Verlag.

IGNEOUS ROCKS

Carmichael, I. S. E.; Turner, F. J.; and Verhoogen, J. 1974. *Igneous petrology.* New York: McGraw-Hill.

Hyndman, D. W. 1972. *Petrology of igneous and metamorphic rocks.* New York: McGraw-Hill.

SEDIMENTARY ROCKS

Blatt, H.; Middleton, G.; and Murray, R. 1980. *Origin of sedimentary rock.* 2d ed. Englewood Cliffs, N.J.: Prentice-Hall.

Friedman, G. M., and Sanders, J. E. 1978. *Principles of sedimentology.* New York: John Wiley & Sons.

METAMORPHIC ROCKS

Miyashiro, A. 1978. *Metamorphism and metamorphic belts.* New York: Halsted Press.

SOILS

Brady, N. C. 1974. *The nature and properties of soils.* 8th ed. New York: Macmillan.

Dixon, J. B., and Weed, S. B., eds. 1977. *Minerals in soil environments.* Madison, Wis.: Soil Science Society of America.

Grim, R. E. 1968. *Clay mineralogy.* 2d ed. New York: McGraw-Hill.

Hunt, C. B. 1972. *Geology of soils: Their evolution, classification, and uses.* San Francisco: W. H. Freeman.

WATER

Cedergren, H. R. 1977. *Seepage, drainage, and flow nets.* 2d ed. New York: John Wiley & Sons.

Davis, S. N., and De Wiest, R. J. M. 1966. *Hydrogeology.* New York: John Wiley & Sons.

Fetter, C. W., Jr. 1980. *Applied hydrogeology.* Columbus, Ohio: Charles E. Merrill.

Freeze, R. A., and Cherry, J. A. 1979. *Groundwater.* Englewood Cliffs, N.J.: Prentice-Hall.

HISTORICAL GEOLOGY

Dott, R. H., Jr., and Batten, R. L. 1980. *Evolution of the earth.* 3d ed. New York: McGraw-Hill.

EROSION

Toy, T. J. 1980. *Erosion: Research techniques, erodibility and sediment delivery.* Norwich, England: Geo Abstracts Ltd.

RIVERS

Crickmay, C. H. 1974. *The work of the river: A critical study of the central aspects of geomorphogeny.* New York: American Elsevier Publishers.

Hjelmfelt, A. T., Jr., and Cassidy, J. J. 1975. *Hydrology for engineers and planners.* Ames: Iowa State University Press.

Linsley, R. K., Jr., and Franzini, J. B. 1972. *Elements of hydraulic engineering: Water resources engineering.* New York: McGraw-Hill.

Morisawa, M. 1968. *Streams: Their dynamics and morphology.* New York: McGraw-Hill.

COASTS

Bascom, W. 1980. *Waves and beaches: The dynamics of the ocean surface.* Rev. ed. Garden City, N.Y.: Anchor Press.

Brahtz, J. F., ed. 1968. *Ocean engineering: Goals, environment, technology.* Huntington, N.Y.: Krieger.

King, C. A. 1972. *Beaches and coasts.* 2d ed. New York: St. Martin's Press.

Quinn, A. D. 1971. *Design and construction of ports and marine structures.* 2d ed. New York: McGraw-Hill.

Shepard, F. P., and Wanless, H. R. 1971. *Our changing coastlines.* New York: McGraw-Hill.

U.S. Army Coastal Engineering Research Center. 1977. *Shore protection manual.* Washington, D.C.: U.S. Army Corps of Engineers.

SLOPES

Hoek, E., and Bray, J. 1977. *Rock slope engineering.* London: The Institution of Mining and Metallurgy.

Schuster, R. L., and Krizek, R. J., eds. 1978. *Landslides: Analysis and control.* Transportation Research Board Special Report 176, National Research Council. Washington, D.C.: National Academy of Sciences.

Záruba, Q., and Mencl, V. 1972. *Landslides and their control.* Amsterdam: Elsevier Scientific Publishing.

TECTONICS

Bolt, B. A. 1980. *Earthquakes and volcanoes: Readings from Scientific American.* San Francisco: W. H. Freeman.

______. 1978. *Earthquakes: A primer.* San Francisco: W. H. Freeman.

Ramsay, J. G. 1967. *Folding and fracturing of rocks.* New York: McGraw-Hill.

Rikitake, T. 1976. *Earthquake prediction.* Amsterdam: Elsevier Scientific Publishing.

Spencer, E. W. 1977. *Introduction to the structure of the earth.* 2d ed. New York: McGraw-Hill.

Wiegel, R. L., et al. 1969. *Earthquake engineering.* Englewood Cliffs, N.J.: Prentice-Hall.

Williams, H., and McBirney, A. R. 1979. *Volcanology.* San Francisco: W. H. Freeman.

ICE

Midland Soil Mechanics and Foundation Engineering Society, Department of Civil Engineering, ed. 1978. *The engineering behaviour of glacial materials.* Proceedings of symposium at the University of Birmingham, April 1975. Norwich, England: Geo Abstracts Ltd.

Washburn, A. L. 1980. *Geocryology: A survey of periglacial processes and environments.* New York: Halsted Press.

SUBSURFACE

Lewis, R. S., and Clark, G. B. 1964. *Elements of mining.* 3d ed. New York: John Wiley & Sons.

Peters, W. C. 1978. *Exploration and mining geology.* New York: John Wiley & Sons.

Thomas, L. J. 1977. *An introduction to mining: Exploration, feasibility, extraction, rock mechanics.* New York: Halsted Press.

Wahlstrom, E. E. 1973. *Tunneling in rock.* Amsterdam: Elsevier Scientific Publishing.

ENGINEERING

Bates, R. L. 1969. *Geology of the industrial rocks and minerals.* New York: Dover Publications.

Bazant, Z. 1979. *Methods of foundation engineering.* Amsterdam: Elsevier Scientific Publishing.

Bolt, B. A.; Horn, W. L.; MacDonald, G. A.; and Scott, R. F. 1977. *Geological hazards: Earthquakes, tsunamis, volcanoes, avalanches, landslides, floods.* New York: Springer-Verlag.

Catanese, A. J., and Snyder, J. C., eds. 1979. *Introduction to urban planning.* New York: McGraw-Hill.

Legget, R. F. 1973. *Cities and geology.* New York: McGraw-Hill.

Peck, R. B.; Hanson, W. E.; and Thornburn, T. H. 1974. *Foundation engineering.* 2d ed. New York: John Wiley & Sons.

Wahlstrom, E. E. 1974. *Dams, dam foundations, and reservoir sites.* Amsterdam: Elsevier Scientific Publishing.

Walters, R. C. S. 1971. *Dam geology.* London: Butterworth & Co.

PROFESSIONAL PRACTICE

American Society of Civil Engineers. *Subsurface exploration for underground excavation and heavy construction.* Proceedings of a Specialty Conference, August 1974, at New England College, Henniker, New Hampshire. New York: American Society of Civil Engineers.

Avery, T. E. 1977. *Interpretation of aerial photographs.* 3d ed. Minneapolis: Burgess Publishing.

Dobrin, M. B. 1976. *Introduction to geophysical prospecting.* 3d ed. New York: McGraw-Hill.

Dowding, C. H., ed. 1979. *Site characterization and exploration.* Proceedings of Specialty Workshop, June 1978, at Northwestern University, Evanston, Illinois. New York: American Society of Civil Engineers.

Griffiths, D. H., and King, R. F. 1980. *Applied geophysics for engineers and geologists.* New York: Pergamon Press.

Hvorslev, M. J. 1965. *Subsurface exploration and sampling of soils for civil engineering purposes.* New York: American Society of Civil Engineers.

Lahee, F. H. 1961. *Field geology.* 6th ed. New York: McGraw-Hill.

Moffitt, F. H., and Mikhail, E. M. 1980. *Photogrammetry.* 3d ed. New York: Harper & Row.

Ray, R. G. 1960. *Aerial photographs in geologic interpretation and mapping.* U.S. Geological Survey Professional Paper 373. Washington, D.C.

Sabins, F. F., Jr. 1978. *Remote sensing: Principles and interpretation.* San Francisco: W. H. Freeman.

Varnes, D. J. 1974. *The logic of geologic maps, with reference to their interpretation and use for engineering purposes.* U.S. Geological Survey Professional Paper 837. Washington, D.C.

Way, D. S. 1978. *Terrain analysis: A guide to site selection using aerial photographic interpretation.* Community Development Series. New York: McGraw-Hill.

GENERAL

American Geological Institute. 1976. *Dictionary of geological terms.* Garden City, N.Y.: Anchor Press.

Bates, R. L., and Jackson, J. A., eds. 1980. *Glossary of geology.* Falls Church, Va.: American Geological Institute.

MECHANICS

Bagnold, R. A. 1966. *An approach to the sediment transport problem from general physics.* U.S. Geological Survey Professional Paper 422–I. Washington, D.C.

Burwell, J. T., and Rabinowicz, E. 1953. The nature of the coefficient of friction. *Journal of Applied Physics* 24:136–39.

ROCK MATERIALS

Adams, W. M. 1970. Quasi-permanent deformation in the elastic zone near explosions. In *Engineering geology case histories,* no. 8, ed. W. M. Adams, pp. 5–9. Boulder, Colo.: Geological Society of America.

Adler, L. 1963. Failure in geologic materials containing planes of weakness. *Transactions of the Society of Mining Engineers of the American Institute of Mining Engineers* 226:88–94.

Bienawski, Z. T. 1967. Mechanism of brittle fracture of rock. *International Journal of Rock Mechanics and Mining Science* 4:395–430.

Brace, W. F. 1960. An extension of Griffith theory of fracture to rocks. *Journal of Geophysics Research* 65:3477–80.

Byerlee, J. D. 1967. Theory of friction based on brittle fracture. *Journal of Applied Physics* 38:2928–34.

Cleaves, A. B. 1963. Engineering geology characteristics of basal till, St. Lawrence Seaway Project. In *Engineering geology case histories,* no. 4, eds. P. D. Trask and G. A. Kiersch, pp. 51–56. Boulder, Colo.: Geological Society of America.

Coates, D. F. 1970. *Rock mechanics principles.* Mines Branch Monograph 874. Ottawa: Department of Energy, Mines and Resources.

Griffith, A. A. 1921. The phenomena of rupture and flow in solids. *Transactions of the Royal Society of London* 221:163–98.

Griggs, D. T., and Handin, J., eds. 1960. *Rock deformation—A symposium.* Geological Society of America Memoir 79. Boulder, Colo.: Geological Society of America.

Hodgson, R. A. 1961. Classification of structures on joint surfaces. *American Journal of Science* 259: 493–502.

Hubbert, M. K., and Rubey, W. W. 1959. Role of fluid pressure in mechanics of overthrust faulting. *Bulletin of the Geological Society of America* 7:115.

Judd, W. R. 1959. Effect of elastic properties on rocks of civil-engineering design. In *Engineering geology case histories,* no. 3, ed. P. D. Trask, pp. 53–76. Boulder, Colo.: Geological Society of America.

Nichols, T. C., Jr.; Lee, F. T.; and Abel, J. F., Jr. 1969. Some influences of geology and mining upon the three-dimensional stress field in a metamorphic rock mass. *Bulletin of the Association of Engineering Geologists* 6:131–43.

Paulding, W. J., Jr. 1970. *Coefficient of friction of natural rock surfaces.* Soil Mechanics and Foundations Division of the American Society of Civil Engineers, Sm2, pp. 385–94. New York.

Sowers, G. E. 1972. Theory of spacing of extension fracture. In *Engineering geology case histories,* no. 9, ed. H. Pincus, pp. 27–54. Boulder, Colo.: Geological Society of America.

SOILS

Abbott, M. B. 1960. One-dimensional consolidation of multi-layered soils. *Geotechnique* 10:151–65.

Aitchison, G. D., and Richards, B. G. 1965. A broad scale study of moisture conditions in pavement subgrade throughout Australia. In *Moisture equilibria and moisture changes in soils beneath covered areas,* ed. G. D. Aitchison, pp. 184–90. Sydney, Australia: Butterworth Scientific Publications.

Andersland, O. B., and Douglas, A. G. 1970. Soil deformation rates and activation energies. *Geotechnique* 20:1.

Bennett, R. H.; Bryant, W. R.; and Keller, G. H. 1977. *Clay fabric and geotechnical properties of selected submarine sediment cores from the Mississippi delta.* National Oceanic and Atmospheric Administration Professional Paper no. 9. Washington, D.C.

Bjerrum, L., and Rosenqvist, I. T. 1956. Some experiments with artificially sedimented clays. *Geotechnique* 6:124.

Bolt, G. H. 1965. Physico-chemical analysis of the compressibility of pure clays. *Geotechnique* 6:86.

Casagrande, A. 1940. The structure of clay and its importance in foundation engineering. In *Contributions to soil mechanics, 1925–1940,* pp. 72–126. Boston Society of Civil Engineers.

Collins, K., and McGown, A. 1974. The form and function of microfabric features in a variety of natural soils. *Geotechnique* 24:223.

Croney, D., and Coleman, J. D. 1961. Pore pressure and suction in soil. In *Pore pressure and suction in soils,* pp. 31–37. London: Butterworth Scientific Publications.

Font, R. G. 1977. Influence of anisotropies on the shear strength and field behavior of heavily over-consolidated, plastic and expansive clay-shales. *Texas Journal of Science* 29:21.

Foster, R. H. 1972. Behavior of kaolin fabric under shear loading at low stress. In *Stress-strain behavior of soils,* ed. R. H. G. Parry, pp. 81–88. Roscoe Memorial Symposium, March 1971, Cambridge, England. Oxfordshire, England: G. T. Foulis & Co.

Haefield, R. 1950. Investigation and measurements of the shear strengths of saturated cohesive soils. *Geotechnique* 2:186.

Hvorslev, M. J. 1960. Physical components of the shear strength of saturated clays. Paper read at Research Conference on Shear Strength of Cohesive Soils, American Society of Civil Engineers, Boulder, Colorado.

Kenney, T. C. 1967. The influence of mineral composition on the residual composition of natural soils. In *Proceedings of the Geotechnical Conference on Shear Strength Properties of Natural Soils and Rocks,* vol. 1, pp. 123–29. Oslo: Norwegian Geotechnical Institute.

Lambe, T. W. 1953. Structure of inorganic soil. In *Proceedings of the American Society of Civil Engineers,* Separate 315, pp. 1–49.

Low, P. F. 1960. Viscosity of water in clay systems. *Clays and Clay Minerals* 8:170.

Low, P. F., and White, J. L. 1970. Hydrogen bonding and polywater in clay-water systems. *Clays and Clay Minerals* 18:63.

Luessink, H., and Kirchenbauer, H. M. 1967. Determination of the shear strength behavior of sliding planes caused by geological means. In *Proceedings of the Geotechnical Conference on Shear Strength Properties of Natural Soils and Rocks,* vol. 1, pp. 131–37. Oslo: Norwegian Geotechnical Institute.

Lyle, W. M., and Smerdon, E. T. 1965. Relation of compaction and other soil properties to erosion resistance of soils. *Transactions of the American Society of Agricultural Engineers* 8:419–22.

Malyshev, M. V. 1967. A generalization of the Mohr-Coulomb strength criterion for soils. In *Proceedings of the Geotechnical Conference on Shear Strength Properties of Natural Soils and Rocks,* vol. 1, pp. 221–24. Oslo: Norwegian Geotechnical Institute.

Meade, R. H. 1964. *Removal of water and rearrangement of particles during the compaction of clayey sediments.* U.S. Geological Survey Professional Paper 497–B. Washington, D.C.

Mesri, G., and Olson, R. E. 1970. Shear strength of montmorillonite. *Geotechnique* 20:261.

Mitchell, J. K. 1964. Shearing resistance of soils as a rate process. *Journal of Soil Mechanics and Foundations* 94:231.

Mitchell, J. K., and McConnell, J. R. 1965. Some characteristics of the elastic and plastic deformation of clay on loading. In *Proceedings of the Sixth International Conference on Soil Mechanics and Foundation Engineering,* vol. 1, pp. 313–17. Toronto: University of Toronto Press.

Morgenstern, N. R., and Tchalenko, J. S. 1967. Microscopic structures in kaolin subject to direct shear. *Geotechnique* 17:309–28.

Rosenqvist, I. T. 1955. Physico-chemical properties of soils—Soil-water systems. *Journal of Soil Mechanics and Foundations* 85:31–53.

Schmertmann, J. H. 1955. The undisturbed consolidation behavior of clay. *Transactions of the American Society of Civil Engineers* 120:1201–27.

Seed, B. H., and Chan, C. K. 1959. Structure and strength characteristics of compacted clays. *Journal of Soil Mechanics and Foundations* 85:87–128.

Skempton, A. W., and Bjerrum, L. 1957. A contribution to settlement analysis of foundations on clay. *Geotechnique* 7:168–78.

Skempton, A. W., and Petley, D. J. 1967. The strength along structural discontinuities in stiff clays. In *Proceedings of the Geotechnical Conference on Shear Strength Properties of Natural Soils and Rocks,* vol. 2, p. 29. Oslo: Norwegian Geotechnical Institute.

Tan, T-K. 1957. Discussion of soil properties and their measurement. In *Proceedings of the Fourth International Conference on Soil Mechanics and Foundation Engineering,* vol. 3, pp. 87–88. London: Butterworth Scientific Publications.

EXPANSIVE SOILS

Aylmore, L. A. G., and Quirk, J. P. 1959. Swelling of clay-water systems. *Nature* 183:1752.

Davits, J. C., and Low, P. F. 1970. Relation between crystal-lattice configuration and swelling montmorillonites. *Clays and Clay Minerals* 18:325.

El Swaify, S. A., and Henderson, D. W. 1967. Water retention by osmotic swelling of certain colloidal clays with varying ionic composition. *Journal of Soil Science* 18:223.

Felt, E. J. 1953. Influence of vegetation on soil moisture contents and resulting soil volume change. In *Proceedings of the Third International Conference on Soil Mechanics and Foundation Engineering,* vol. 1, pp. 24–27. Zurich: The Organizing Committee.

Foster, M. D. 1955. The relation between composition and swelling in clays. *Clays and Clay Minerals* 3:205.

Franzmeier, D. P., and Ross, S. J. 1968. Soil swelling: Laboratory measurement and relation to other soil properties. *Proceedings of the Soil Science Society of America* 32:573.

Gibbs, H. J. 1959. *Expansive clay properties and problems.* U.S. Bureau of Reclamation Lab Report no. Em–568. Washington, D.C.

Gibbs, H. J., and Holtz, W. G. 1956. The engineering properties of expansive clay. *Transactions of the American Society of Civil Engineers* 121:669.

Gromko, G. J. 1974. Review of expansive soils. *Journal of the Geotechnical Division, American Society of Civil Engineers* 6:667–87.

Jennings, J. E. 1953. The heave of buildings on desiccated clay. In *Proceedings of the Third International Conference on Soil Mechanics and Foundation Engineering,* vol. 1, p. 390. Zurich: The Organizing Committee.

Jones, E. C., and Holtz, W. G. 1973. Expansive soils—The hidden disaster. *Civil Engineering* 8:49–51.

Lambe, T. W. 1960. *The character and identification of expansive soils.* Federal Housing Administration Technical Publication 701. Washington, D.C.

Lytton, R. L., and Woodburn, I. A. 1973. Design and performance of mat foundations on expansive clays. In *Proceedings of the Third International Conference on Expansive Clay Soils,* vol. 1, p. 301. Haifa, Israel: Jerusalem Academic Press.

Mathewson, C. C.; Castleberry, J. P.; and Lytton, R. L. 1975. Analysis and modeling of the performance of home foundations on expansive soils in central Texas. *Bulletin of the Association of Engineering Geologists* 12:275–302.

Sids, G., and Borden, L. 1971. The microstructure of dispersed and flocculated samples of kaolinite, illite and montmorillonite. *Canadian Geotechnical Journal* 8:391.

Warkentin, B. P.; Bolt, G. H.; and Miller, R. D. 1957. Swelling pressure of montmorillonite. *Proceedings of the Soil Science Society of America* 21:495–97.

WATER

Adams, J. K.; Boggess, D. H.; and Davis, C. F. 1964. *Water-table, surface-drainage, and engineering soils map of the Dover quadrangle, Delaware.* U.S. Geological Survey Hydrologic Investigation Atlas HA–139. Washington, D.C.

Arnold, A. B. 1964. Relief wells on the Garrison Dam and Snake Creek Embankment, North Dakota. In *Engineering geology case histories,* no. 5, ed. G. A. Kiersch, pp. 45–52. Boulder, Colo.: Geological Society of America.

Beard, D. C., and Weyl, P. D. 1973. Influence of texture on porosity and permeability of unconsolidated sand. *Bulletin of the American Association of Petroleum Geologists* 57:349–69.

Berg, R. R. 1970. Method for determining permeability from reservoir rock properties. *Transactions of the Gulf Coast Association Geological Society* 20:303–17.

Beyer, J. L. 1974. Global summary of human response to natural hazards—Floods. In *Natural hazards: Local, national, global,* ed. G. F. White, pp. 265–73. New York: Oxford University Press.

Bue, C. D. 1967. *Flood information for floodplain planning.* U.S. Geological Survey Circular 539. Washington, D.C.

Davis, G. H.; Small, J. D.; and Counts, H. B. 1963. Land subsidence related to decline of artesian pressure in the Ocala limestone at Savannah, Georgia. In *Engineering geology case histories,* no. 4, eds. P. D. Trask and G. A. Kiersch, pp. 1–8. Boulder, Colo.: Geological Society of America.

Foose, R. M. 1969. Mine dewatering and recharge in carbonate rocks near Hershey, Pennsylvania. In *Engineering geology case histories,* no. 7, eds. G. A. Kiersch and A. B. Cleaves, pp. 45–60. Boulder, Colo.: Geological Society of America.

Gabrysch, R. K., and Bonnet, C. W. 1975. *Land-surface subsidence in the Houston-Galveston region, Texas.* Texas Water Development Board Report 188. Austin: Texas Water Development Board.

Geraghty, J. J.; Miller, D. W.; van der Leaden, F.; and Troise, F. L. 1973. *Water atlas of the United States.* Port Washington, N.Y.: Water Information Center.

Graton, L. C., and Fraser, H. J. 1935. Systematic packing of spheres with particular relation to porosity and permeability. *Journal of Geology* 38:786–943.

Hubbert, M. K. 1940. The theory of groundwater motion. *Journal of Geology* 48:785–944.

Laverty, F. B. 1958. Recharging groundwater with reclaimed sewage effluent. *Civil Engineering* 28:585–87.

Leopold, L. B. 1968. *Hydrology for urban land planning—A guidebook on the hydrologic effects of urban land use.* U.S. Geological Survey Circular 554. Washington, D.C.

Mann, J. F., Jr. 1969. Groundwater management in the Raymond Basin, California. In *Engineering geology case histories,* no. 7, eds. G. A. Kiersch and A. B. Cleaves, pp. 61–74. Boulder, Colo.: Geological Society of America.

Meinzer, O. E. 1950. Geology and engineering in the production and control of groundwater. In *Application of geology to engineering practice* (Berkey Volume), pp. 151–80. New York: Geological Society of America.

Moline, N. T. 1974. Perception research and local planning: Floods on the Rock River, Illinois. In *Natural hazards: Local, national, global,* ed. G. F. White, pp. 52–59. New York: Oxford University Press.

Poland, J. F. 1958. Land subsidence due to groundwater development. *Proceedings of the American Society of Civil Engineers* 84:11.

Richter, R. C. 1958. Geologic considerations in artificial recharge of groundwater reservoirs in California. In *Engineering geology case histories,* no. 2, ed. P. D. Trask, pp. 25–32. Boulder, Colo.: Geological Society of America.

Sheaffer, J. R., and Associates. 1967. *Introduction to flood proofing.* Chicago: Center for Urban Studies, University of Chicago.

Theis, C. V. 1938. The significance and nature of the cone of depression in groundwater bodies. *Economic Geology* 33:889–902.

Tolman, C. F. 1937. *Groundwater.* New York: McGraw-Hill.

U.S. Army Corps of Engineers. 1967. Guidelines for reducing flood damages. In *Floodplain—Handle with care* (Pamphlet EP1105–2–4), p. 33. Mississippi River Commission, Washington, D.C.

Wiitala, S. W.; Jetter, K. R.; and Sommerville, A. J. 1961. *Hydraulic and hydrologic aspects of floodplain planning.* U.S. Geological Survey Water Supply Paper 1526. Washington, D.C.

EROSION

Swanson, D. N., and Swanson, F. J. 1976. Timber harvesting, mass erosion and steep land forest geomorphology in the Pacific Northwest. In *Geomorphology and engineering,* ed. D. R. Coates, pp. 199–221. Stroudsburg, Pa.: Dowden, Hutchinson & Ross.

RIVERS

Emerson, J. W. 1971. Channelization: A case study. *Science* 173:325–26.

Fisk, H. N. 1952. Mississippi River Valley geology relation to river regime. *Transactions of the American Society of Civil Engineers* 117:667–82.

Gibbs, H. J. 1962. *A study of erosion and tractive force characteristics in relation to soil mechanics properties.* U.S. Bureau of Reclamation Lab Report no. Em–643. Washington, D.C.

Graf, W. H. 1971. *Hydraulics of sediment transport.* New York: McGraw-Hill.

Gunn, C. A.; Reed, D. J.; and Couch, R. E. 1972. *Cultural benefits from metropolitan river recreation—San Antonio prototype.* Texas Water Resources Institute Technical Report no. 43. College Station, Texas.

Haffen, M. 1963. Grouting deep alluvial fill in the Durance River Valley, Serre Ponçon Dam, France. In *Engineering geology case histories,* no. 4, eds. P. D. Trask and G. A. Kiersch, pp. 33–44. Boulder, Colo.: Geological Society of America.

Hjulstrom, F. 1935. The morphological activity of rivers as illustrated by River Fyris. *Bulletin of the Geological Institute, Uppsala* 25:221.

———. 1939. Transportation of detritus by running water. In *Recent marine sediments,* ed. P. D. Trask, pp. 5–29. New York: Dover Publications.

Kolb, C. R. 1958. Geologic aspects of control of the Mississippi-Atchafalaya diversion, Louisiana. In *Engineering geology case histories,* no. 2, ed. P. D. Trask, pp. 13–15. Boulder, Colo.: Geological Society of America.

Leopold, L. B., and Maddock, T., Jr. 1953. *The hydraulic geometry of stream channels and some physiographic implications.* U.S. Geological Survey Professional Paper 252. Washington, D.C.

Leopold, L. B., and Wolman, M. G. 1957a. *River channel patterns—Braided, meandering and straight.* U.S. Geological Survey Professional Paper 282–B. Washington, D.C.

______. 1957b. *River flood plains: Some observations on their formation.* U.S. Geological Survey Professional Paper 282–C. Washington, D.C.

______. 1960. River meanders. *Bulletin of the Geological Society of America* 71:769–94.

Schuum, S. A. 1963. Sinuosity of alluvial rivers on the Great Plains. *Bulletin of the Geological Society of America* 74:1089–99.

Simons, D. B., and Richardson, E. V. 1961. Forms of bed roughness in alluvial channels. *Proceedings of the American Society of Civil Engineering* 87:87–105.

Smerdon, E. T., and Beasley, R. P. 1959. *The tractive force theory applied to stability of open channels in cohesive soils.* University of Missouri Agricultural Experiment Station Research Bulletin 715. Columbia, Missouri.

Sutherland, A. J. 1967. Proposed mechanism for sediment entrainment by turbulent flows. *Journal of Geophysical Research* 72:6183–94.

Treasher, R. C. 1958. Russian River reservoir project, Coyote Valley, California. In *Engineering geology case histories,* no. 2, ed. P. D. Trask, pp. 33–36. Boulder, Colo.: Geological Society of America.

Winkler, M. 1958. Hydro-electric power system of Tauern Kaprun, Austria. In *Engineering geology case histories,* no. 2, ed. P. D. Trask, pp. 37–40. Boulder, Colo.: Geological Society of America.

COASTS

Bosco, D. R.; Bouma, A. H.; and Dunlap, W. A. 1974. *Assessment of the factors of dredged material deposited in unconfined subaqueous disposal areas.* U.S. Army Corps of Engineers Dredged Material Research Program Contract Report no. D–74–8. Vicksburg, Mississippi.

Boyd, M. B.; Saucier, R. T.; Kedey, J. W.; Montgomery, R. I.; Brown, R. D.; Mathis, D. B.; and Buice, C. S. 1972. *Disposal of dredge spoil—Problem identification and assessment and research program development.* U.S. Army Engineer Waterways Experiment Station Technical Report H–72–8. Vicksburg, Mississippi.

Bruun, P., and Gerritsen, F. 1960. *Stability of coastal inlets.* Amsterdam: North-Holland Publishing.

Davis, A. B., Jr. 1962. *Design of hurricane flood protection on the upper Texas coast.* Galveston, Tex.: U.S. Army Corps of Engineers.

Dolmange, V. 1964. Removal of ripple rocks, Seymour Narrows, British Columbia, Canada. In *Engineering geology case histories,* no. 5, ed. G. A. Kiersch, pp. 7–13. Boulder, Colo.: Geological Society of America.

Ginsburg, R. N., and Lowenstam, H. A. 1958. The influence of marine bottom communities on the depositional environment of sediments. *Journal of Geology* 66:310–18.

Hayes, M. O. 1967. *Hurricanes as geologic agents: Case studies of hurricanes Carla, 1961, and Cindy, 1963.* Report of Investigations no. 61. Austin: Bureau of Economic Geology, University of Texas.

Herbick, J. R. 1975. *Coastal and deep ocean dredging.* Houston: Gulf Publishing.

Herron, W. J., and Harris, R. L. 1966. Littoral by-passing and beach restoration in the vicinity of Port Hueneme, California. *Proceedings of the Tenth Conference on Coastal Engineering,* pp. 651–75. New York: American Society of Civil Engineers.

Koman, P. D. 1976. The transport of cohesionless sediments on continental shelves. In *Marine sediment transport and environmental management,* ed. D. J. Stanley, pp. 107–25. New York: John Wiley & Sons.

Krumbein, W. C. 1950. Geological aspects of beach engineering. In *Application of geology to engineering practice* (Berkey Volume), pp. 195–223. New York: Geological Society of America.

Mathewson, C. C.; Clary, J. H.; and Stinson, J. E., II. 1975. Dynamic physical processes on a south Texas barrier island—Impact on construction and maintenance. In *Proceedings of the Institute of Electrical and Electronics Engineers, Conference 75,* ed. E. Herz, pp. 327–30. San Diego, Calif.: Institute of Electrical and Electronics Engineers, Inc.

McGowen, J. H., and Garner, L. E. 1970. Physiographic features and stratification types of coarse-grained point bars: Modern and ancient examples. *Sedimentology* 14:77–111.

Meccia, R. M.; Allanach, W. C.; and Griffis, F. H. 1975. Engineering challenges of dredged material disposal. *Journal of the Waterways and Harbors Division of the American Society of Civil Engineers* 101:1–13.

Pierce, J. W. 1970. Tidal inlets and washover fans. *Journal of Geology* 78:230–34.

Price, W. A. 1947. Equilibrium of form and forces in tidal basins of the coast of Texas and Louisiana. *Bulletin of the American Association of Petroleum Geologists* 31:1619–63.

______. 1952. Reduction of maintenance by proper orientation of ship channels through tidal inlets. In *Proceedings of the Second Conference on Coastal Engineering,* ed. J. W. Johnson, pp. 243–45. Houston: American Society of Civil Engineering.

Smith, D. D. 1975. Dredging and spoil disposal—Major geologic processes in San Diego Bay, California, sedimentary processes. Paper presented at the Third International Estuarine Research Foundation Conference: Recent Advances in Estuarine Research, 7–9 October 1975, at Galveston, Texas.

Watts, G. M. 1962. Mechanical by-passing of littoral drift at inlets. *Journal of the Waterways and Harbors Division of the American Society of Civil Engineers* 88:83–89.

SLOPES

Anderson, J. G. 1906. Solifluction, a component of sub-aerial denudation. *Journal of Geology* 14:91–112.

Baker, R. F. 1954. Analysis of corrective actions for highway landslides. *Transactions of the American Society of Civil Engineers* 119:665.

Bjerrum, L. 1967. Mechanism of progressive failure in slopes of over-consolidated plastic clays and clay shales. *Journal of the Soil Mechanics and Foundations Division of the American Society of Civil Engineers* 93:1.

Blanc, R. P., and Cleveland, G. B. 1968. *Natural slope stability as related to geology, San Clemente area, Orange and San Diego counties.* California Division of Mines and Geology Special Report 98. Sacramento, California.

Brawner, C. O., and Milligan, V., eds. 1971. *Stability in open pit mining.* First International Conference on Stability in Open Pit Mining. New York: Society of Mining Engineers of the American Institute of Mining and Petroleum Engineers.

Cadman, J. D., and Goodman, R. E. 1967. Landslide noise. *Science* 158:1182–84.

Cleaves, A. B. 1969. Flagg's Restaurant area landslide, Pacific Palisades, California. In *Engineering geology case histories,* no. 7. eds. G. A. Kiersch and A. B. Cleaves, p. 75. Boulder, Colo.: Geological Society of America.

Coates, D. R., ed. 1977. *Landslides.* Boulder, Colo.: Geological Society of America.

Crawford, C. B., and Eden, W. J. 1964. Nicolet landslide of November, 1955, Quebec, Canada. In *Engineering geology case histories,* no. 4, eds. P. D. Trask and G. A. Kiersch, pp. 229–34. Boulder, Colo.: Geological Society of America.

Fleming, R. W., and Johnson, A. M. 1975. Rates of seasonal creep of silty clay soil. *Quarterly Journal of Engineering Geology* 8:1.

Fox, P. P. 1957. Geology, exploration, and drainage of the Serra Slide, Santos, Brazil. In *Engineering geology case histories,* no. 1, ed. P. D. Trask, pp. 17–24. Boulder, Colo.: Geological Society of America.

Hill, R. A. 1934. Clay stratum dried out to prevent landslips. *Civil Engineering* 4:403–7.

Hoek, E. 1970. *Estimating the stability of excavated slopes in open cast mines.* London: Institute of Mining and Metallurgy.

Hogg, A. D. 1959. Some engineering studies of rock movement in the Niagara area. In *Engineering geology case histories,* no. 3, ed. P. D. Trask, pp. 1–12. Boulder, Colo.: Geological Society of America.

Hunter, J. H., and Schuster, R. L. 1971. Chart solution for analysis of earth slopes. *Highway Research Record* 345:77.

Hutchinson, J. N., and Bhandara, R. K. 1971. Undrained loading, a fundamental mechanism of mudflows and other mass movements. *Geotechnique* 21:353–58.

Jahn, R. H. 1958. Residential ills of the Heartbreak Hills of Southern California. *Engineering and Science, Caltech Alumni Magazine* 22:13–20.

Janbu, N. 1973. Slope stability computations. In *Embankment-dam engineering*

(Casagrande Volume), eds. R. C. Hirschfeld and S. J. Poulos, pp. 47–86. New York: John Wiley & Sons.

Kiersch, G. A. 1964. Vaiont Reservoir disaster. *Civil Engineering* 3:32–39.

Kirkby, M. J. 1967. Measurement and theory of soil creep. *Journal of Geology* 75:359–78.

Lo, K. Y. 1965. Stability of slopes in anisotropic soils. *Journal of the Soil Mechanics and Foundations Division of the American Society of Civil Engineers* 91:85–106.

Long, A. E. 1964. Problems in designing stable open-pit mine slopes. *Canadian Mining and Metallurgical Bulletin* 57:741–46.

Morgenstern, N. R., and Price, V. E. 1965. The analysis of the suitability of general slip surfaces. *Geotechnique* 15:79–93.

Müller, L. 1964. The rock slide in the Vaiont Valley. *Rock Mechanics and Engineering Geology* 4:148–212.

Peck, R. B. 1967. Stability of natural slopes. *Journal of the Soil Mechanics and Foundations Division of the American Society of Civil Engineers* 93:403–17.

Prior, D. B., and Stephens, N. 1972. Some movement patterns of temperate mudflows: Examples from northeastern Ireland. *Bulletin of the Geological Society of America* 33:2533–44.

Prior, D. B., and Suhayda, J. N. 1979. Submarine mudslide morphology and development mechanisms, Mississippi Delta. *Offshore Technology Conference,* vol. 2, pp. 1055–62. Houston: Marine Technical Society.

Seed, H. B. 1973. Landslides caused by soil liquefaction: The great Alaska earthquake of 1964. In *Engineering Volume of the National Academy of Sciences,* pp. 73–119. Washington, D.C.

Shreve, R. L. 1968. *The Blackhawk landslide.* Geological Society of America Special Paper 108. Boulder, Colorado.

Skempton, A. W. 1964. Long-term stability of clay slopes. *Geotechnique* 14:77–102.

Snow, D. T. 1964. Landslide of Cerro Condor-Sencca, Department of Avacucho, Peru. In *Engineering geology case histories,* no. 5, ed. G. A. Kiersch, pp. 1–6. Boulder, Colo.: Geological Society of America.

Staples, L. W. 1957. Landslide at north abutment of Lookout Point Dam, Oregon. In *Engineering geology case histories,* no. 1, ed. P. D. Trask, p. 43. Boulder, Colo.: Geological Society of America.

Stout, M. L. 1969. *Radiocarbon dating of landslides in southern California and engineering geology implications.* Geological Society of America Special Paper 123, pp. 167–79. Boulder, Colorado.

Supp, C. W. A.; Whikeheart, R. E.; and Obear, G. H. 1957. Occurrence, investigation, and treatment of embankment foundation failure on Ohio Turnpike Project No. 1. In *Engineering geology case histories,* no. 1, ed. P. D. Trask, pp. 57–66. Boulder, Colo.: Geological Society of America.

Terzaghi, K. 1950. Mechanics of landslides. In *Application of geology to engineering practice* (Berkey Volume), pp. 83–125. New York: Geological Society of America.

______. 1962. Stability of steep slopes on hard unweathered rock. *Geotechnique* 12:251–70.

Wahrhaftig, C., and Cox, A. 1959. Rock glaciers in the Alaska Range. *Bulletin of the Geological Society of America* 70:383–436.

Williams, P. J. 1959. An investigation into processes occurring in solifluction. *American Journal of Science* 257:481–90.

Woods, H. D. 1958. Causes of the Sear's Point landslide, Sonoma County, California. In *Engineering geology case histories,* no. 2, ed. P. D. Trask, pp. 41–43. Boulder, Colo.: Geological Society of America.

Youd, L. T. 1974. *Liquefaction.* U.S. Geological Survey Circular 733. Washington, D.C.

TECTONICS

Adams, W. M. 1970. Tsunami effects and risk at Kahuku Point, Oahu, Hawaii. In *Engineering geology case histories,* no. 8, ed. W. M. Adams, pp. 63–70. Boulder, Colo.: Geological Society of America.

Bardwell, G. E. 1970. Some statistical features of the relationship between Rocky Mountain Arsenal waste disposal and frequency of earthquakes. In *Engineering geology case histories,* no. 8, ed. W. M. Adams, pp. 33–38. Boulder, Colo.: Geological Society of America.

Carder, D. S. 1970. Seismic ground effects from nuclear explosions. In *Engineering geology case histories,* no. 8, ed. W. M. Adams, pp. 11–18. Boulder, Colo.: Geological Society of America.

______. 1970. Reservoir loading and local earthquakes. In *Engineering geology case histories,* no. 8., ed. W. M. Adams, pp. 51–62. Boulder, Colo.: Geological Society of America.

Evans, D. M. 1970. The Denver area earthquakes and the Rocky Mountain Arsenal disposal well. In *Engineering geology case histories,* no. 8, ed. W. M. Adams, pp. 25–32. Boulder, Colo.: Geological Society of America.

Logan, J.; Iwasaki, T.; Friedman, M.; and Kling, S. A. 1972. Experimental investigation of sliding friction in multilithologic specimens. In *Engineering geology case histories,* no. 9, ed. H. Pincus, pp. 55–68. Boulder, Colo.: Geological Society of America.

Munson, R. C. 1970. Relationship of effect of water flooding of the Rangely Oil Field on seismicity. In *Engineering geology case histories,* no. 8, ed. W. M. Adams, pp. 39–50. Boulder, Colo.: Geological Society of America.

Scopel, L. J. 1970. Pressure injection disposal well, Rocky Mountain Arsenal, Denver, Colorado. In *Engineering geology case histories,* no. 8, ed. W. M. Adams, pp. 19–24. Boulder, Colo.: Geological Society of America.

Scott, R. F. 1971. *Earthquake effects on utilities—Van Norman Dams: Engineering features of the San Fernando earthquake of February 9, 1971.* Caltech Earthquake Engineering Research Lab Report no. 71–02, pp. 302–10. Pasadena, California.

ICE

Anderson, J. G. 1906. Solifluction, a component of sub-aerial denudation. *Journal of Geology* 14:91–112.

Chandler, R. J. 1972. Periglacial mudslides in Vestspitbergen and their bearing on the origin of fossil "solifluction" shears in low-angled clay slopes. *Quarterly Journal of Engineering Geology* 5:223–41.

Davison, C. 1889. On the creeping of the soil-cap through the action of frost. *Geological Magazine* 6:255.

Embleton, C., and King, C. A. M. 1975. *Periglacial geomorphology.* New York: John Wiley & Sons.

French, H. M. 1976. *The periglacial environment.* New York: Longman.

Harris, C. 1977. Engineering properties, groundwater conditions, and the nature of soil movement on a solifluction slope in North Norway. *Quarterly Journal of Engineering Geology* 10:27–43.

SUBSURFACE

Abel, J. F. 1970. Tunnel support—Where does the geology come in? *Civil Engineering* 40:69–71.

Bohman, R. A. 1958. Tunnel construction in the Mancos Shale Formation on U.S. Bureau of Public Roads Project 1-F, Mesa Verde National Park, Colorado. In *Engineering geology case histories,* no. 2, ed. P. D. Trask, pp. 1–2. Boulder, Colo.: Geological Society of America.

Content, C. S. 1958. Summary of six Rocky Mountain case histories. In *Engineering geology case histories,* no. 2, ed. P. D. Trask, pp. 3–5. Boulder, Colo.: Geological Society of America.

Fluhr, T. W. 1957a. Geology of the Queens Midtown Tunnel, New York. In *Engineering geology case histories,* no. 1, ed. P. D. Trask, pp. 1–12. Boulder, Colo.: Geological Society of America.

______. 1957b. Geologic engineering features of the West Delaware Tunnel. In *Engineering geology case histories,* no. 1, ed. P. D. Trask, pp. 11–16. Boulder, Colo.: Geological Society of America.

Lachel, D. 1972. The engineering geologist's role in hard rock tunnel machine selection. In *Engineering geology case histories,* no. 9, ed. H. Pincus, pp. 5–10. Boulder, Colo.: Geological Society of America.

Lofgren, B. E. 1963. *Land subsidence in the Arvin-Maricopa area, San Joaquin Valley, California.* U.S. Geological Survey Professional Paper 475–B, Art. 47, pp. B171–B175. Washington, D.C.

Moody, W. T. 1959. Importance of geological information as a factor in tunnel-lining design. In *Engineering geology case histories,* no. 3, ed. P. D. Trask, pp. 45–52. Boulder, Colo.: Geological Society of America.

Sherman, W. F. 1963. Engineering geology of Cody Highway Tunnels, Park County, Wyoming. In *Engineering geology case histories,* no. 4, eds. P. D. Trask and G. A. Kiersch, pp. 27–32. Boulder, Colo.: Geological Society of America.

Trask, P. D., ed. 1959. Rock mechanics in the investigation and construction of T. 1 underground power station, Snowy Mountains, Australia. In *Engineering geology case histories,* no. 3, ed. P. D. Trask, p. 13. Boulder, Colo.: Geological Society of America.

Zeevaert, L. 1963. *Foundation problems related to ground surface subsidence in Mexico City.* American Society of Testing and Materials Special Technical Paper 322. Philadelphia, Pa.: American Society for Testing and Materials.

ENGINEERING: ENGINEERING GEOLOGY MAPPING

Coates, D. R., ed. 1976a. *Geomorphology and engineering.* Stroudsburg, Pa.: Dowden, Hutchinson & Ross.

______. 1976b. *Urban geomorphology.* Geological Society of America Special Paper 174. Boulder, Colorado.

Maberry, J. O. 1972. *Folio of the Parder Quadrangle.* Colorado Map I–770–C (A–K Series). U.S. Geological Survey. Washington, D.C.

Montgomery, H. B. 1974. What kinds of geologic maps for what purposes? In *Engineering geology case histories,* no. 10, ed. H. F. Ferguson, pp. 1–8. Boulder, Colo.: Geological Society of America.

Nichols, D. R., and Campbell, C. C., eds. 1971. Environmental planning and geology. In *Symposium on engineering geology in the urban environment,* p. 204. Washington, D.C.

Olson, R. A., and Wallace, M. M., eds. 1969. *Geologic hazards and public problems.* Conference Proceedings, Office of Emergency Preparedness. Santa Rosa, California.

Rockaway, J. D., Jr., and Lutzen, E. E. 1970. *Engineering geology of the Creve Coeur Quadrangle, St. Louis County, Missouri.* Engineering Geology Series no. 2. Rolla: Missouri Geological Survey and Water Resources.

U.S. Geological Survey. 1953. *Interpreting geologic maps for engineering purposes, Hollidaysburg Quadrangle, Pennsylvania.* Prepared by members of the Engineering Geology and Groundwater branches. Washington, D.C.

Varnes, D. J. 1974. *The logic of geological maps, with reference to their interpretation and use for engineering purposes.* U.S. Geological Survey Professional Paper 837. Washington, D.C.

ENGINEERING: LAND USE

Artim, E. R. 1973. *Geology in land use planning.* Information Circular 47. Olympia, Wash.: Department of Natural Resources, Division of Mines and Geology.

Bolt, B. A.; Horn, H. L.; MacDonald, G. A.; and Scott, R. F. 1977. *Geological hazards: Earthquakes, tsunamis, volcanoes, avalanches, landslides, floods.* New York: Springer-Verlag.

Burton, I., and Kates, R. W. 1964. The perception of natural hazards in resource management. *Natural Resources Journal* 3:412–41.

Dacy, D. C., and Dunreuther, H. 1969. *Economics of natural disasters.* New York: The Free Press.

Jacobs, A. M. 1971. *Geology for planning in St. Clair County, Illinois.* Springfield: Illinois State Geological Survey, Department of Registration and Education.

Legget, R. F. 1974. Engineering geological maps for urban development. In *Engineering geology case histories,* no. 10, ed. H. F. Ferguson, pp. 19–21. Boulder, Colo.: Geological Society of America.

Mathewson, C. C., and Font, R. G. 1974. Geologic environment: Forgotten aspect in the land use planning process. In *Engineering geology case histories,* no. 10, ed. H. F. Ferguson, pp. 22–28. Boulder, Colo.: Geological Society of America.

Meltz, S. J. 1974. Geonatural resources planning. In *Engineering geology case histories,* no. 10, ed. H. F. Ferguson, pp. 13–16. Boulder, Colo.: Geological Society of America.

Meyers, C. R., Jr. 1974. Regional land use analysis and simulation models: A step forward in the planning process. In *Engineering geology case histories,* no. 10, ed. H. F. Ferguson, pp. 9–12. Boulder, Colo.: Geological Society of America.

Montgomery, H. G. 1974. Environmental analysis in local development planning. In *Engineering geology case histories,* no. 10, ed. H. F. Ferguson, pp. 29–40. Boulder, Colo.: Geological Society of America.

Rober, W. P.; Ladwid, L. R.; Hornbaker, A. L.; Schwochow, S. D.; Hart, S. S.; Shelton, D. C.; Scroggs, D. L.; and Soule, J. M. 1974. *Guidelines and criteria for identification and land-use controls of geologic hazard and mineral resource areas.* Special Publication no. 6. Denver: Colorado Geological Survey, Department of Natural Resources.

ENGINEERING: CONSTRUCTION

Clark, B. E. 1963. Geologic exploration and foundation treatment, Barkley Lock, Cumberland River, Kentucky. In *Engineering geology case histories,* no. 4, eds. P. D. Trask and G. A. Kiersch, pp. 19–26. Boulder, Colo.: Geological Society of America.

Croney, D.; Coleman, J. D.; and Black, W. P. 1958. *The movement and distribution of water in soil in relation to highway design and performance.* Highway Research Board Special Report no. 40. Washington, D.C.

Gullixson, R. L. 1958. Foundation grouting, McNary Dam, Oregon and Washington. In *Engineering geology case histories,* no. 2, ed. P. D. Trask, pp. 5–8. Boulder, Colo.: Geological Society of America.

Hatheway, W. W., and McClure, C. R., Jr., eds. 1979. *Geology in the siting of nuclear power plants.* Reviews in Engineering Geology, vol. 4, Boulder, Colo.: Geological Society of America.

McGill, J. T. 1954. Residential building-site problems in Los Angeles, California. *California Division of Mines and Geology Bulletin* 170:11–18.

National Academy of Sciences. 1968. *Criteria for selection and design of residential slabs-on-ground.* F.H.A. Report no. 33. Washington, D.C.

Russam, L., and Coleman, J. D. 1961. The effect of climatic factors on subgrade conditions. *Geotechnique* 11:22.

Supp. C. W. A. 1957. Engineering geology of the Chesapeake Bay Bridge. In *Engineering geology case histories,* no. 1, ed. P. D. Trask, pp. 49–56. Boulder, Colo.: Geological Society of America.

Terzaghi, K. 1925. Modern concepts concerning foundation engineering. *Journal of the Boston Society of Civil Engineers* 12:1.

ENGINEERING: DAMS

American Society of Civil Engineers. 1974. *Inspection, maintenance, and rehabilitation of old dams.* New York: American Society of Civil Engineers.

Burwell, E. B., and Moneymaker, B. C. 1950. Geology in dam construction. In *Application of geology to engineering practice* (Berkey Volume), pp. 11–43. New York: Geological Society of America.

Castle, R. O.; Yerkes, R. F.; and Youd, T. L. 1973. Ground rupture in the Baldwin Hills—An alternative explanation. *Association of Engineering Geologists* 10:21–43.

Fluhr, T. W. 1964. Earth dams in glacial terrain, Catskill Mountain Region, New York. In *Engineering geology case histories,* no. 5, ed. G. A. Kiersch, pp. 15–30. Boulder, Colo.: Geological Society of America.

Hill, L. C.; Evens, H. W.; and Fowler, R. H. 1929. Essential facts concerning the failure of St. Francis Dam. *Proceedings of the American Society of Civil Engineers* 55:2147–63.

Holdredge, C. P. 1957. Geologic reports on damsites in the John Day Basin, Oregon. In *Engineering geology case histories,* no. 1, ed. P. D. Trask, pp. 25–32. Boulder, Colo.: Geological Society of America.

Honkala, A. U. 1964. Engineering geology of the Demirkopru Dam Site, Salihli, Turkey. In *Engineering geology case histories,* no. 5, ed. G. A. Kiersch, pp. 71–76. Boulder, Colo.: Geological Society of America.

James, L. B. 1968. Failure of Baldwin Hills Reservoir, Los Angeles, California. In *Engineering geology case histories,* no. 6, ed. G. A. Kiersch, pp. 1–11. Boulder, Colo.: Geological Society of America.

Kaye, C. A. 1958. Antonio Lucchetti Dam, Puerto Rico. In *Engineering geology case histories,* no. 2, ed. P. D. Trask, pp. 9–12. Boulder, Colo.: Geological Society of America.

Mollard, J. D. 1958. Dam-site studies from aerial photographs. In *Engineering geology case histories,* no. 2, ed. P. D. Trask, pp. 21–24. Boulder, Colo.: Geological Society of America.

Monohan, C. G. 1957. Geologic features at McNary Dam, Oregon. In *Engineering geology case histories,* no. 1, ed. P. D. Trask, pp. 33–38. Boulder, Colo.: Geological Society of America.

Outland, C. F. 1963. *Man-made disaster: The story of St. Francis Dam.* Glendale, Calif.: A. H. Clark.

Ransome, F. L. 1928. Geology of the St. Francis Dam site. *Economic Geology* 23:553–63.

Terzaghi, K. 1929. Effect of minor geologic details on the safety of dams. In *Bulletin of the American Institute of Mining Engineering,* Technical Publication 215, Class 1, Mining Geology, no. 26, pp. 31–46.

U.S. Department of the Interior. 1976. *Failure of Teton Dam.* Report by Independent Panel to Review Cause of Failure. Idaho Falls, Idaho.

Wahlstrom, E. E. 1963. Geology and foundation treatment, Williams Fork Dam, Power Plant, and West Dike, Colorado. In *Engineering geology case histories,* no. 4, eds. P. D. Trask and G. A. Kiersch, pp. 9–18. Boulder, Colo.: Geological Society of America.

ENGINEERING: EXCAVATION

Hamburger, R. 1972. Geologic factors in rapid excavation with nuclear explosives. In *Engineering geology case histories,* no. 9, ed. H. Pincus, pp. 17–26. Boulder, Colo.: Geological Society of America.

Obert, L. A. 1972. Rapid excavation and the role of engineering geology. In *Engineering geology case histories,* no. 9, ed. H. Pincus, pp. 1–4. Boulder, Colo.: Geological Society of America.

Pugliese, H. 1972. Some geological structural influences in quarrying limestone and dolomite. In *Engineering geology case histories,* no. 9, ed. H. Pincus, pp. 11–16. Boulder, Colo.: Geological Society of America.

Roach, C. H. 1972. Total systems approach to rapid excavation and its geological requirements. In *Engineering geology case histories,* no. 9, ed. H. Pincus, pp. 69–78. Boulder, Colo.: Geological Society of America.

ENGINEERING: WASTE DISPOSAL

Barnes, H. 1974. Geologic and hydrologic background for selecting site of pilot-plant repository for radioactive waste. *American Engineering Geology Bulletin* 11:83–92.

Brunner, D. R.; Hubbard, S. J.; Keller, D. J.; and Newton, J. L. 1971. *Closing open dumps.* Washington, D. C.: Environmental Protection Agency Solid Waste Management Office.

Cartwright, D., and McComas, M. R. 1968. Geophysical surveys in the vicinity of sanitary landfills in northeastern Illinois. *Groundwater* 6:8.

Environmental Protection Agency. 1971. *Recommended standards for sanitary landfill design, construction and evaluation and model sanitary landfill operation agreement.* Washington, D.C.

Galley, J. E., ed. 1968. *Subsurface disposal in geologic basins.* American Association of Petroleum Geologists Memoir 10. Tulsa, Oklahoma.

Hughes, G. M. 1972. *Hydrogeologic considerations in the siting and design for landfills.* Environmental Geology Note no. 61. Springfield: Illinois State Geological Survey.

Hughes, G. M., and Cartwright, K. 1972. Scientific and administrative criteria for shallow waste disposal, civil engineering. *Transactions of the American Society of Civil Engineers* 42:70–73.

McMichael, F. C., and McKee, J. E. 1965. *Report of research on waste-water reclamation at Whittier Narrows.* Pasadena: California Institute of Technology.

Merz, R. C., and Stone, R. 1964. Gas production in a sanitary landfill. *Public Works* 95:84.

Newcomb, R. C. 1973. Requirements of an optimum site for subsurface storage or disposal of radioactive wastes. *Association of Engineering Geologists Bulletin* 10:219–30.

Piper, A. M. 1969. *Disposal of liquid wastes by injection underground—Neither myth nor millennium.* U.S. Geological Survey Circular 631. Washington, D.C.

Reichert, S. O., and Fenimore, J. W. 1964. Lithology and hydrology of radioactive waste-disposal sites, Savannah River plant, South Carolina. In *Engineering geology case histories,* no. 5, ed. G. A. Kiersch, pp. 53–70. Boulder, Colo.: Geological Society of America.

Riccio, J. F., and Hyde, L. W. 1971. *Hydrogeology of sanitary landfill sites in Alabama—A preliminary appraisal.* Geological Survey of Alabama Circular 71. University, Ala.: Division of Water Resources.

Schneider, W. J. 1970. *Hydrologic implication of solid-waste disposal.* U.S. Geological Survey Circular 601–F. Washington, D.C.

______. 1973. Hydrologic implications of solid waste disposal. In *Focus on environmental geology,* ed. R. Tank, pp. 412–25. London: University Press.

Turk, L. J. 1970. Disposal of solid wastes—Acceptable practice or geological nightmare? In *American Geological Institute Short Course, Lecture Notes on Environmental Geology,* 9–10 November 1970, Milwaukee, Wisconsin. Washington, D.C.: American Geological Institute.

White, W. A., and Kyriazis, M. D. 1968. *Effects of waste effluents on the plasticity of earth materials.* Environmental Geology Notes no. 23. Springfield: Illinois State Geological Survey.

ENGINEERING: MATERIALS

Mather, K. 1958. Cement-aggregate reaction—What is the problem? In *Engineering geology case histories,* no. 2, ed. P. D. Trask, pp. 83–85. Boulder, Colo.: Geological Society of America.

McConnell, D.; Mielenz, R. D.; Holland, W. Y.; and Greene, K. T. 1950. Petrology of concrete affected by cement-aggregate reaction. In *Application of geology to engineering practice* (Berkey Volume), pp. 225–50. New York: Geological Society of America.

Treasher, R. C. 1964. Geologic investigations for sources of large rubble. In *Engineering geology case histories,* no. 5, ed. G. A. Kiersch, pp. 31–44. Boulder, Colo.: Geological Society of America.

PROFESSIONAL PRACTICES

Ackerman, A. J. 1961. The engineer's obligation to disclose all facts. *Civil Engineering* 31:60–63.

Berkey, C. P. 1929. *Responsibilities of the geologist in engineering projects.* American Institute of Mining, Metallurgical, and Petroleum Engineers Technical Publication 215, pp. 4–9. Littleton, Colorado.

Burwell, E. B., and Roberts, G. D. 1950. The geologist in the engineering organization. *Application of geology to engineering practice* (Berkey Volume), p. 327. New York: Geological Society of America.

Clark, J. R. 1961. *Concerning some legal responsibilities in the practice of architecture and engineering and some recent changes in contract documents.* New York: Engineering Joint Council.

Ehrlich, M.; Scharon, H. L.; and Mateker, E. J., Jr. 1969. Vibration analysis and construction blasting. In *Engineering geology case histories,* no. 7, eds. G. A. Kiersch and A. B. Cleaves, pp. 13–20. Boulder, Colo.: Geological Society of America.

Gaskins, J. W., and Cleaves, A. B. 1969. Jim Woodruff Dam—Changed conditions claim. In *Engineering geology case histories,* no. 7, eds. G. A. Kiersch and A. B. Cleaves, pp. 7–12. Boulder, Colo.: Geological Society of America.

Kiersch, G. A. 1969a. The geologists and legal cases—Preparation, responsibility, and the expert witness. In *Engineering geology case histories,* no. 7, eds. G. A. Kiersch and A. B. Cleaves, pp. 1–6. Boulder, Colo.: Geological Society of America.

______. 1969b. Pfeiffer versus General Insurance Corporation—Landslide damage to insured dwelling, Orinda, California, and relevant cases. In *Engineering geology case histories,* no. 7, eds. G. A. Kiersch and A. B. Cleaves, pp. 81–94. Boulder, Colo.: Geological Society of America.

Miles, K. R. 1969. Geological conditions and arbitration of dredged channel and turning basin, Bluff, New Zealand. In *Engineering geology case histories,* no. 7, eds. G. A. Kiersch and A. B. Cleaves, pp. 21–32. Boulder, Colo.: Geological Society of America.

Sidrer, P. 1957. Engineers' problems in estimating quantity of bedrock to be removed in estimating cost of highway construction jobs. In *Engineering geology case histories,* no. 1, ed. P. D. Trask, pp. 39–42. Boulder, Colo.: Geological Society of America.

Supp, W. W. A. 1969. Maryland State Roads Commission versus Champion Brick Company—Condemnation of clay-bearing lands. In *Engineering geology case histories,* no. 7, eds. G. A. Kiersch and A. B. Cleaves, pp. 95–112. Boulder, Colo.: Geological Society of America.

Waggoner, E. B.; Sherard, J. L.; and Clevenger, W. A. 1969. Geological conditions and construction claims on earth and rock-fill dams and related structures. In *Engineering geology case histories,* no. 7, eds. G. A. Kiersch and A. B. Cleaves, pp. 33–44. Boulder, Colo.: Geological Society of America.

5: Professional Societies

American Association of Petroleum Geologists, 1444 S. Boulder Avenue, P.O. Box 979, Tulsa, OK 74101.

American Geological Institute, 5205 Leesburg Pike, Falls Church, VA 22041.

American Geophysical Union, 2000 Florida Avenue, N.W. Washington, D.C. 20009.

American Institute of Mining, Metallurgical and Petroleum Engineers, Inc., Caller No. D, Littleton, CO 80127.

American Society of Civil Engineers, 345 East 47th Street, New York, NY 10017.

The American Society of Photogrammetry, 105 N. Virginia Avenue, Falls Church, VA 22046.

American Society for Testing and Materials, 1916 Race Street, Philadelphia, PA 19103.

American Water Resources Association, St. Anthony Falls Hydraulic Laboratory, Mississippi River at 3rd Avenue S.E., Minneapolis, MN 55414.

Association of Engineering Geologists, 8310 San Fernando Way, Dallas, TX 75218.

Boston Society of Civil Engineers, 80 Boylston, Boston, MA 02116.

British Mineralogic Society, 41 Queen's Gate, London, England SW7 5HR.

Clay Mineral Society, P.O. Box 2295, Bloomington, IN 47402.

Geological Association of Canada, Department of Earth Sciences, University of Waterloo, Waterloo, Ontario, N2L 3G1.

The Geological Society, London, Burlington House, Piccadilly, London, England WIV OJU.

Geological Society of America, 3300 Penrose Place, Boulder, CO 80301.

International Association of Engineering Geology, Geologisches Landesamt NW, de-Greiff-StraBe 195, Postfach, 1080, D-4150 Krefeld 1, Federal Republic of Germany.

International Association of Scientific Hydrology, 19 Rue Eugene Carriere, F-75018, Paris, France.

International Society for Rock Mechanics, c/o Laboratorio Nacional De Engenharia Civil, Av. Do Brazil, Lisbon-5, Portugal.

International Society for Soil Mechanics and Foundation Engineering (Association of Soil and Foundation Engineers), 8811 Colesville Road, Suite 225, Silver Spring, MD 20910.

National Research Council of Canada, Ottawa, Canada K1A OR6.

National Water Well Association, 500 W. Wilson Bridge Road, Worthington, OH 43085.

Seismological Society of America, 2620 Telegraph Avenue, Berkeley, CA 94704.

Society of Economic Geologists, 185 Estes Street, Lakewood, CO 80226.

Society of Economic Paleontologists and Mineralogists, P.O. Box 4756, Tulsa, OK 74104.

Society of Exploration Geophysicists, Box 3098, Tulsa, OK 74101.

Soil Science Society of America, 677 South Segoe Road, Madison, WI 53711.

index